CHROMATIN AND EPIGENETICS

Edited by **Colin Logie**
and **Tobias Aurelius Knoch**

Chromatin and Epigenetics
http://dx.doi.org/10.5772/intechopen.71387
Edited by Colin Logie and Tobias Aurelius Knoch

Contributors

Metin Budak, Jin-Wook Kim, Yujing Li, Pengfei Ji, Nina Xie, Cuili Ma, Hiroshi Yasuda, Yoshiyuki Watanabe, Ritsuko Oikawa, Fumio Itoh, Paul Enríquez, Yoshiaki Onishi, Tatsunosuke Tomita, Kinan Drak Alsibai, Didier Meseure, Hugo Van Ingen, Velten Horn, Tobias Aurelius Knoch, Ray Kit Ng, Molly Pui Man Wong, Judith Miné-Hattab, Xavier Darzacq, Sergio Angel, Laura Vanagas, Susana Marisol Contreras

First published in London, United Kingdom, 2020 by IntechOpen
IntechOpen is the global imprint of INTECHOPEN LIMITED, registered in England and Wales, registration number: 11086078, 7th floor, 10 Lower Thames Street, London,
EC3R 6AF, United Kingdom
Printed in Croatia

British Library Cataloguing-in-Publication Data
A catalogue record for this book is available from the British Library

Additional hard and PDF copies can be obtained from orders@intechopen.com

Chromatin and Epigenetics, Edited by Colin Logie and Tobias Aurelius Knoch
p. cm.
Print ISBN 978-1-78984-492-4
Online ISBN 978-1-78984-493-1
eBook (PDF) ISBN 978-1-78984-982-0

For EU product safety concerns:
IN TECH d.o.o., Prolaz Marije Krucifikse Kozulić 3, 51000 Rijeka, Croatia,
info@intechopen.com or visit our website at intechopen.com.

Meet the editors

Born in Brussels, Colin Logie studied Genetics and Molecular Biology at Glasgow University, graduating in 1991. For his PhD, he built the first ligand-regulated site-specific DNA recombinases with A. Francis Stewart at the EMBL, Heidelberg, Germany, in 1995. He went on to engineer nucleosome remodelling assays during his post-doc with Craig L. Peterson at the UMASS, Worcester, MA, USA. Since 1999 he has conducted academic research on chromatin remodelling complexes at the Radboud University Science Faculty Molecular Biology Department (RIMLS), Nijmegen, The Netherlands, which is a joint venture with the Medical Faculty of the Radboud University. For the past seven years he has focused on human blood monocytes, and the cell types they can differentiate into such as macrophages, dendritic cells, and osteoclasts as an experimental research system. He is interested in the systems-wide integration of nuclear receptor action in health and disease. Through the COST-funded Gene Regulation Ensemble Effort for the Knowledge Commons (GREEKC) he became a member of the Gene Ontology Consortium. There he participates in setting the stage for computer-readable gene annotation for the 21st century, with a focus on 'breaking through the DNA wall' so as to enable automated reasoning through cellular signal transduction all the way to the trans-acting sequence-specific DNA binding transcription factors and their co-factors that act in concert on cis-acting chromosomal DNA sequences to permit cell type- and environmentally-conditioned gene transcription regulation. This includes the concepts of epigenetic memory embodied by post-replicational DNA modifications such as DNA methylation, nucleosome-borne epigenetic information in the form of histone post-translational modifications, nucleosome remodelling, and the positioning of nucleosomes along the length of chromosomes. Also chromatin domain boundaries have been a focus of his recent research. Over the years, his research has been funded by EMBL, Human Frontiers Science Program (HFSP), the Dutch Research Organisation NWO, the European Science Foundation (ESF), and indirectly by the European Research Council (ERC) with Hendrik G. Stunnenberg and Stefano Ceri.

Born in the Rhein-Neckar region Mannheim/Heidelberg, Germany, Dr. Knoch studied Physics, Mathematics, and Biology at the University of Heidelberg. In 1998, he graduated in (bio-)physics with "Three-Dimensional Organization of Chromosomes domains in Simulation and Experiment", followed by his dissertation "Approaching the Three-Dimensional Organization of the Human Genome" both at the German Cancer Research Center (DKFZ), Heidelberg, in 2002. In 2002/2004 Dr. Knoch founded his group Biophysical Genomics located at the Kirchhoff Institute for Physics, University of Heidelberg, and until today at the Cell Biology Department, Erasmus Medical Center, Rotterdam, The Netherlands. His work is focusing on the determination and understanding of genome organization from the DNA sequence level to the entire nuclear morphology. Therefore, approaches from theoretical physics have been combined with molecular biology in highly interdisciplinary projects ranging from advanced DNA sequence analyses, parallel high-performance computer modelling of genomic architectures, and new image analysis methods, to advanced fluorescence in situ hybridization and high-resolution chromatin conformation interaction genome mapping. Major achievements have been: an artefact-free in vivo labelling method of nuclear chromatin, the first system-biological genome browser (GLOBE 3D Genome Browser), the set-up of one of the largest desktop computing grids, and last but not least the final determination of the general structural organization of higher mammalian genomes leading to a consistent systems genomics view of genomes from genotype to phenotype. All this has resulted in patents, publications, the foundation/coordination of international interdisciplinary cooperative networks, and consortia. He also (co-)founded many initiatives improving institutional/university study and management performance including the science outreach to the public and industry. Besides, he also conducts environmental and human ecology research, has achieved law-changing contributions in the human-rights sector, is an increasingly recognized artist in the fine arts, and last but not least has founded and is running two companies in the renewable energy production and development sector. His achievements have resulted in prestigious scholarships, awards, and prices as early as 1983.

Contents

Preface IX

Chapter 1 **Logic of Epigenetics and Investigation of Potential Gene Regions 1**
Metin Budak

Chapter 2 **Recognition of Nucleosomes by Chromatin Factors: Lessons from Data-Driven Docking-Based Structures of Nucleosome-Protein Complexes 21**
Velten Horn and Hugo van Ingen

Chapter 3 **Chromatin Dynamics upon DNA Damage 47**
Judith Miné-Hattab and Xavier Darzacq

Chapter 4 **A Consistent Systems Mechanics Model of the 3D Architecture and Dynamics of Genomes 67**
Tobias A. Knoch

Chapter 5 **Apicomplexa and Histone Variants: What's New? 95**
Laura Vanagas, Susana Marisol Contreras and Sergio Oscar Angel

Chapter 6 **Epigenetic Modulation of Circadian Rhythms: Bmal1 Gene Regulation 125**
Tatsunosuke Tomita and Yoshiaki Onishi

Chapter 7 **Epigenome Editing 145**
Paul Enríquez

Chapter 8 **Resetting Cell Fate by Epigenetic Reprogramming 165**
Molly Pui Man Wong and Ray Kit Ng

Chapter 9 **Cytosine Modifications and Distinct Functions of TET1 on Tumorigenesis 187**
Cuili Ma, Pengfei Ji, Nina Xie and Yujing Li

Chapter 10 **Role of COX-2 Promoter Methylation and Helicobacter pylori Infection in Impaired Gastric Ulcer Healing 207**
Hiroshi Yasuda, Yoshiyuki Watanabe, Ritsuko Oikawa and Fumio Itoh

Chapter 11 **Epigenetic Regulation of Hepatitis B Virus Replication 223**
In Young Moon and Jin-Wook Kim

Chapter 12 **Part 1: The PIWI-piRNA Pathway Is an Immune-Like Surveillance Process That Controls Genome Integrity by Silencing Transposable Elements 233**
Didier Meseure and Kinan Drak Alsibai

Chapter 13 **Part 2: Deregulated Expressions of PIWI Proteins and piRNAs as New Candidate Biomarkers and Potential Therapeutic Tools in Cancer 263**
Didier Meseure and Kinan Drak Alsibai

Preface

Since Charles Darwin put forward his famous bottom-up theory of natural selection in "The Origin of Species" in 1859, genetics and nowadays genomics increasingly gained world-wide attention [1]. The frequently debated cartoons depicting the descent of man from apes mark the transition from ancient mystical beliefs to a scientifically falsifiable modern theory of the origin of life and mankind. Jean-Baptiste Lamarck had already put forward a top-down hypothesis in 1809 in his famous book "Philosophie Zoologique" symbolized by the elongating giraffe-neck longing for leaves at tree-tops [2]. It achieved public attention due to the 'motivated change' paradigm that remains simple to grasp, actually to this day. The discovery of the DNA double helix and the four base code culminated in the publicly much anticipated race to decipher the sequence of the human genome in 2002 by the public Human Genome Project and the private initiative of Craig Venter. Now everything would be known about humans, their lives from birth to death as well as all species, the origin of life and its future - so resounded the promise of some and the happy belief of many.

But are 3 billion A, T, G, C bases and the ensuing code for ~20,000 proteins really sufficient explanation for the holy grail of human life, as would be suggested by monozygotic twin physiologic and physiognomic identity? Or is the real situation much more complex, potentially involving a multitude of interactions of the genotype ranging from its own internal physicochemical foundation, via the phenotype it is creating, up to the entire ecosystem, which is itself made up of many genomes? There is little doubt that a purely reductionistic approach cannot explore the full extent of the interaction-networks of genomes. While reductionism can formally identify biological variables that are rate-limiting *under* the experimental conditions that are tested, there may always be biological signals that are not observed because they were not solicited during the experiment.

Since the initial investigations in the late 18th century much progress has been made, achieving a picture of the notoriously hard-to-access eukaryotic cell nucleus and the chromatin it contains [3]. This revealed a layered organization of the physical genome comprising DNA, nucleosomes, chromatin quasi-fibres, chromatin loops, loop aggregates/rosettes, as well as chromosome arms, entire chromosomes, and their position within the cell nucleus as outlined in Chapter 4 of this book.

Intriguingly, it also has become apparent relatively early already that there are heritable phenotype changes that do not involve alterations in the nucleotide sequence. This led to the concept of epigenetic coding initially introduced by Conrad Hall Waddington in 1942 [4] to explain why almost every cell of an organism harbours the same DNA but does not express the same parts of the genetic information, enabling cell fate determination and cell lineage differentiation to yield all the cells of an organism, as explained in Chapter 1.

This book addresses current issues in the fields of epigenetics and chromatin ranging from more theoretical overviews in the first four chapters to much more detailed methodologies and insights into diagnostics and treatments in the other chapters.

In the elaborate Chapter 1 *"Logic of Epigenetics and Investigation of Potential Gene Regions"*, an overview is presented on epigenetic coding and the regulation of genes as well as what kind of modifications appear and how this can be understood within a broader scope. Chapter 2 *"Recognition of Nucleosomes by Chromatin Factors: Lessons from Data-Driven Docking-Based Structures of Nucleosome-Protein Complexes"* focuses on the intricate molecular interactions and thus how epigenetics actually works, i.e. the involved proteins and their structure. In Chapter 3 *"Chromatin Dynamics Upon DNA Damage"*, fingers are laid deep into the chromatin dynamics after DNA damage and its repair, thus showing how a genome reacts to the inevitable reality of life and consequent DNA damaging events. Finally, this first part of the book is rounded-off by Chapter 4 *"A Consistent Systems Mechanics Model of the 3D Architecture and Dynamics of Genomes"*, in which for the first time a broad systems genomics model based on the complete determination of genomic layers is presented leading to new perspectives on genome evolution and general complex systems development in nature.

As far as much more detailed methodologies and insights are concerned, in Chapter 5 *"Apicomplexan and Histone Variants: What's new?"*, the clinically relevant histone variants in Plasmodium spp. and Toxoplasma gondii are investigated in depth. Furthermore, in the far-reaching Chapter 6 *"Epigenetic Modulation of Circadian Rhythms: Bmal1 Gene Regulation"*, regulation of complex metabolic effects and phenotypic modulations are considered within a disease-relevant context. In Chapter 7 *"Epigenome Editing"*, we dive deeply into an overview of modern epigenome editing systems locally and in genome-wide contexts, of which the CRISPR-Cas9 system has gained prominent public attention. This leads immediately to Chapter 8 *"Resetting Cell Fate by Epigenetic Reprogramming"*, which details phenomenally the setting of epigenetic patterns during differentiation and different reprogramming strategies that can be applied. In Chapter 9 *"Cytosine Modifications and Distinct Functions of TET1 on Tumorigenesis"*, very practical aspects of epigenetics during tumorigenesis are investigated. Other important clinical aspects are considered in Chapter 10 *"Role of COX-2 Promoter Methylation and Helicobacter pylori Infection in Impaired Gastric Ulcer Healing"*, where a treatment of epigenetic inhibition is shown. In the broad Chapter 11 *"Epigenetic Regulation of Hepatitis B Virus Replication"*, the importance of epigenetics for the regulation of the Hepatitis B virus is described, thus shedding light on how epigenetic control of transcription can be used for therapeutic strategies. The book ends with a great double: In Chapter 12 *"Part 1: The PIWI-piRNA Pathway is an Immune-Like Surveillance Process that Controls Genome Integrity by Silencing Transposable Elements"*, we get an in-depth introduction into the pivotal roles of the PIWI pathway focusing on origin, properties, and functions in the germ line and somatic tissues. This is followed by Chapter 13 *"Part 2: Deregulated Expressions of PIWI Proteins and piRNAs as New Candidate Biomarkers and Potential Therapeutic Tools in Cancer"*, where we dive into how such epigenetic-based research as in the case of PIWI can lead to biomarkers which could be used as potential therapeutic tools in the perhaps emotionally most feared disease, namely cancer.

Consequently, the general development of genomics, which these chapters on epigenetics and chromatin illustrate in both their depth as well as broadness, shows clearly that genetic information is stored on all structural and dynamical levels within the nucleus with corresponding modifications. Only an integrative systems approach allows to understand and manipulate the consequently holistic interplay of geno- and phenotype creating functional genomes. All this opens the door to a concrete grasp of life as well as in more general terms complex systems as a whole. Hence, genomics and the entire biology are driven into the future by broad complex physical and mathematical methodologies and approaches. This not only opens

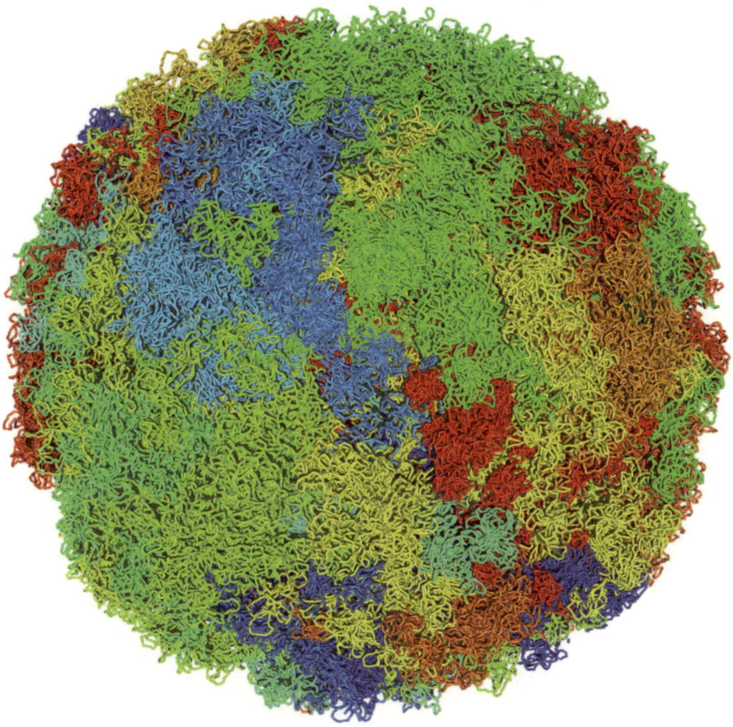

Figure 1.
Visions make the world go round: The simulation of an entire human genome with all chromosomes (different colours; [5]) depicting a chromatin quasi-fibre folding into stable loop aggregates/rosettes (Multi-Loop Subcompartment Model) has opened huge opportunities for an immediate intuitive understanding of the dynamic genome organization and function not only as, i.e. concerned spatial stable proximities, accessibilities, and replicability are concerned, but also in respect to the bigger genotype-phenotype evolutionary systems genomics entanglement.

opportunities for a true universal view of genetic information, but also is the key for a general understanding of genomes, their function, as well as life and evolution in general. Furthermore, these insights pave the path for diagnostics and disease treatment, for future genome manipulation and engineering efforts, and ultimately also for *de novo* created life forms. This will leave much room for new ample opportunities and we are sure that the most interesting times are yet to come.

Thus, in practical terms and in the formulation of T.A.K., to ultimately approach Sustainable Development Goals as health/disease, and death, i.e. to achieve "eternal" as well as artificial intelligence and life, demands the following: i) R&D *must* work inter-/trans-disciplinarily in an open innovative network! Here, *THE* keys are new virtual paper tools representing and seamlessly visualizing, integrating, and manipulating the complexity of systems wholeness (Figure 1, 2; Movie 1). Beyond, as foundation, ii) broad humanistic education (the baroquian *Bildung* ideal) *must* be achieved with inter-/trans-disciplinary curricula of *ALL* sciences, arts, and professional crafts to efficiently exploit the opportunities of systems complexity. Learning for its own sake beyond mere training on the job *must* be the final mantra. Lastly, of ultimate importance, iii) society as a whole *must* epitomize an overall integrative thinking and operation, i.e. living an internalized Human Ecology autopoietic systems perspective [10]. Hence, *ALL* this *must* be represented in a humanistic systems vision of terrestrial, interplanetary, and artificial intelligence/life, being for everybody *ad hoc* graspable in a playful "Glass Bead Game" [11] manner, both for the detailed daily practice as well as for a general "enlightened" understanding of existence.

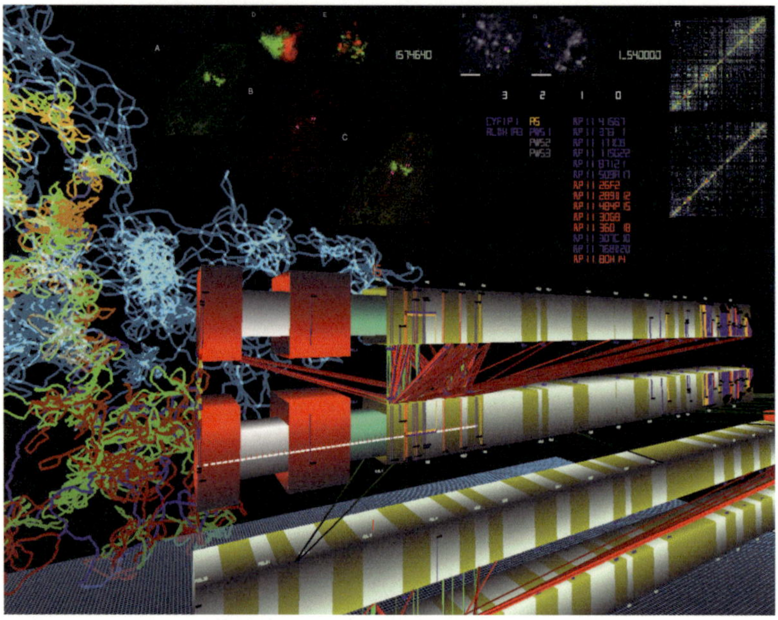

Figure 2, Movie 1.
Practical systemic virtual paper tools make seeing is believing accessible: The GLOBE 3D Genome Platform [6] creates a virtual desktop environment for genomics combining i) visual data representation, ii) data access/ management, and iii) data analysis/creation. This allows, i.e. to combine 3D genome simulations (background left [5]) with complex DNA similarity and syndrome analysis of a linear ideogram chromosome representation (foreground) in respect to actual data (microscopy/FISH of two chromosome loci: PWLS/AS region (A-C; [5, 7]), chromosome arms (D; courtesy S. Dietzel), subchromosomal domains (E; courtesy D. Zink) [5], IgH region (F, G; [8]; T2C chromosomal interaction maps, (H, I; [9]), allowing complexity to be intuitively approached systemically.

Last but not least, we would like to point out the unusually long history of this book which initially started as two separate titles: "Chromatin" has been edited by Dr. Colin Logie since autumn 2017 and "Epigenetics" edited originally by Dr. Hasibe Cingilli Vural, from the Necmettin Erbakan University Meram Medicine Faculty, Turkey, in October 2017. Dr. Vural collected a phenomenal set of chapters, but due to circumstances not in her hands had to resign from her editorship. With everything on stake, Dr. Tobias A. Knoch (already a contributor) offered to take over the editorship of Dr. Vural. Over the course of time by the odds of life, Dr. Tobias A. Knoch, also contributor to the "Chromatin" book, and Dr. Colin Logie decided to join forces and combine the books into *"Chromatin and Epigenetics"*, of which you now hold the result in your hands.

Thus, we also would like to thank all the authors for their cooperation and patience.

During this project, it was a pleasure for us to work with IntechOpen Publisher. We would like to express our appreciation to all members of this team participating in this effort, especially Author Service Managers Danijela Sakić, Romina Rovan and Edi Lipović, as well as Commissioning Editor Andrea Korić.

Dr. Colin Logie and Dr. Tobias A. Knoch
Nijmwegen/Rotterdam, The Netherlands, November 2019

Video

Video file available from: https://bit.ly/37KBTgU

Cover image

Simulation of an entire human genome with all chromosomes (different colours) depicting a chromatin quasi-fibre folding into stable loop aggregates/rosettes (Multi-Loop Subcompartment Model); with courtesy of Dr. Tobias A. Knoch (for more details see Figure 1 of the Preface and there literature [5], or Chapter 5 and references therein).

References

[1] Darwin CR. On the origin of species. 1859

[2] Lamarck J-B. Philosophie Zoologique. 1809

[3] Cremer T, Cremer C. Centennial of Wilhelm Waldeyer's introduction of the term "chromosome" in 1888. Cytogenetics and Cell Genetics.1988;**48**(2):65-67

[4] Waddington CH. The epigenotype. Endeavour. 1942;**1**:18-20

[5] Knoch TA. Approaching the three-dimensional organization of the human genome: Structural-, scaling- and dynamic properties in the simulation of interphase chromosomes and cell nuclei, long-range correlations in complete genomes, *in vivo* quantification of the chromatin distribution, construct conversions in simultaneous co-transfections. Mannheim, Germany: TAKPress; 2002. ISBN 3-00-009959-X

[6] Knoch TA, Lesnussa M, Kepper FN, Eussen HB, Grosveld FG. The GLOBE 3D Genome Platform—Towards a novel system-biological paper tool to integrate the huge complexity of genome organization and function. Studies in Health Technology and Informatics. 2009;**147**:105-116

[7] Rauch J, Knoch TA, Solovei I, Teller K, Stein S, Buiting K, et al. Lightoptical precision measurements of the Prader-Willi/Angelman Syndrome imprinting locus in human cell nuclei indicate maximum condensation changes in the few hundred nanometer range. Differentiation. 2008;**76**(1):66-82

[8] Jhunjhunwala S, van Zelm MC, Peak MM, Cutchin S, Riblet R, van Dongen JJM, et al. The 3D-structure of the immunoglobulin heavy chain locus: implications for long-range genomic interactions. Cell. 2008;**133**(2):265-279

[9] Knoch TA, Wachsmuth M, Kepper N, Lesnussa M, Abuseiris A, Imam AMA, et al. The detailed 3D multi-loop aggregate/rosette chromatin architecture and functional dynamic organization of the human and mouse genomes. Epigenetics & Chromatin. 2016;**58**(9):1-22

[10] Knoch TA. Sustained renewability: Approached by systems theory and human ecology. In M. Nayeripour M, Keshti M, editors. Renewable Energy Sources. Vol. 2. IntechOpen. 2011. pp. 21-48. ISBN 978-953-307-573-0. Available at: https://www.intechopen.com/books/sustainable-growth-and-applications-in-renewable-energy-sources/sustainedrenewability-approached-by-systems-theory-and-human-ecology

[11] Hesse H. The Glass Bead Game. Holt, Rinehart and Winston, New York; 1943

Chapter 1

Logic of Epigenetics and Investigation of Potential Gene Regions

Metin Budak

Additional information is available at the end of the chapter

http://dx.doi.org/10.5772/intechopen.81739

Abstract

In living organisms, all molecular structures are formed, and all events are carried out by specific DNA sequences referred to as 'genes'. However, these genes need to be governed and controlled to function properly. In this way, they can participate in biological processes at the optimal time. Genes are actively controlled by other genes and specific proteins called 'transcription factors'. In addition, there is another mechanism that determines gene expression, which can be transmitted from generation to generation and from cell to cell. This mechanism is referred to as the 'epigenetic code'. The DNA sequence does not undergo any changes during the formation of this code, but the relevant part of DNA fragment becomes no longer meaningful. While histone modifications control expression of DNA in chromosome structure, methylation modifications at the gene level are quite effective in controlling expressions the cytosine-end methylation seen in mammalian genome often occurs in the nucleotide pairs which are also called the CpG dinucleotides. The most common epigenetic modifications are the changes in histone proteins and DNA methylation, and the most widely studied and the most well-established epigenetic mechanism is the latter.

Keywords: epigenetic, methylation, mitochondria, methylation-specific PCR, analysis

1. Introduction

Epigenetic changes in living organisms can basically be grouped under two headings. One is protein acetylation, which is an epigenetic modification at the protein level. The other is DNA methylation, an epigenetic modification that occurs at the DNA level. In living species, all macro-molecular structures are determined by nucleotide sequences in the genome, and there is a different mechanism that can be transferred to cell from cell, which has inherent ability to determine gene expression. It is called epigenetic code. During creation of this code, DNA sequence does not undergo any change. The genetic and epigenetic alterations mentioned above result in the activation of oncogenes or the inactivation of tumour suppressor genes. Methylation may occur in any living organism from bacteria to complex species such as humans. The most common type of methylation is the methylation of gene promoters. This is followed by exon methylation, intron methylation and exon-intron methylations, which may be observed quite frequently.

Methylation-specific PCR (MSP) and methylation-sensitive restriction fragment length polymorphism (MS-RFLP) are the two most widely utilized methods in DNA methylation studies. Also, modified DNA sequencing with bisulphite treatment, known as bisulphite sequencing, may also be employed to investigate the conformation of the region of interest. These 3 are considerably successful methods. With the advances in technology and reduced costs, methylation-specific DNA sequencing has become a frequently used method to investigate the methylated regions identified by means of these methods. Whether a methylation region affects the expression of the gene of interest is another aspect to take into account as some genes may not yield any products although they are not methylated. In that case, one should consider that the gene in question may be activated by other mechanisms. With a better understanding of such histone and DNA modifications, they now attract attention as therapeutic targets in cancer and various diseases. They have started to create new alternatives especially in cancer treatments. Various computer programs have begun to be developed for methylation analysis. This section discusses all of the aforementioned conditions separately.

1.1. Histone modifications

The most basic unit of the structure called chromatin is nucleosomes. A nucleosome is a unit of 146 bp stretch of DNA over H2A, H2B, H3 and H4 central histone proteins and binding of H1 protein to the structure as a lock. In addition, these constructs provide necessary packaging for the DNA molecule, which is quite large, to fit in a small area (**Figure 1**).

Covalent changes in amino acids are found in tail parts of central histone proteins form the epigenetic code. As a result of these changes, chromosome structure constitutes expression control constructs in DNA by acquiring heterochromatin (expressionally inactive) or by forming regions euchromatin (expressionally active regions). Histone modifications can be classified as acetylation, methylation and phosphorylation. The modifications are mostly visible and reversible at the amino (NH_3-) and carboxyl (COO-) ends of central histone proteins. Each of these

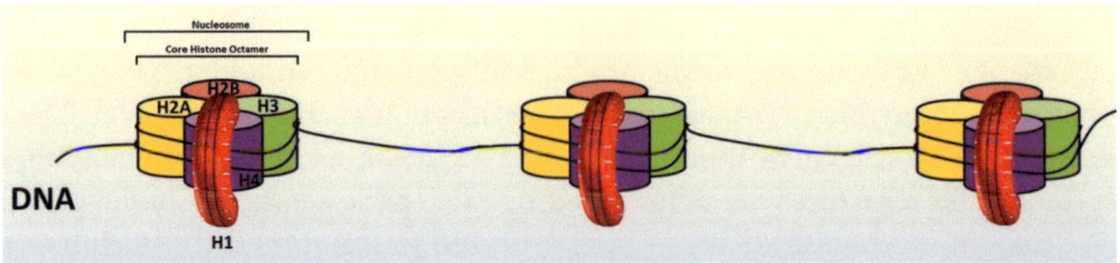

Figure 1. Core histone and nucleosome structure.

http://dx.doi.org/10.5772/intechopen.81739

changes to square may cause differentiation by altering histone function. For example, it can be seen that one, two and three bases of methyl group are added in methylation of arginine [1]. Histone changes are carried out by various enzymes. These include histone acetyl transferases (HATs), histone deacetylases (HDACs) and histone methyl transferases. The equilibrium in activity of these enzymes and their associated proteins is important when they can perform functions of normal cells. These equilibrium distortions can cause problems that can occur from loss of cell function to cancer formation.

1.2. DNA modifications—methylations

The underlying transcriptional silencing mechanism of DNA methylation is based on the overmethylation of cytosine in CpG-rich islands in the promoter region of a gene. This mechanism cooperates with histone deacetylation to suppress the chromatin structure. GC-rich DNA sequences in the human genome are often found in the promoter region and exon 1 of about 50% of all genes [2]. DNA methylation is the main underlying mechanism that regulates gene expression in mammalian cells, as it happens to be one of the major mechanisms for the silencing of genes involved in cell cycle as well as cell growth and death [3].

The most widely studied and the most well-established epigenetic mechanism is DNA methylation. It is an enzymatic change where cytosines are converted to 5′-methylcytosine. The cytosine-end methylation seen in mammalian genome often occurs at the 5′-CpG-3′ dinucleotides, which are also called CpG dinucleotides [4].

Methylation occurs by means of DNA methyltransferase (DNMT) enzymes. The DNMT family consists of four members, namely DNMT1, DNMT2, DNMT3A and DNMT3B. These enzymes are stratified into two groups: those that protect the methylated region and the ones that add new methyl groups. About 70% of all CpG dinucleotides of the human genome are methylated [5]. The remaining are the CpG-rich promoter regions of about 200 base pairs or are the first exons of genes. These regions are also called CpG islands and are found in 60% of all genes [6]. CpG methylation is programmed during the early embryonic period and preserved in later periods. CpG methylation is highly important with regard to normal functions of a given cell, as it affects the regulation of gene expression. For example, DNA methylation plays an important role in gene silencing of the inactive X chromosome as well as the regulation of age-related or tissue-specific gene expression [7].

Although the structural changes that occur in DNA are usually termed as mutations, not every alteration is actually a mutation. A mutation refers to any change at base level such as purine-to-pyrimidine (G-A) or pyrimidine-to-pyrimidine (C-T) changes; single or multiple alterations; insertions, deletions and even single nucleotide polymorphisms (SNP). Yet, SNPs differ from mutations due to their structure. When methylation is compared with other changes in DNA, the methylation process may be considered as another type of mutation, with a change in the structure of the base resulting from a chemical change in DNA. However, mutations are rare changes compared to methylation, and they may or may not be repaired by DNA repair mechanisms [8]. They can be inherited from any ancestor or parent, and they may also occur as germline changes. On the other hand, SNPs can be called DNA alterations, which are more common in the population and which manifest themselves as susceptibility to disease, rather than resulting in a direct disease phenotype. At this point, methylation is not considered as a mutation, despite the fact that it prevents cytosine behaviour by adding a methyl group from CpG dinucleotides to cytosine [9, 10].

The most appropriate means of this option are the CpG sequences within the DNA, which are bound to their conjugates through an enzymatic process in a stronger manner compared to the A-T pairs. This is because ApTs are bound to their complementary pairs in the corresponding chain by means of two hydrogen bonds, while CpGs are bound with three. Such binding characteristics are expected to provide stability to CpGs compared to ApTs. This may explain the greater frequency of methylation in CpGs rather than ApTs in the organism.

Methylation usually occurs through the addition of a methyl group to CpG sequence or to the C base in these CpG islands. Although such a change normally appears as a mutation, it is understood that, unlike mutations, this change is a highly functional mechanism in terms of cellular development and quite common across living organisms from bacteria to highly complex multicellular species. In this way, the organism can adapt to environmental changes by changing the activation of the desired genes in response to external influences when necessary, thereby maintaining vitality and survival. During a methylation reaction, 5-methylcytosine is formed with the addition of a methyl group to the fifth carbon of the cytosine in CpG base pairs by the DNA methyltransferase enzyme (**Figure 1**). Potentially, any CpG base pair or island may undergo methylation. In addition, the fourth nitrogen of cytosine and sixth nitrogen of adenine, which are usually not found in multicellular organisms, may also be methylated in addition to the 5-methylcytosine formation in bacteria [11].

Genomic imprinting is another example of DNA methylation that is involved in single-allele gene expression. Approximately 80 loci are suppressed in this way. The tissue-specific and condition-specific expressions of these genes occur through the regulation of methylation [12]. At the end of 1970s, a decrease in methylcytosine numbers was observed in the genome of tumour cells [13]. This was referred to as hypomethylation of DNA and was demonstrated in benign and malignant tumours [14]. Hypomethylation of DNA may also activate oncogenes. Studies have shown hypomethylation in SI00A4, a metastasis-associated gene in colorectal cancers and the genes, cyclin-D2 and maspin, in gastric carcinomas [14, 15]. Hypomethylation may cause loss of imprinting (LOI), thereby promoting cell proliferation. One of the best examples of this process is the loss of imprinting in the IGF2/H19 region, which is seen in about 40% of colorectal cancers [1].

2. Bacterial epigenetic mechanisms

Modulation of chromosome organisation is one of the host defence mechanisms against bacterial attacks in eukaryotes. The host cell can often resist bacteria through these highly special and successful defence mechanisms. However, bacteria also have mechanisms that are developed against this system. Some bacteria may contain eukaryote-like proteins and eukaryotic histone translation proteins, which target the chromosomal machinery. In this way, they can activate appropriate enzymes in the host [16].

Several bacteria contain an N4-methylcytosine base whose function has not been fully characterised. There are studies indicating that these N4C modifications affect global gene expression in *Helicobacter pylori*, an example of carcinogenic bacteria [17].

Methylation in bacteria is different from eukaryotes in that it is seen in the fourth carbon of the cytosine as well as methylated adenine (N6-methyladenine) in addition to the fourth carbon of cytosine. DNA methylation occurs in bacteria by methyl binding to cytosine C-5 or N-4 and N6-adenine on the DNA methyltransferase enzyme side. N6-methyladenine is found only in bacteria and in less complex eukaryotes, and not in vertebrates [11]. Interestingly, bacteria also contain a restriction modification system that digests DNA methylase to provide protection against foreign DNA. These consist of the restriction enzyme systems called DcM, which recognises the 5-C cytosine, and Dam, which recognises methylated adenine. Of these, the Dam family is the most well-known protein group. The functional domain of Dam is a DNA MTase with an alpha molecule consisting of a polypeptide of 10 amino acids [18].

Similar to eukaryotes, bacteria also have rRNA methylation. The most important aspect of this methylation is that it creates targets for bacterial infections that can cause infection in humans. While promoter methylation is associated with negative expression, this may not always be the case for exon methylation. Still, sometimes exon methylation shows no effect on gene expression. Investigation on genetic mechanisms affecting cardiomyocyte differentiation includes some studies, which show that intragenic methylations create cellular memory through this mechanism, particularly in pluripotent cells [11].

3. Mitochondrial methylation

In all eukaryotic cells, mitochondrion is the most important organelle for cellular energy and the only organelle containing genomic material apart from the nucleus. Owing to its unique and small genome, this organelle exerts certain proteins and RNAs needed for respiratory reactions and cell growth. Together with the nucleus, it is one of the two genetic systems found in the cell. Mitochondrial DNA (mtDNA) has a circular structure and is located inside mitochondrial matrix, bound to the internal membrane. The mtDNA consists of 16,569 base pairs in a loop form, containing a heavy chain (H) and a light chain. This chain structure contains 2 rRNA molecules, 22 tRNA molecules, and 13 genes necessary for oxidative phosphorylation and electron transport (**Figure 2**). A healthy mitochondrion exerts adequate functions by means of certain proteins that are present in the mechanism of oxidative phosphorylation. This genome is about 16.5 kb in humans, and 13 proteins and rRNAs are synthesised from the mitochondrial genome in mammals [19, 20]. Therefore, the slightest change in mitochondrial genome can potentially affect the life of the cell, and thus the organism [21].

As is the case with mutations, methylation is a mechanism that alters the way the genes work together with the diet, drugs and oxidative stress. Methylation profile of human mtDNA starts from the intrauterine period. With the aid of foetal thyroid hormones, mtDNA copy number and mtDNA methylation are regulated by a thyroid-dependent pathway [23]. In addition, mtDNA is also affected by airway pollutants. The elemental carbon present in benzene and exhaust gas in traffic may influence the number of mtDNA copies by means of ribosomal RNA methylation [24].

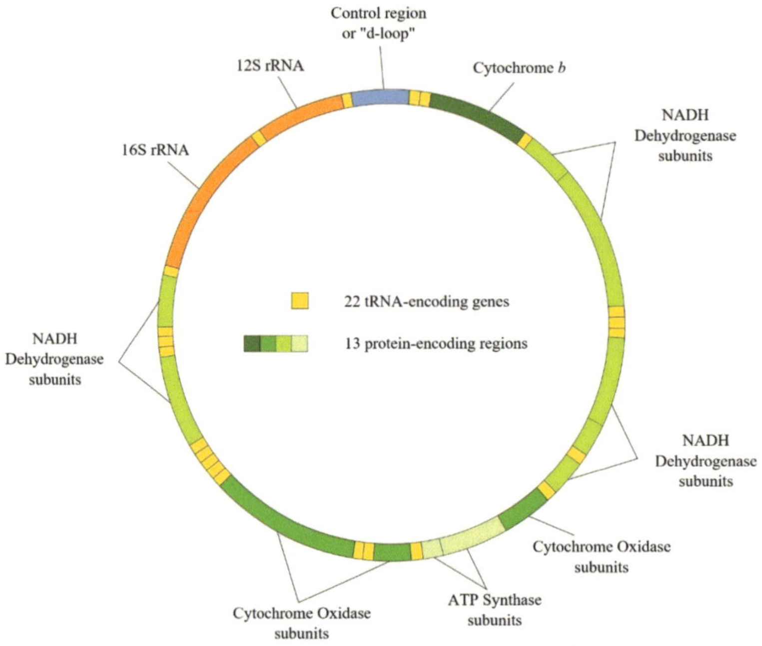

Figure 2. Gene structure of mitochondrial genome [22].

Despite the understanding of these methylation changes in mitochondrial genome, the function of methylated mtDNA has not been fully understood; however, Monique et al. have revealed a different situation. Contrary to what is expected with the methylation of CpGs and GpCs in mtDNA, they have demonstrated that methylation of CpG base pairs had no effect on expression while methylated GpCs were associated with decreased expression [25].

Furthermore, since the mitochondria in humans are entirely of maternal origin, life style of the mother may also have effects at mitochondrion level. Habitual behaviour of the mother, her diet and excessive consumption of fats and sugars trigger obesity, which may affect epigenetics, including that of mitochondria in subsequent generations. A study conducted in mice revealed increased methylation leading to alterations in gene expression and suppression, particularly in the respiratory tract of the offspring of mice that were fed high-fat diets [26]. Furthermore, because the structure of mitochondrion is highly similar to that of bacteria, some genetic factors and structures may also be the same.

4. Detection of the methylation region

Any DNA region containing a CpG sequence may potentially undergo methylation. For this reason, any gene may be subjected to methylation; however, methylation most commonly occurs in the promoter region of genes. That is quite reasonable given the fact that the promoter region is the recognition site for RNA polymerases and therefore of critical importance

http://dx.doi.org/10.5772/intechopen.81739

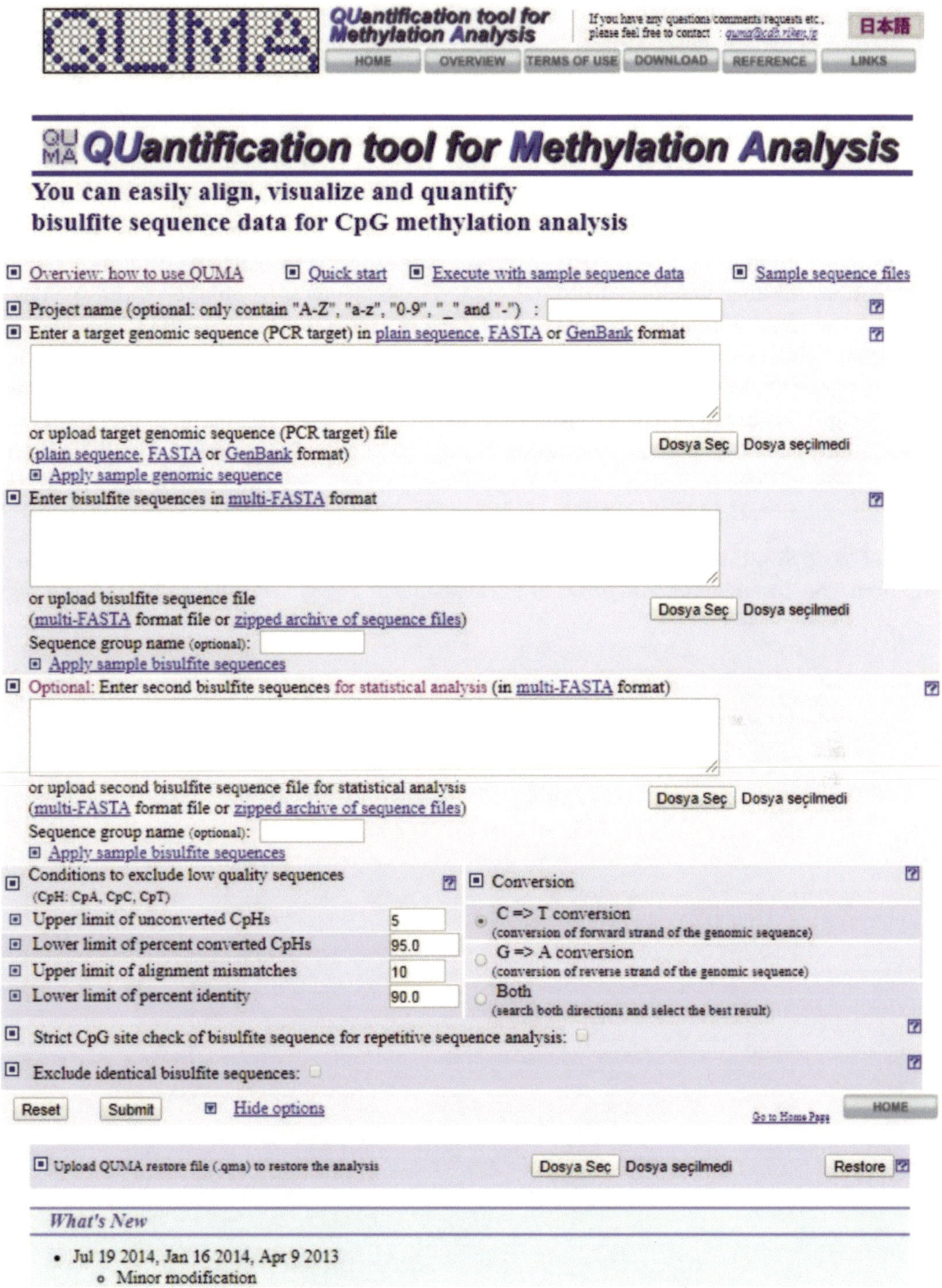

Figure 3. QUMA Web-based program interface.

for gene expression. Although more rarely, methylation may also be observed in exon 1 and other exons of certain genes. One of the ways of finding out whether a gene or DNA region may undergo methylation is to investigate that region with sequence analysis programs specifically developed for this task. There are paid programs by companies such as Fermentas (Methyl Primer Express® Software v1.0) serving this purpose as well as Web-based free access programs such as 'MethPrimer' developed by Li LC and Dahiya R. (available at http://www.urogene.org/methprimer/) [27]. There is also a Web-based program 'DiseaseMeth' (available at http://202.97.205.78/diseasemeth/Analyze.html#form3), which offers researchers who is interested in using a study of which disease or cancer is associated with a desired gene methylation. Another Web-based program is SMS (Sequence Manipulation Suite; available at http://www.bioinformatics.org/sms2/) [28]. One of many useful programs at this address is program which shows CpG islands in a desired DNA region. Another program is a Web-based program QUMA (Quantification tool for Methylation Analysis; available at http://quma.cdb.riken.jp/) (**Figure 3**) [29]. These programs allow methylation region mapping, designing methylation-specific primers, determining bisulphite sequencing primers, identification of CpG islands and determination of DNA sequences that are altered or newly formed due to bisulphite modifications.

In order to do this, the initial step should be obtaining the FASTA sequence of the sequence in question. The Ensembl genome browser 92, available at https://www.ensembl.org/, is a very

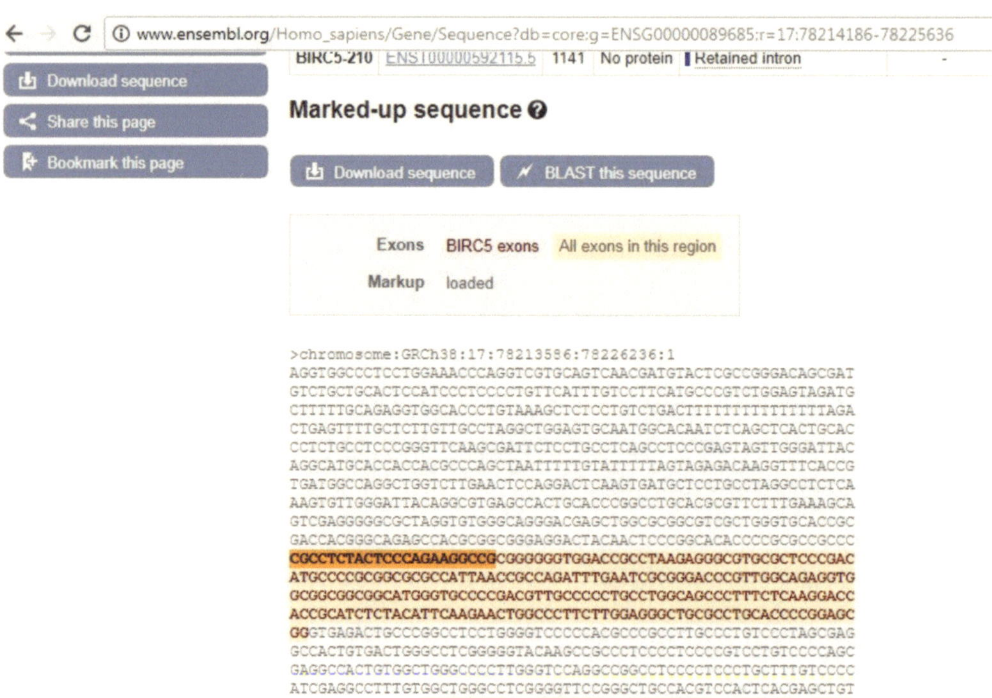

Figure 4. Survivin gene sequence on Ensembl genome browser 92.

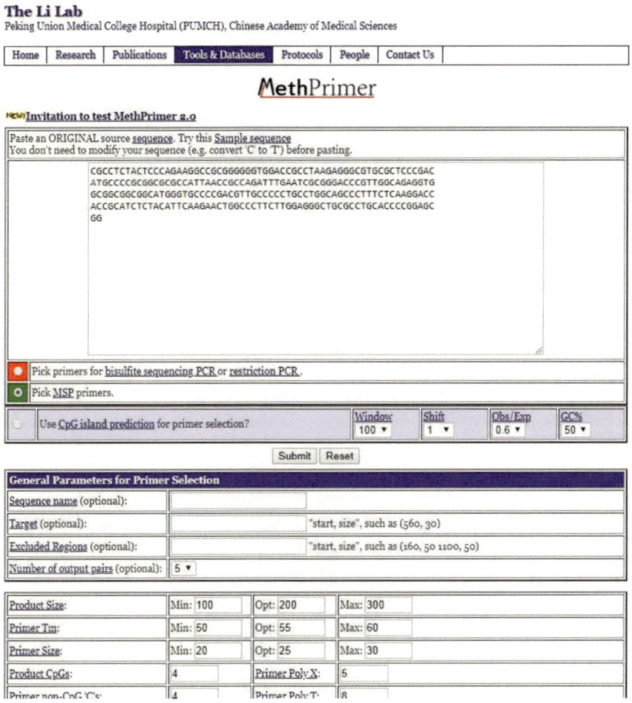

Figure 5. Enter the sequence into Methyl Primer Express. The selection is made for bisulphite sequestration or MSP.

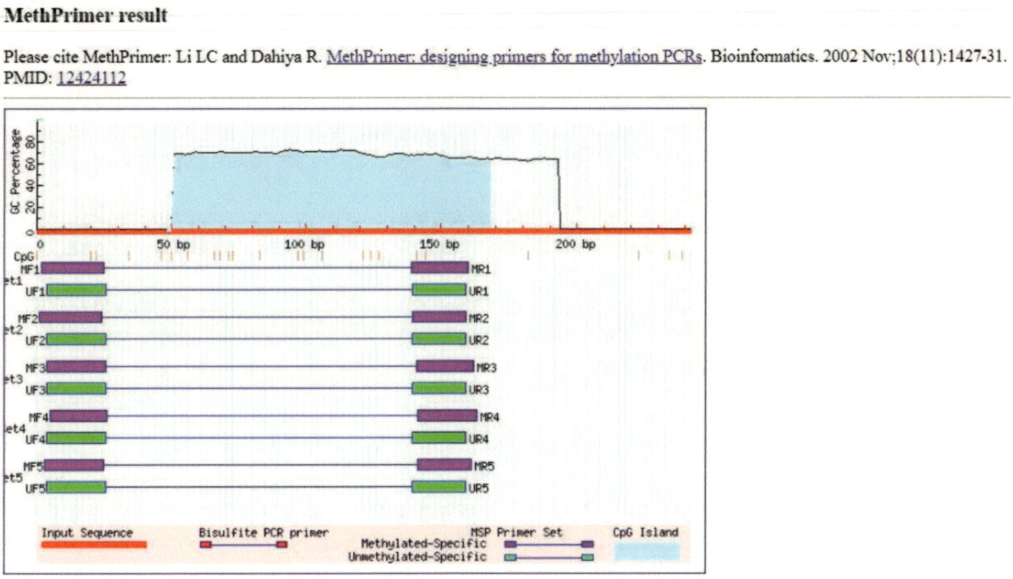

Figure 6. The possible gene region resulting from the analysis.

```
Sequence Name:
Sequence Length: 242

CpG island prediction results
Criteria used: Island size > 100, GC Percent > 50.0, Obs/Exp > 0.60
 1 CpG island(s) were found in your sequence
           Size        (Start - End)
 Island 1  121 bp       (48 - 168)

Primer picking results for methylation specific PCR (MSP)

   Primer             Start Size  Tm      GC%    'C's  Sequence
1  Left  M primer        2   23   58.51   69.57   8   GTTTTTATTTTTAGAAGGTCGCG
   Right M primer      160   21   58.76   76.19   9   CAAACAAAAAACAACGTCGAA
   Product size: 159, Tm: 76.0
   Left  U primer        4   22   53.26   68.18   7   TTTTATTTTTAGAAGGTTGTGG
   Right U primer      159   20   50.93   75.00   9   AAACAAAAAACAACATCAAA
   Product size: 156, Tm: 71.0
2  Left  M primer        1   23   58.51   69.57   8   CGTTTTTATTTTTAGAAGGTCGC
   Right M primer      159   20   55.94   75.00   9   AAACAAAAAACAACGTCGAA
   Product size: 159, Tm: 76.0
   Left  U primer        4   22   53.26   68.18   7   TTTTATTTTTAGAAGGTTGTGG
   Right U primer      159   20   50.93   75.00   9   AAACAAAAAACAACATCAAA
   Product size: 156, Tm: 71.0
3  Left  M primer        4   22   59.65   68.18   7   TTTTATTTTTAGAAGGTCGCGG
   Right M primer      162   21   59.93   76.19   8   ACCAAACAAAAAACAACGTCG
   Product size: 159, Tm: 76.0
   Left  U primer        4   22   53.26   68.18   7   TTTTATTTTTAGAAGGTTGTGG
   Right U primer      159   20   50.93   75.00   9   AAACAAAAAACAACATCAAA
   Product size: 156, Tm: 71.0
4  Left  M primer        5   21   58.45   66.67   6   TTTATTTTTAGAAGGTCGCGG
   Right M primer      163   22   59.94   72.73   8   TACCAAACAAAAAACAACGTCG
   Product size: 159, Tm: 76.0
   Left  U primer        4   22   53.26   68.18   7   TTTTATTTTTAGAAGGTTGTGG
   Right U primer      159   20   50.93   75.00   9   AAACAAAAAACAACATCAAA
   Product size: 156, Tm: 71.0
5  Left  M primer        3   22   57.75   68.18   8   TTTTTATTTTTAGAAGGTCGCG
   Right M primer      161   20   59.10   75.00   7   CCAAACAAAAAACAACGTCG
   Product size: 159, Tm: 76.0
   Left  U primer        4   22   53.26   68.18   7   TTTTATTTTTAGAAGGTTGTGG
   Right U primer      159   20   50.93   75.00   9   AAACAAAAAACAACATCAAA
   Product size: 156, Tm: 71.0
```

Figure 7. Methyl and unmethyl primer sequences and product length for the gene region of interest.

good resource that can be utilised for this step. Ensembl is a database that allows access to DNA sequences in formats such as BLAST and BLAT for comparative genomic studies, evolution studies, sequence variants and transcriptional variants across vertebrate genomes. The relevant DNA sequence obtained from such a database is added to the methylation primer program, the sites of interest are labelled, and the program is run (**Figures 4–8**).

http://dx.doi.org/10.5772/intechopen.81739

```
  1 CGCCTCTACTCCCAGAAGGCCGCGGGGGGTGGACCGCCTAAGAGGGCGTGCGCTCCCGAC
    ++::|:||:|:::||||||:++++|||||||||:++::||||||||++||++:|::++|:
  1 CGTTTTTATTTTTAGAAGGTCGCGGGGGGTGGATCGTTTAAGAGGGCGTGCGTTTTCGAT
    M>>>>>>>>>>>>>>>>>>>>>>>
      U>>>>>>>>>>>>>>>>>>>>>>

 61 ATGCCCCGCGGCGCGCCATTAACCGCCAGATTTGAATCGCGGGACCCGTTGGCAGAGGTG
    |||:::++++|++++::|||||:++::||||||||||++++|||::++||||:|||||||
 61 ATGTTTCGCGGCGCGTTATTAATCGTTAGATTTGAATCGCGGGATTCGTTGGTAGAGGTG

121 GCGGCGGCGGCATGGGTGCCCCGACGTTGCCCCCTGCCTGGCAGCCCTTTCTCAAGGACC
    |++|++|++|:|||||||:::++|++|||:::::||::|||:||:::|||:|:|||||::
121 GCGGCGGCGGTATGGGTGTTTCGACGTTGTTTTTGTTTGGTAGTTTTTTTTAAGGATT
                      M<<<<<<<<<<<<<<<<<<<<<
                      U<<<<<<<<<<<<<<<<<<<<

181 ACCGCATCTCTACATTCAAGAACTGGCCCTTCTTGGAGGGCTGCGCCTGCACCCCGGAGC
    |:++:||:|:||:|||:|||||:|||:::||:||||||||:||++::||:|:::++|||+
181 ATCGTATTTTTATATTTAAGAATTGGTTTTTTTTGGAGGGTTGCGTTTGTATTTCGGAGC

241 GG
    +|
241 GG
     M
```

Figure 8. Upper row: original sequence. Lower row: bisulphite modified sequence. (For display, assume all CpG sites are methylated) ++, CpG sites; ::::, non-CpG 'C' converted to 'T'; M>>>>>>, left methylated-specific primer; M<<<<<<, right methylated-specific primer; U>>>>>>, left unmethylated-specific primer; U<<<<<<, right unmethylated-specific primer.

5. Methylation-specific PCR (MSP)

MSP is an established and the most commonly utilised method to determine the presence or absence of methylation in a gene region of interest as well as the extent of methylation, if any [30, 31].

In this method, DNA is initially subjected to total bisulphite treatment. In this way, all of the unmethylated cytosines in the DNA sequence are transformed into thymine. However, the methylated cytosines remain unchanged (**Figure 5**). This results in a motif change in the methylated region. Subsequently, spectrophotometric DNA quantification is conducted with DNA samples. For this, measurements are made at wavelengths of 260/280 nm and multiplied by the dilution factor to determine the amount of DNA as nanogram per microlitre; and according to these amounts, MS-PCR is performed taking care to include equal amounts of DNA samples in the PCR. In this type of PCR, the PCR is conducted in two separate tubes for each sample. While the primer specific for the methylated region is added into one tube, the other tube contains the primer for the unmethylated region, and PCR is performed as 35–40 cycles. PCR samples are then analysed by visualisation with ethidium bromide agarose gel imaging systems.

5.1. Sample reaction: (for the prestin gene promoter region in guinea pigs)

Methylation-specific polymerase chain reaction: DNA purity was measured at wavelengths of 260 and 280 nm on spectrophotometry, and DNA quantification was performed using the DNA (μg/mL) = A260 × Dilution Factor × 50 (coefficient) formula at 260 nm UV. Subsequently, 10 μL of DNA was taken from each sample and bisulphite modification was carried out for DNA with Millipore CpGenome modification kit according to the manufacturer's instructions. This modification converts the cytosines of the unmethylated region to thymines. For the region thought to be altered in this manner, CpG sequences in exon 1 of the prestin gene were detected using the MethPrimer V1.1 beta program [30]. MSP was conducted according to the following protocols to investigate methylation utilising the PCR primers for exon 1 of the prestin gene stated below, and PCR conditions for the methylated and unmethylated regions are as follows:

Methylated region (M: methylated):

M-Forward: ATGTTGAAGAAAATGAAATTTTCGT,

M-Reverse: ACTTATCCCCGATAAAATCCG,

PCR product: 164 base pairs (bp).

Unmethylated region (UnM: unmethylated):

UnM-Forward: TTTATTTTTAGAAGGTTGTGG,

UnM-Reverse: AAACTACCAAACAAAAAACAACATC,

PCR product: 163 bp.

PCR conditions:

PCR buffer 1×, MgCl: 2 2.5 mM, DMSO: 5% (v/v), dNTP: 12.5 mM, Primer Forward: 10 pmol, Primer Reverse: 10 pmol, Taq Polymerase: 1 U (5 U/μL), Template DNA: 100 ng and dH2O were used to obtain a total of 50 μL, and PCR thermal cycling procedure was as follows:

95°C for 10′ (Initiation),

95°C for 45″,
59°C for 30″, } 35 Cycles
72°C for 30″,

72°C for 10″ (Termination).

This investigation allows determining whether the region of interest is methylated and quantifying methylation by measuring the band intensity with any gel analysis system. If desired, results may also be obtained while real-time PCR is performed with SYBR-green or Taqman or similar fluorescent probes (the probe must be designed according to the bisulphite DNA sequence).

5.2. Evaluation of agarose gel imaging

The resulting primers were stained with ethidium bromide on 2% agarose gel, and agarose gel findings were evaluated by examination under ultraviolet light (**Figures 7–10**).

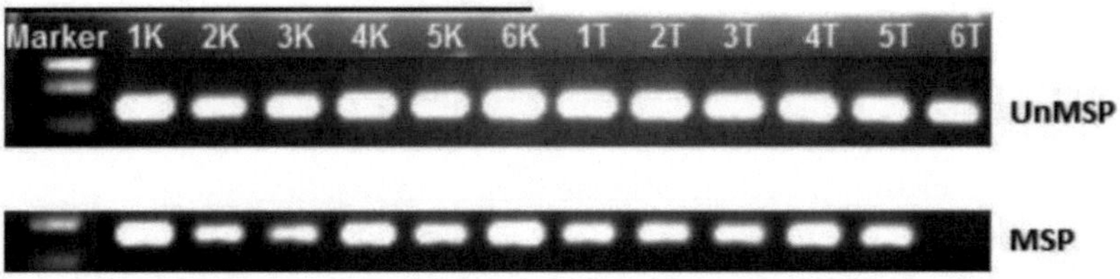

Figure 9. Results of polymerase chain reaction in all study groups at the time point of second week for prestin (K1–K3: Group 1; K3–K6: Group 2; T1–T6: Group 3). Complete methylation is seen in sample 6T, while heterozygous methylation is observed in other samples [32].

Example run:

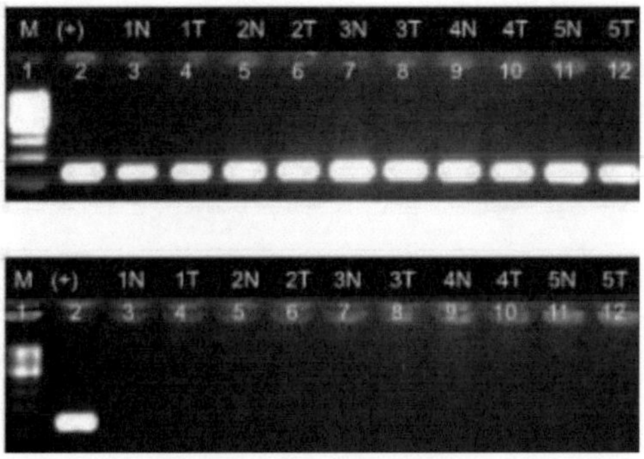

Figure 10. Imaging of MSP and un-MSP PCRs of survivin exon 1. Well 1: marker, 100 bp marker; well 2: positive control; 3–4, 5–6, 7–8, 9–10, 11–12, while unmethylated PCR results are found to be at the same (+) intensity in MSP and un-MSP sample wells of the same cases, the following methylated PCR results were not observed except the control DNA (+), and all samples were accepted as unmethylated. Well 13 is the negative control of the reaction (**Figure 10**).

5.3. MS RFLP

Another method used to detect any DNA methylation is the methylation-specific restriction fragment length polymorphism (MS-RFLP) method, which produces methylation-specific digestion. This method employs restriction enzymes obtained from bacteria, which recognise

5′	C↓	C	G	G	3′
3′	G	G	C↑	C	5′

and cut only specific methylated regions. The most commonly used restriction enzymes for this purpose are HpaII and MspI.

This process digests the methylated CpG (if present) in the DNA region of interest. If that region is not methylated, digestion occurs via HpaII, an isomer of MspI. Concurrent use of these two enzymes provides insight on the methylation status of the region in question.

6. COBRA (combined bisulphite restriction analysis)

COBRA is a method developed by combined use of bisulphite modification and RFLP methods. In this combined method, DNA is first differentiated in the methylation-dependent DNA sequence with sodium bisulphite. As mentioned earlier, in practice, methylase cytosines are not affected, while unmethylated cytosines are converted into uracil. PCR is performed with primers designed specifically for these new DNA sequences obtained by bisulphite method. Unlike MS-PCR, primers used in this PCR step should not contain CpG sequences. After this step, digestion step of restriction comes. At this step, PCR products are treated with two restriction enzymes with TaqI (TCGA) and BstUI (CGCG). These enzymes form a methylation profile by cutting off DNA fragments to properties of whether residues of cytosine are methylated or not (**Figure 11**).

Methylated regions in DNA fragments digested BstUI and unmethylated homologue of the same region digested TaqI enzyme. These new fragments, which are formed according to methylation state, can be calculated as percentages of methylation rate for region investigated according to patterns and density of band formed by conducting the polyacrylamide gel electrophoresis. It should be noted here that band densities should be determined by a photo

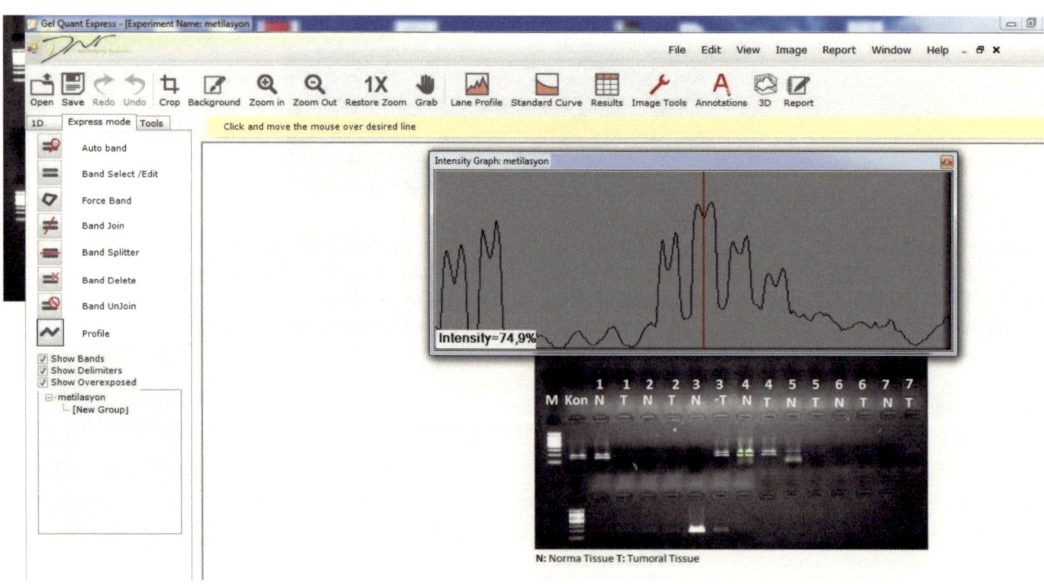

Figure 11. Calculation of band intensity in survivin exon 1 with 'Gel Quant Express'.

http://dx.doi.org/10.5772/intechopen.81739

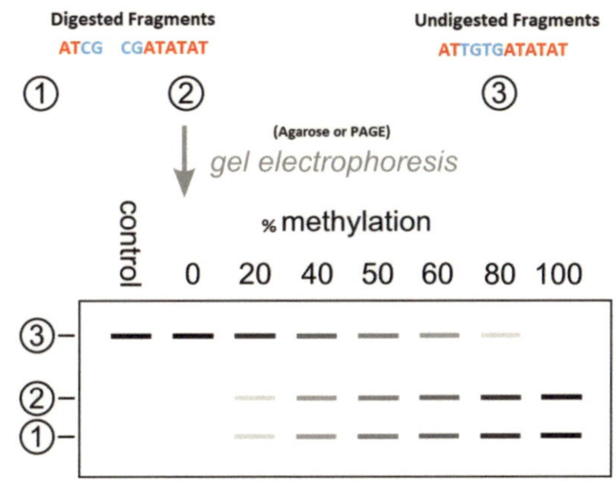

Figure 12. The image of COBRA method and calculation step.

image analysis program (e.g., **Figure 12**). Once these values have been obtained, percentage of methylation can be calculated by substituting the following formula [33, 34] (**Figure 13**).

6.1. Advantages

It is simple, cheap and fast. DNA methylation levels can be shown without needing for any extra bisulphite sequencing.

Due to its high specificity, it is very successful even in DNA material, which is obtained from paraffin blocks.

It is quite quantitative compared to methylation-specific PCR, which is a qualitative method.

In MS-PCR, only locus-specific methylation information can be obtained, whereas in this method, entire region within locus is examined.

6.2. Disadvantages

This method is limited by existing restriction regions in region being investigated.

In addition, incomplete digestions of restriction enzymes can have misleading results for the amount of methylation.

Due to cell-type heterogeneity in different cell complexes, methylated CG sequence may be transformed into other sequences, such as CA or CT, leading to a change in restriction sites.

Considering all these advantages and disadvantages, the COBRA method emerges as an effective method for determining a highly effective level of methylation.

$$\%\ \text{Methylation} = 100 \times \frac{\text{Digested Fragments}}{\text{Digested Fragments} + \text{Undigested Fragments}}$$

Figure 13. Mathematical calculation of percentage methylation rates in COBRA method.

7. Conclusion

Many studies after the first discovery of epigenetic changes have shown that epigenetic modifications are quite important in natural flow of life and that many genes are mechanisms used for expression and inactivation when needed.

Furthermore, as these mechanisms are understood better, they have been associated with many pathological conditions, from cancer to mental retardations such as fragile X and Prader-Willi Angelman, to chromosomal instability. In this regard, such diseases have emerged in new therapeutic targets. Chemical agents such as 5-azacytidine and 5-aza-2′-deoxycytidine, which inhibit methylation by binding to and inhibiting DNMT enzyme, are now being tested in phase II-III studies [35–38]. Furthermore, use of oligonucleotides that bind to promoter regions at specific gene level and perform gene inhibition is seen as approaches that may contribute to cancer treatments [39–41].

In addition, histone acetyltransferase inhibitors, which inhibit formation of epigenetic modifications at histone level, have emerged as novel cancer treatment agents. For example, H3-H4 of a soy protein Lunasin has been found to exhibit anticancer properties in mammals by suppressing histone acetylation. Wenyi et al. have shown that YEAST domain, an acetyl lysine-binding module, is effective in the development of cancer, and this domain appears to be the target for anticancer therapies. It seems that such approaches in the future will start to give more successful results [42].

In light of the information presented above, one may conclude that methylation is a highly important genomic mechanism for the cell from unicellular organisms to multicellular organisms. This mechanism is seen in bacteria, mitochondria and all eukaryotic cells in proportion to the complexity and development level of the organism. The identification of methylated regions is as important as the methylation process itself, as this may allow identifying several potential novel targets related to subject matters such as the development mechanism of diseases, certain roles in cancer development and bacterial resistance.

Conflict of interest

There is no conflict of interest.

Author details

Metin Budak

Address all correspondence to: genomicdna2@yahoo.com

Biophysics Department, Faculty of Medicine, Trakya University, Edirne, Turkey

References

[1] Wong JJL, Hawkins NJ, Ward RL. Colorectal cancer: A model for epigenetic tumorigenesis. Gut. 2007;**56**(1):140-148

[2] Roman-Gomez J, Jimenez-Velasco A, Agirre X, Prosper F, Heiniger A, Torres A. Lack of CpG island methylator phenotype defines a clinical subtype of T-cell acute lymphoblastic leukemia associated with good prognosis. Journal of Clinical Oncology. 2005;**23**(28): 7043-7049

[3] Chim C, Pang R, Liang R. Epigenetic dysregulation of the Wnt signalling pathway in chronic lymphocytic leukaemia. Journal of Clinical Pathology. 2008;**61**(11):1214-1219

[4] Ramsahoye BH, Biniszkiewicz D, Lyko F, Clark V, Bird AP, Jaenisch R. Non-CpG methylation is prevalent in embryonic stem cells and may be mediated by DNA methyltransferase 3a. Proceedings of the National Academy of Sciences. 2000;**97**(10):5237-5242

[5] Ehrlich M, Gama-Sosa MA, Huang L-H, Midgett RM, Kuo KC, McCune RA, et al. Amount and distribution of 5-methylcytosine in human DNA from different types of tissues or cells. Nucleic Acids Research. 1982;**10**(8):2709-2721

[6] Antequera F, Bird A. Number of CpG islands and genes in human and mouse. Proceedings of the National Academy of Sciences. 1993;**90**(24):11995-11999

[7] Plass C, Soloway PD. DNA methylation, imprinting and cancer. European Journal of Human Genetics. 2002;**10**(1):6

[8] Cinemre F, Serinkan B, Cinemre H, Yucel A, Degirmencioglu S, Tuten A, et al. The role of selenoprotein P and selenium in the etiopathogenesis of gestational diabetes mellitus: Association with selenoprotein P1 gene (rs3877899) polymorphism. Trace Elements and Electrolytes. 2018;**35**(4):174-182

[9] Karki R, Pandya D, Elston RC, Ferlini C. Defining "mutation" and "polymorphism" in the era of personal genomics. BMC Medical Genomics. 2015;**8**(1):37

[10] Nei M. The new mutation theory of phenotypic evolution. Proceedings of the National Academy of Sciences. 2007;**104**(30):12235-12242

[11] Sánchez-Romero MA, Cota I, Casadesús J. DNA methylation in bacteria: From the methyl group to the methylome. Current Opinion in Microbiology. 2015;**25**:9-16

[12] Robertson KD. DNA methylation and human disease. Nature Reviews Genetics. 2005; **6**(8):597

[13] Feinberg AP, Vogelstein B. Hypomethylation distinguishes genes of some human cancers from their normal counterparts. Nature. 1983;**301**(5895):89

[14] Oshimo Y, Nakayama H, Ito R, Kitadai Y, Yoshida K, Chayama K, et al. Promoter methylation of cyclin D2 gene in gastric carcinoma. International Journal of Oncology. 2003;**23**(6): 1663-1670

[15] Akiyama Y, Maesawa C, Ogasawara S, Terashima M, Masuda T. Cell-type-specific repression of the maspin gene is disrupted frequently by demethylation at the promoter region in gastric intestinal metaplasia and cancer cells. The American Journal of Pathology. 2003;**163**(5):1911-1919

[16] Rolando M, Gomez-Valero L, Buchrieser C. Bacterial remodelling of the host epigenome: Functional role and evolution of effectors methylating host histones. Cellular Microbiology. 2015;**17**(8):1098-1107

[17] Kumar S, Karmakar BC, Nagarajan D, Mukhopadhyay AK, Morgan RD, Rao DN. N4-cytosine DNA methylation regulates transcription and pathogenesis in Helicobacter pylori. Nucleic Acids Research. 2018;**46**(7):3429-3445

[18] Low DA, Weyand NJ, Mahan MJ. Roles of DNA adenine methylation in regulating bacterial gene expression and virulence. Infection and Immunity. 2001;**69**(12):7197-7204

[19] Anderson S, Bankier AT, Barrell BG, de Bruijn MH, Coulson AR, Drouin J, et al. Sequence and organization of the human mitochondrial genome. Nature. 1981;**290**(5806):457

[20] Bogenhagen DF. Biochemical isolation of mtDNA nucleoids from animal cells. Mitochondrial DNA. 2009:3-14. Springer

[21] Sun H, Shi W, Wang X. How far can mitochondrial DNA drive the disease? Mitochondrial DNA and Diseases. 2017:1-8. Springer

[22] Falkenberg M, Larsson N-G, Gustafsson CM. DNA replication and transcription in mammalian mitochondria. Annual Review of Biochemistry. 2007;**76**:679-699

[23] Janssen BG, Byun H-M, Roels HA, Gyselaers W, Penders J, Baccarelli AA, et al. Regulating role of fetal thyroid hormones on placental mitochondrial DNA methylation: Epidemiological evidence from the ENVIRONAGE birth cohort study. Clinical Epigenetics. 2017;**9**(1):66

[24] Byun H-M, Panni T, Motta V, Hou L, Nordio F, Apostoli P, et al. Effects of airborne pollutants on mitochondrial DNA methylation. Particle and Fibre Toxicology. 2013;**10**(1):18

[25] van der Wijst MG, van Tilburg AY, Ruiters MH, Rots MG. Experimental mitochondria-targeted DNA methylation identifies GpC methylation, not CpG methylation, as potential regulator of mitochondrial gene expression. Scientific Reports. 2017;**7**(1):177

[26] Keleher MR, Zaidi R, Shah S, Oakley ME, Pavlatos C, El Idrissi S, et al. Maternal high-fat diet associated with altered gene expression, DNA methylation, and obesity risk in mouse offspring. PLoS One. 2018;**13**(2):e0192606

[27] Li L-C, Dahiya R. MethPrimer: Designing primers for methylation PCRs. Bioinformatics. 2002;**18**(11):1427-1431

[28] Stothard P. The sequence manipulation suite: JavaScript programs for analyzing and formatting protein and DNA sequences. 2000

[29] Kumaki Y, Oda M, Okano M. QUMA: Quantification tool for methylation analysis. Nucleic Acids Research. 2008;**36**(suppl_2):W170-W175

[30] Yalcin O, Budak M. Un-methylation of the survivin gene has no effect on immunohistochemical expression of survivin protein in lung cancer patients with squamous cell carcinoma. Bratislavské Lekárske Listy. 2017;**118**(3):160-163

[31] Herman JG, Graff JR, Myöhänen S, Nelkin BD, Baylin SB. Methylation-specific PCR: A novel PCR assay for methylation status of CpG islands. Proceedings of the National Academy of Sciences. 1996;**93**(18):9821-9826

[32] Bulut E, Budak M, Öztürk L, Türkmen MT, Uzun C, Sipahi T. DNA methylation of the prestin gene and outer hair cell electromotileresponse of the cochlea in salicylate administration. Turkish Journal of Medical Sciences. 2017;**47**(5):1626-1633

[33] Bilichak A, Kovalchuk I. The combined bisulfite restriction analysis (COBRA) assay for the analysis of locus-specific changes in methylation patterns. Plant Epigenetics. 2017:63-71. Springer

[34] Varinli H, Statham AL, Clark SJ, Molloy PL, Ross JP. COBRA-Seq: Sensitive and quantitative methylome profiling. Genes. 2015;**6**(4):1140-1163

[35] Bauman J, Verschraegen C, Belinsky S, Muller C, Rutledge T, Fekrazad M, et al. A phase I study of 5-azacytidine and erlotinib in advanced solid tumor malignancies. Cancer Chemo therapy and Pharmacology. 2012;**69**(2):547-554

[36] Lin J, Gilbert J, Rudek MA, Zwiebel JA, Gore S, Jiemjit A, et al. A phase I dose-finding study of 5-azacytidine in combination with sodium phenylbutyrate in patients with refractory solid tumors. Clinical Cancer Research. 2009;**15**(19):6241-6249

[37] Levis M, Faderl SH, Garcia-Manero G, Jabbour E, Daver NG, Dellasala SE, et al. Final report of phase II study of sorafenib and 5-azacytidine in patients with relapsed or untreated acute myeloid leukemia and FLT3-ITD mutation. American Society of Hematology. 2013;**122**(21):3934

[38] Kuendgen A, Müller-Thomas C, Lauseker M, Haferlach T, Urbaniak P, Schroeder T, et al. Efficacy of azacitidine is independent of molecular and clinical characteristics-an analysis of 128 patients with myelodysplastic syndromes or acute myeloid leukemia and a review of the literature. Oncotarget. 2018;**9**(45):27882

[39] Urdinguio RG, Sanchez-Mut JV, Esteller M. Epigenetic mechanisms in neurological diseases: Genes, syndromes, and therapies. The Lancet Neurology. 2009;**8**(11):1056-1072

[40] Hua C-D, Bian E-B, Chen E-F, Yang Z-H, Tang F, Wang H-L, et al. Repression of Dok7 expression mediated by DNMT1 promotes glioma cells proliferation. Biomedicine & Pharmacotherapy. 2018;**106**:678-685

[41] Chen M, Nie J, Liu Y, Li X, Zhang Y, Brock MV, et al. Phase Ib/II study of safety and efficacy of low-dose decitabine-primed chemoimmunotherapy in patients with drug-resistant relapsed/refractory alimentary tract cancer. International Journal of Cancer. 2018

[42] de Lumen BO. Lunasin: A cancer-preventive soy peptide. Nutrition Reviews. 2005;**63**(1): 16-21

Recognition of Nucleosomes by Chromatin Factors: Lessons from Data-Driven Docking-Based Structures of Nucleosome-Protein Complexes

Velten Horn and Hugo van Ingen

Additional information is available at the end of the chapter

http://dx.doi.org/10.5772/intechopen.81016

Abstract

The function of chromatin ultimately depends on the many chromatin-associated proteins and protein complexes that regulate all DNA-templated processes such as transcription, repair and replication. As the molecular docking platform for these proteins, the nucleosome is the essential gatekeeper to the genome. As such, the nucleosome-binding activity of a myriad of proteins is essential for a healthy cell. Here, we review the molecular basis of nucleosome-protein interactions and classify the different binding modes available. The structural data needed for such studies not only come from traditional sources such as X-Ray crystallography but also increasingly from other sources. In particular, we highlight how partial interaction data, derived from for example NMR or mutagenesis, are used in data-driven docking to drive the modeling of the complex into an atomistic structure. This approach has opened up detailed insights for several nucleosome-protein complexes that were intractable or recalcitrant to traditional methods. These structures guide the formation of new hypotheses and advance our understanding of chromatin function at the molecular level.

Keywords: nucleosome, protein interactions, epigenetics, chromatin binding, acidic patch, histone tails, post-translational modifications, data-driven docking, NMR spectroscopy, XL-MS, crystallography, cryo-EM, structural models

1. Introduction

The packaging of DNA into chromatin represents one of the most fundamental layers of the biology of the cell. It provides the required structural compaction of DNA to fit in the nucleus and plays crucial roles in controlling cell fate and protecting genome integrity. The fundamental unit of chromatin is the nucleosome in which 147 base pairs (bp) of DNA are wrapped around an octameric protein complex composed of two copies of histone proteins H2A, H2B, H3 and H4 [1–3]. Nucleosomes are arranged as beads-on-a-string forming 10 nanometer (nm) wide fiber that subsequently condense into higher order structures [4]. Nucleosomes as the basis of chromatin are responsible for its dynamics. Chromatin state and changes in DNA accessibility are determined at the nucleosome level. These changes are mediated through interactions of histone proteins and nucleosomal DNA alike with a wide range of protein complexes that control the structure of chromatin. They interpret, write and erase post-translational modifications or act as ATP-dependent nucleosome remodelers. This allows changes in the functional state of chromatin and regulation of DNA-templated processes. While promoting a large variety of effects on chromatin structure, nucleosome-interacting proteins share the molecular basis of recognizing and binding the nucleosome. Understanding the basis of chromatin dynamics therefore demands understanding the molecular basis of nucleosome-protein interactions.

In particular, insights into the molecular mechanistic basis of how histone-modifying enzymes install or remove post-translational modifications (writers and erasers, respectively) and how these modifications are recognized by effector proteins (readers) are of immense interest, especially in drug development. Deregulation of these proteins is strongly connected to pathological outcome, including cardiovascular diseases, neurological disorders, metabolic disorders and cancer [5]. So-called epigenetic drugs that target the nucleosome interaction of these chromatin factors offer new therapeutic potential [6–9]. A selection of epigenetic drugs including those currently undergoing clinical trial is described in detail elsewhere [10]. Advancement in their development requires insights into the underlying molecular mechanism of nucleosome recognition, enabling control over subsequent modification of the chromatin state.

In the following, we will review the molecular basis of nucleosome-protein interactions, focusing on the different binding epitopes presented by the nucleosome. After an overview of the nucleosome-protein structures determined by crystallography or cryo-electron microscopy (cryo-EM), we highlight several studies in which experimental data from nuclear magnetic resonance spectroscopy (NMR), cross-link-based mass spectrometry (XL-MS) or mutational analysis were used to build atomistic structural models of nucleosome complexes. Throughout, we emphasize the role of these data-driven models in deepening our understanding of nucleosome recognition.

2. Nucleosome-binding epitopes

Consisting of DNA and histone proteins, the nucleosome offers a selection of distinct interaction surfaces for binding of effector proteins with high levels of specificity (**Figure 1**).

Histone proteins possess a globular tertiary structure with exposed, disordered N-terminal tails. Histone tails are known to carry a wide range of covalent, post-translational side chain

http://dx.doi.org/10.5772/intechopen.81016

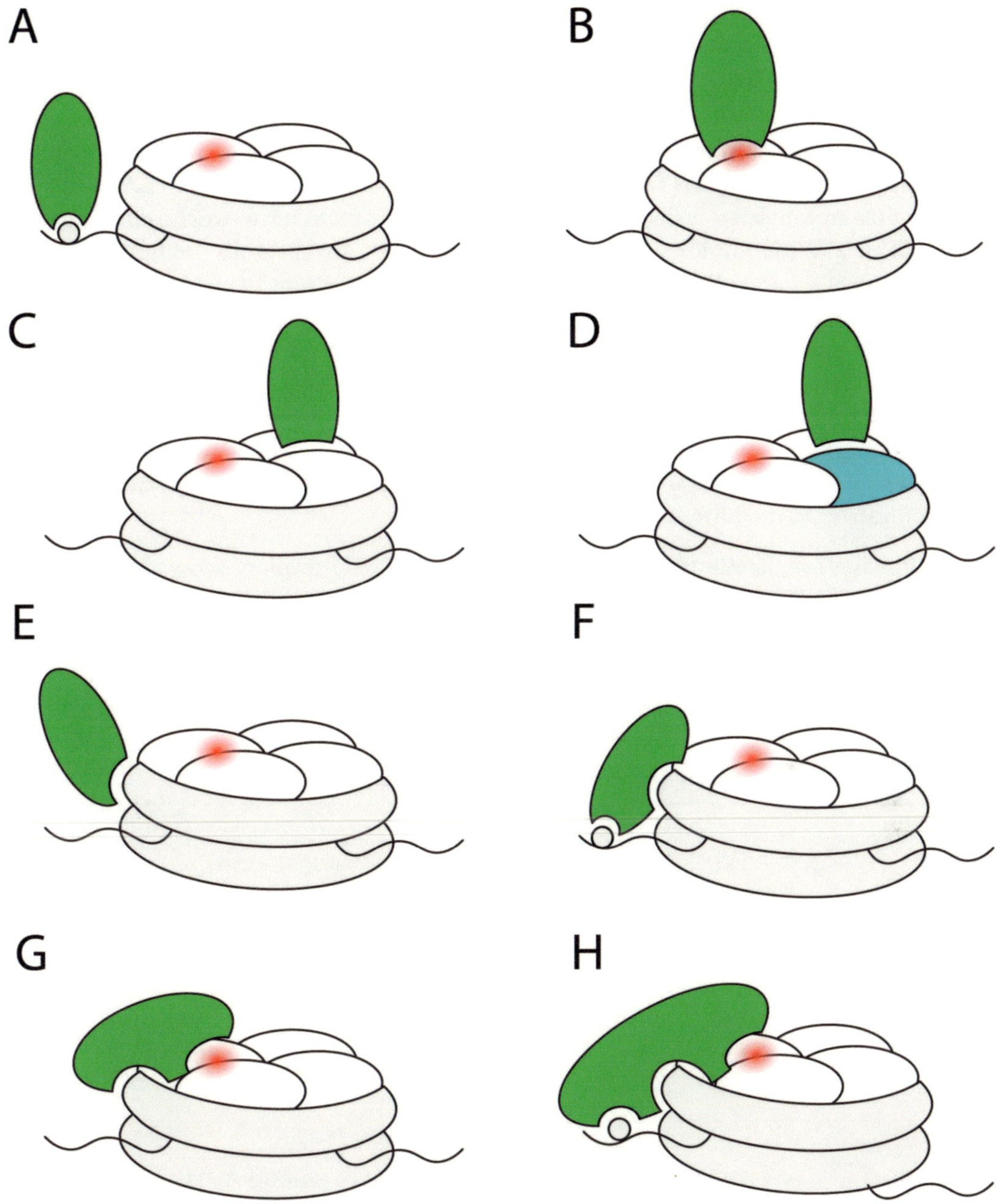

Figure 1. A schematic depiction of different modes of nucleosome recognition. Reported types of epitopes are histone tails including PTMs (A), the H2A-H2B acidic patch (B), the canonical histone surface (C), specific surface motifs formed by histone variants (D) or nucleosomal DNA (E). A manifold of synergetic combinations of binding epitopes are known, such as histone mark and DNA (F), acidic patch and DNA (G) or all three epitopes (H).

modifications (PTMs) such as, mono-, di- and trimethylation (Lys, Arg); acetylation (Lys); phosphorylation (Ser, Thr) and ubiquitination (Lys) [11, 12]. This cosmos of modifications maintains a dynamic nature through the reversibility of the covalent modifications. Modified histones are recognized by so-called reader protein domains specific for the respective modification (**Figure 1A**). Interestingly, nucleosome-interacting proteins can possess more than one reader domain which allows cross talk between different post-translational modifications. Examples of PTM reader domains are Chromo, Tudor, PHD and MBT domains for methylated lysine residues, bromodomains for acetylated lysine residues and 14–3-3 proteins for phosphorylated serine [11, 13] (**Table 1**). The most recent addition to the list is YEATS domains that recognize crotonylated lysine [14–16]. Reader domains often have structurally conserved motifs that are able to complex a specific modification. The "Royal Family" of reader domains is in this respect a particularly instructive example. This superfamily includes the Chromo, MBT, PWWP and plant Agenet domains that bind methylated lysine (Tudor, Chromo, MBT, PWWP, plant Agenet) or arginine (Tudor) residues. Most domains of this family contain a barrel-shaped structure formed by 3–5 antiparallel β-strands that holds a cluster of aromatic residues that form the so-called aromatic cage [17]. The aromatic cage presents an electron-rich yet hydrophobic surface that is ideally suited to bind methylated lysines through cation-π interactions [18]. The structural features and similarities, as well as their substrate specificity, have been subject to literature reviews [19–21].

Reader domains can, in addition to the post translational modification, show specificity for a defined amino acid sequence motif around the epigenetic mark that supports complex formation. For example, the WD40 domain of the EED (embryonic ectoderm development) protein selectively reads out trimethylated lysine in a A-R-K-S sequence motif (as for H3K27me3) but not in a R-T-K-Q motif (as for H3K4me3) [37].

(sub)Domain	Modification	Protein	Function
Royal family			
Tudor	Kme1, Kme2, Kme3, Rme2	53BP1	DNA damage response [24]
		TDRD3	Transcription activation [25]
MBT	Kme1, Kme2	L3MBTL1	Transcriptional repression [26, 27]
PWWP	Kme3	PSIP1	Transcriptional co-activation, DNA repair [28, 29]
Chromo	Kme, Kme2, Kme3	CHD1	Chromatin remodeling [30, 31]
		HP1	Heterochromatin [32]
		MRG15	Splicing [33]
Plant Agenet	Kme, Kme2, Kme3	FMRP	DNA damage response [34]
Bromodomain			
	KAc	BRD2/3	Transcriptional regulation [35]
14-3-3			
	Sph	14–3-3ζ	Transcriptional activation [36]

[a]See Refs. [21–23] for more in-depth discussion.

Table 1. Overview of selected reader domains for post-translational modifications[a].

Next to histone tails, the nucleosome also possesses intrinsic docking platforms on its histone surface. The most prominent of these is composed of histones H2A and H2B. While the histone octamer is overall highly positively charged, there is a patch on the H2A-H2B dimer surface formed by acidic residues with negative surface charge. This structural feature is named the acidic patch and engages in a manifold of interactions with specific binding domains (**Figure 1**), including the tail of histone H4 of adjacent nucleosomes that promotes chromatin compaction. A common feature observed for acidic patch-interacting proteins is a positively charged arginine residue that interacts with a triad of acidic residues on H2A (Glu61, Asp90, Glu92). This is referred to as the arginine anchor [38]. It is often supported by surrounding positively charged residues interacting with acidic H2A/H2B interface residues.

Other parts of the histone core surface may also mediate protein-nucleosome interactions (**Figure 1C**). First, a solvent exposed cleft between H4 and H2B was shown to be involved in binding interactions with Sir3 or 53BP1 [39, 40]. Interestingly, these proteins bind simultaneously to both the H4-H2B cleft and the acidic patch using one nucleosome-binding domain for each epitope. Second, incorporation of non-canonical histones in nucleosomes introduces specific interaction surfaces that allow histone variant-specific nucleosome binding (**Figure 1D**). An example hereof are CENP-N and CENP-C that recognize the incorporated histone H3 variant CENP-A [41, 42].

Finally, the nucleosomal DNA is a major protein interaction site. First, it forms the binding site of linker histone H1 [43–45] (see also Section 4.9). Second, it is often involved in additional synergistic interactions to nucleosome-binding domains (**Figure 1E**). Finally, recent studies have identified transcription factor proteins that primarily bind to nucleosomal DNA. These so-called pioneer factors bind their DNA target sites while embedded in the nucleosome [46–48]. The structural details of these are however still lacking.

Throughout the advances in studies on nucleosome binding, it has become clear that binding of effector proteins in many cases involves interactions of nucleosome-binding domains to multiple nucleosome epitopes (**Figure 1G, H**). However, due to their size and complexity as well as the stability and dynamics of complex formation, the nucleosome is a challenging system for structural biology.

3. Crystal clear: lessons from crystallography and single particles

A key role in the research of protein interactions are high-resolution three-dimensional structures of the complexes, typically obtained by crystallography and, increasingly, cryo-electron microscopy. These structures enable the identification of binding sites and intermolecular interactions, offering a guided approach to design binding-deficient mutants or competitive binders. The history of nucleosome structural biology peaked with the publication of the high-resolution crystal structure of the nucleosome in 1997 [3]. Luger *et al.* achieved crystallization of the nucleosome together with a palindromic version of human α-satellite DNA [49]. This milestone study provided the foundation to also study the structures of nucleosomes together with chromatin factors in complexes. **Table 2** lists the structures of nucleosome-protein complexes solved to date by crystallography and cryo-electron microscopy [39, 50–58]. The most recent addition to this ever-growing list is the spectacular structures of the complex between the INO80 chromatin remodeler and the nucleosome [59, 60]. Below, we discuss a few cardinal studies to highlight the different nucleosomal binding modes of effector proteins.

3.1. The first crystal structure of a nucleosome complex (LANA)

The first high-resolution structure of a nucleosome-protein complex was the crystal structure of a peptide model of Kaposi's sarcoma-associated herpesvirus LANA N-terminal region bound to the nucleosome [61]. The binding site identified in this study was the acidic patch. The atomistic resolution allowed to identify intermolecular side chain interactions including the arginine anchor bound to the acidic triad. Ever since, the LANA-nucleosome has become a golden standard for comparisons with other acidic patch interactions [50, 55]. Importantly, LANA is used to investigate the acidic patch binding ability of other proteins by competitive binding [62–64]. Interestingly, this exact epitope happened to be the binding interface also for the first full protein domain that was crystalized in its nucleosome-bound state.

Name	PDB-id	Role	Year	Technique	Reference	Resolution [Å]
Proteins						
RCC1	3MVD	Ran recruitment	2010	X-Ray	[50]	2.9
Sir3 BAH	4JJN, 3TU4, 4LD9, 4KUD	Chromatin compaction	2011, 2011, 2013, 2013	X-Ray	[39, 51–53]	3.0 – 3.3
CENP-C	4X23	H3 variant binding	2013	X-Ray	[54]	3.5
Ring1B	4R8P	E3 ligase	2014	X-Ray	[55]	3.3
53BP1	5KGF	Reader	2016	EM	[40]	4.5
SAGA/DUB	4ZUX	Eraser	2016	X-Ray	[56]	3.8
Set8	5HQ2	Writer	2016	X-Ray	[57]	4.5
Chd1	5O9G	Remodeler	2017	X-Ray	[74]	4.8
Snf2	5X0X, 5X0Y	Remodeler	2017	EM	[71]	4.0
CENP-N	6BUZ, 6C0W	H3 variant binder	2017, 2018	EM	[72, 73]	3.9/4.0
H1	4QLC, 5NL0	Linker histone	2015, 2017	X-Ray	[45, 75]	3.5
INO80	6FML, 6ETX	Remodeling complex	2018, 2018	EM	[59, 60]	4.4/4.8
Peptides						
LANA	1ZLA, 5GTC	Viral protein	2006, 2017	X-Ray	[61, 76]	2.9/2.7
IE1	5E5A	Viral protein	2016	X-Ray	[77]	2.8
GAG	5MLU	Synthetic acetylation system	2017	X-Ray	[78]	2.8

Table 2. Structures of nucleosome-protein or nucleosome-peptide complexes deposited in the RCSB protein databank PDB.

3.2. The first crystal structure of a nucleosome-bound protein domain (RCC1)

The first structure of a protein bound to the nucleosome was the RCC1-nucleosome complex published by the Tan lab in 2010. RCC1 (regulator of chromosome condensation) is essential during mitosis by recruiting Ran GTPase, which plays a role in nucleus reorganization, to the nucleosome [65, 66]. A comparison with LANA highlighted the crucial and conserved interaction of arginine residues with the acidic patch triad [50]. Strikingly, RCC1 binds to the acidic patch using the canonical arginine anchor, here contained in a loop, and also binds the nucleosomal DNA through its N-terminal tail. Such synergetic interactions have been observed later in many other nucleosome-binding proteins [50, 55, 67–70]. This study was the first to show such complexity of nucleosomes as interaction platforms. It also highlights the importance of properly defining the boundaries of binding domains to capture all binding epitopes in order to reveal possible synergetic interactions and fully understand complex formation and subsequent effects on chromatin structure.

3.3. Specificity of effector protein orientation in nucleosome complex formation (PRC1)

Besides determining the binding mode, synergetic interactions can also provide the structural basis for specificity of effector protein activity. This was shown in the crystal structure, also from the Tan lab, of the polycomb repressive complex 1 (PRC1) that ubiquitinates H2A K119 in a highly specific manner [55]. On its surface, the nucleosome displays various lysine residues that can be ubiquitinated by the respective writer proteins. However, the downstream response wildly differs depending on the position of the ubiquitinated lysine. Thus, target specificity is of high importance for ubiquitin writer proteins. In case of PRC1, this is based on two distinct binding processes. For one, there is the interaction between acidic patch and the arginine anchor of the Ring1B/Bmi1 subunit. In addition, the E2 subunit UbcH5c engages the nucleosomal DNA. Combined, both contributions are responsible for exact positioning of the catalytic center of the ubiquitin carrying E2 to the target H2A K119 (**Figure 2B**).

Besides LANA, RCC1 and PRC1, other crystal structures of nucleosome complexes offered further insights into nucleosome recognition. In particular, the structure of the nucleosome complex of the SAGA DUB deubiquitination module showed a non-canonical acidic patch binding. Morgan *et al.* found that the SAGA nucleosome-binding DUB module possesses three equally crucial arginine residues distributed over an α-helix [56] (**Figure 2A**). This perhaps points towards yet other acidic patch interaction modes.

Recently, also cryo-EM-derived structures of nucleosome-protein complexes have been published. The first structure, solved in 2016, yielded the structure of the complex with 53BP1, a reader protein for post-translational histone modifications [40]. Subsequently, the structures of Snf2 and CENP-N were solved and published [71–73].

Since the first crystal structure two decades ago, the list of nucleosome complexes deposited in the RCSB PDB protein databank is continuously growing. Still, the 12 high-resolution structures solved to date only encompass a fraction of all nucleosome-protein interactions. This discrepancy highlights the need for alternative techniques in chromatin structural biology.

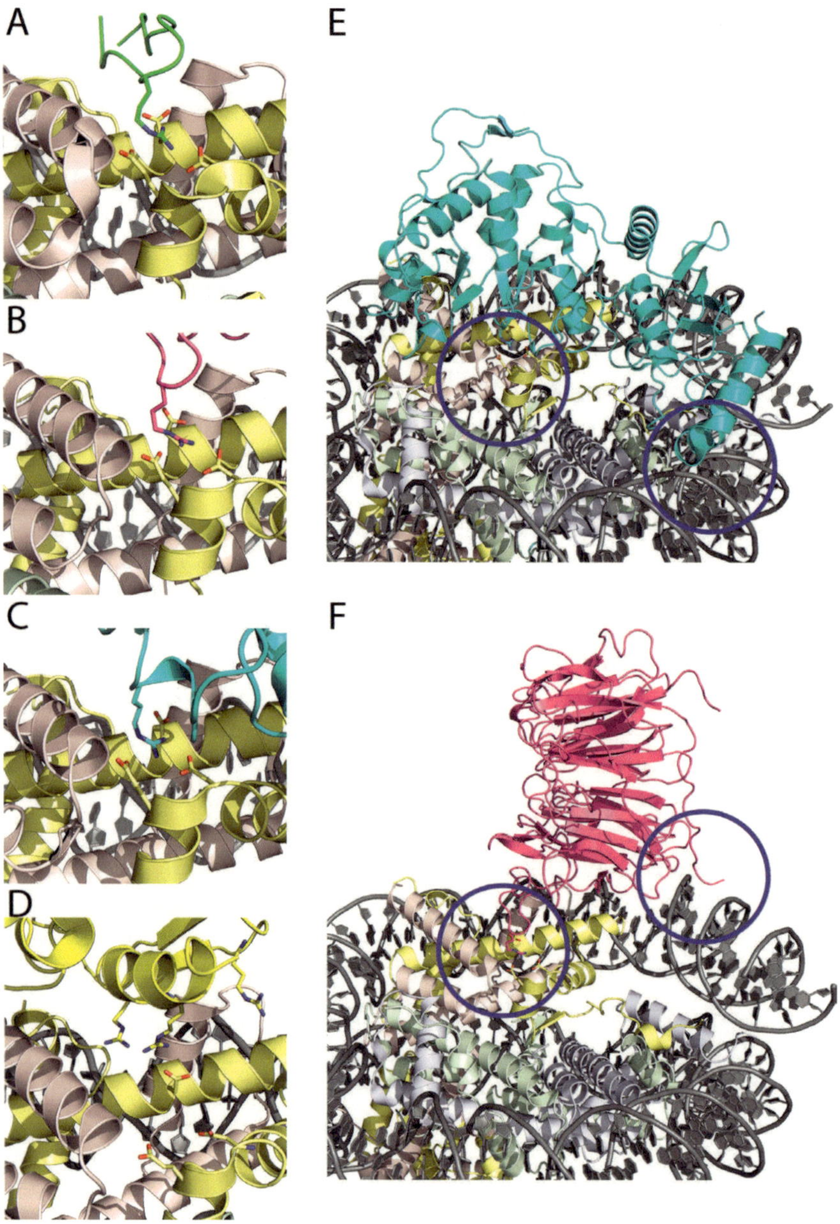

Figure 2. A detailed depiction of the acidic patch region and the triad of H2A that complexes the canonical arginine anchor. (A) LANA was the first peptide model of a binding domain to be crystalized (green, pdb: 1zla). (B) The first crystalized protein domain was RCC1 (pink, pdb: 3mvd). (C) They all share the acidic patch as a common binding epitope for the arginine anchor residue, as is also the case for the PRC1 complex with its acidic patch binding RING domain (cyan, 4r8p). (D) Interestingly, acidic patch binding is not necessarily limited to one single arginine anchor residue. As for the nucleosome-bound structure of the deubiquitinase complex SAGA-DUB (yellow, pdb: 4zux), three arginine residues are essential part of the acidic patch binding, of which none occupies the position in the center of the acidic triad. Combination of nucleosomal DNA and acidic patch binding is shown in the structure of RCC1 and PRC1. (E) PRC1 (cyan, pdb: 4r8p) that besides the acidic patch also engages DNA with its UbcH5c subunit. (F) The same holds true for RCC1 (pink, pdb: 3mvd) that contacts to DNA with the unstructured tail region.

4. Data-driven modeling

An attractive alternative to traditional structure determination methods is the modeling of structures of complexes based on some sort of experimental information on the interaction [79, 80]. In such data-driven modeling of a complex structure, the two interaction partners are docked together, guided by the experimental data, and respecting their biophysical properties. The exact binding interface and relative orientation of the binding partners are typically refined over several steps. Prerequisite for this approach is the availability of the 3D structures of the interacting partners. Several molecular docking programs allow the incorporation and use of experimental data and so increase the accuracy of resulting structures [81]. Hence, data from diverse biophysical techniques are translated into restraints guiding the docking process [82–84]. The type of information includes interaction interface, distances or shape of the complex and its subunits. Techniques that can provide these information are listed in **Table 3**.

Interestingly, all three classes of information can be provided by NMR spectroscopy. It is possible to gather data on intermolecular distances and shape by paramagnetic relaxation enhancement (PRE) and the nuclear Overhauser effect (NOE) as well as information on binding interfaces and binding affinity through chemical shift perturbation (CSP). The use of these NMR methods in docking studies is reviewed in detail elsewhere [79]. An overview of publications that used data-driven docking to investigate nucleosome-protein complexes is listed in **Table 4**.

4.1. Bringing data-driven modeling to nucleosome complexes (LSD1-CoREST)

A pioneer study for data-driven modeling of a nucleosome complex was successfully applied for the lysine-specific demethylase 1 and CoREST complex [86]. Both proteins cooperate in the demethylation of mono- and dimethylated H3K4. While it was possible to solve the crystal structure of LSD1-CoREST, their nucleosome-bound state remains elusive. Yang et al. gained insight into the molecular basis of LSD1-CoREST interaction by identifying point mutations that interfere with the LSD1-CoREST ability to demethylate a methylated peptide model of the histone H3 tail. Since it was previously shown that LSD1 recognizes a specific stretch of the H3 tail [94], it was possible to employ modeling to identify intermolecular interactions between the peptide and both the LSD1 active site and the LSD1-CoREST interface (**Figure 3B**). Lastly, NMR titration experiments of the CoREST SANT2 domain with DNA revealed a DNA-binding interface on SANT2. These pieces of interaction data were used to

Interaction interface	Distances	Shape
Mutagenesis	XL-MS	Cryo-EM
H/D exchange	Forster resonance energy transfer (FRET)	Small angle X-ray or neutron scattering (SAXS/SANS)
	Electron paramagnetic resonance (EPR)	Ion-mobility mass spectrometry (IM-MS)

Table 3. Biochemical and biophysical techniques for structural analysis of protein complexes.

guide a docking approach resulting in a complete structural model of the LSD1-CoREST-nucleosome complex (**Figure 3A**). With the lack of experimental data on the nucleosome interaction, this is a prime example of combining crystal structures, mutagenesis and NMR data to overcome limitations of the separate techniques.

Protein	**Role**	**Data source**	**Reference**
PSIP1-PWWP	Trimethyl lysine reader H3K36	NMR	[67, 68, 85]
CoREST/LSD1	Demethylase	Crystallography/NMR	[86]
Rad6-Bre1	Ubiquitin ligase	XL-MS	[70]
LANA	Viral protein	ssNMR	[87]
NSD1	Methyltransferase H3K36	Mutagenesis	[88]
RNF169	Ubiquitin reader	NMR, SAXS	[69, 89]
H1	Linker histone	NMR	[43, 90]
ISW2	Chromatin remodeler	XL-MS	[91]
Rad18	DNA repair factor	NMR	[89]
RCC1	Ran-recruitment	Crystallography	[62]
PHF1 Tudor	Trimethyl lysine reader H3K36	Crystallography/NMR	[92]
HMGN2	Chromatin decompaction	NMR	[93]

Table 4. Structural models of nucleosome-protein complexes based on biophysical data.

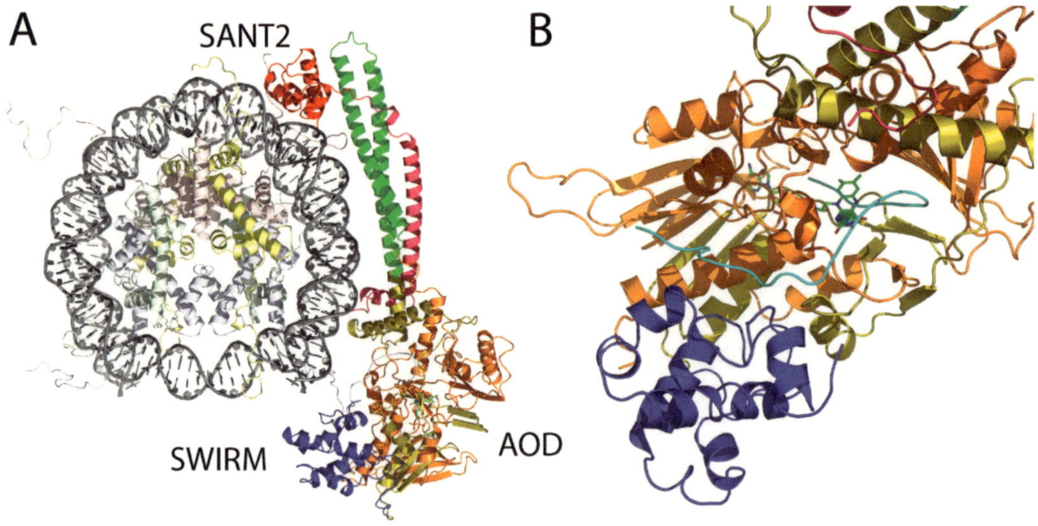

Figure 3. (A) Structural model of LSD1-CoREST bound to the nucleosome. The DNA binding of the SANT2 domain was elucidated by NMR spectroscopy. A previously identified binding motif in the H3 tail sequence was docked onto the interface of amine oxidase (AOD) and SWIRM domain revealing a second binding epitope. (B) The resulting model of the model peptide binding to AOD-SWIRM is shown as a close-up, highlighting how the tail is positioned on the interface of both domains. Figure generated using the author-provided PDB file [86].

4.2. NMR-based structural biology of nucleosome-protein complexes

Over recent years, several studies have demonstrated that state-of-the-art solution NMR can offer high-resolution and site-specific characterization of the structures and dynamics of nucleosome-protein complexes. NMR has the particular advantage of its sensitivity to dynamics and the ease with which interactions can be studied, allowing detailed insights into molecular recognition processes. NMR allows studies when systems are dynamic, or (partially) disordered, while this typically hampers high-resolution structure determination by crystallography and cryo-EM.

The molecular size of nucleosomes, and even more so of complexes with effector proteins, poses a challenge to traditional NMR methods. However, this challenge can be overcome through the use of methodologies designed for high-molecular weight systems. This method, methyl group-based transverse-relaxation-optimized spectroscopy (methyl-TROSY), relies on the highly sensitive observation of NMR signals of protein methyl groups [95]. Here, a specific isotope-labeling scheme is used, which typically results in observation of isoleucine, leucine, valine (ILV) methyl groups. The methyl-TROSY NMR spectra can subsequently be used to delineate binding sites of effector proteins on the nucleosome surface and vice versa [68, 69, 93, 96]. Extracting more detailed structural information is possible through the use of so-called spin-labels that can generate long-range distance restraints between the interaction partners [97, 98]. Whichever way used, NMR-based interaction data are of unique value in the modeling of nucleosome-protein complexes.

4.3. Expanding data sources for nucleosome complex models to NMR (HMGN2)

Kato *et al.* were the first to use the methyl-TROSY approach for the study of nucleosome-protein interactions [93]. Importantly, they reported the NMR signal assignments of the ILV-methyl groups for all histones in the nucleosomes. These assignments are essential in determining protein-binding sites on the nucleosome surface. The approach was demonstrated using high mobility group nucleosomal protein 2 (HMGN2), which regulates a variety of chromatin functions. HMGN2 was found to bind both the acidic patch and nucleosomal DNA. Based on these NMR data, supported by mutagenesis, it was possible to determine a structural model of the complex (**Figure 4A**). HMGN2 binds to the nucleosome as a staple, using two main interaction sites. On one side, HMGN2 is anchored to the acidic patch using a canonical arginine anchor in the N-terminal region of the binding domain, while the lysine-rich motif in its C-terminal region binds to nucleosomal DNA (**Figure 4B**). This binding mode provided a structural basis for the antagonistic function of HMGN2 towards linker histone H1 for nucleosome binding.

4.4. Latest applications of NMR to investigate structures of nucleosome complexes (RNF169 & Rad18)

Two recent studies relied on methyl-TROSY NMR-derived binding data to elucidate the recognition of ubiquitinated nucleosomes. Both focused on the interaction between ubiquitylated H2A K13/15 and the DNA repair factor RNF169. The work of Kitevski-LeBlanc et al.

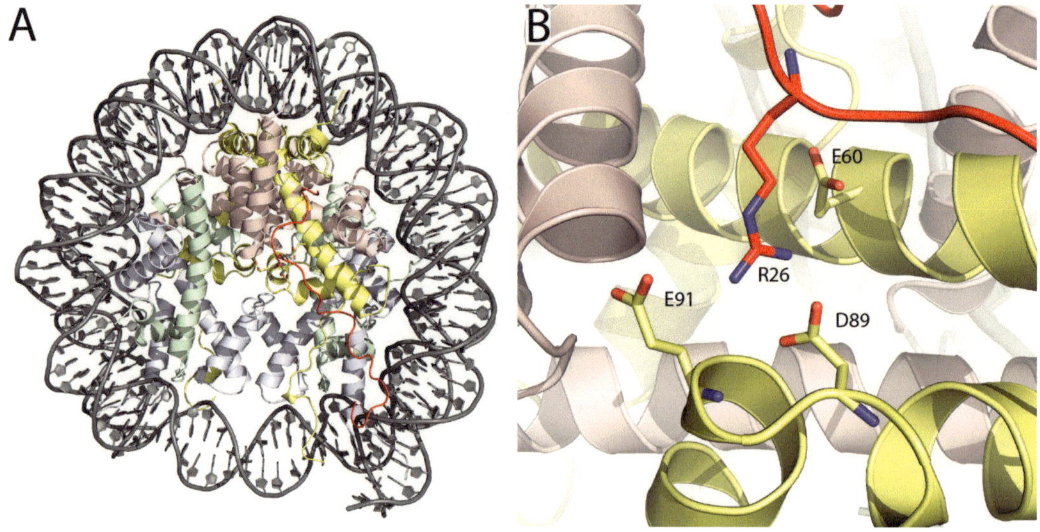

Figure 4. (A) Structural model of HMGN2 (red) bound to the nucleosome. The binding occurs along the nucleosome surface and is driven by interactions with the acidic patch and nucleosomal DNA, resulting in HMGN2 competing with H1 for nucleosome binding. (B) Close view on the acidic patch binding N-terminal HMGN2 region depicting the canonical arginine anchor R26 surrounded by the Glu 91, Asp 89, Glu 60 acidic triad motif of H2A. Figure generated using the author-provided PDB file [93].

established the molecular basis of this interaction. The α-helical MIU2 (motif interacting with ubiquitin) domain binds to a hydrophobic patch on the K13/15-conjugated ubiquitin while a disordered region anchors RNF169 on the nucleosome by binding to the acidic patch. They subsequently reconstructed a model structure that presents both epitopes in their nucleosome-bound state (**Figure 5A**). The work of Hu *et al.* combined traditional NOESY-based structure determination at the level of histone-dimers with interaction studies at the nucleosome level and complemented these with SAXS data into a final model [89]. The authors also extended their findings to an NMR-based structural model for the complex with DNA repair factor Rad18. Both RNF169 and Rad18 are known to interfere with the binding of 53BP1 to nucleosomes ubiquitinated at H2A K13/15. These NMR-based structural models have allowed to hypothesize on the molecular mechanism for this interference.

4.5. Importance of the nucleosomal context in epigenetic read-out (PSIP1-PWWP & PHF1-Tudor)

The complexity of nucleosome recognition by reader proteins is well illustrated by the NMR-based studies on the recognition of H3K36me-nucleosomes by the PWWP domain of PSIP1(Ledgf). NMR studies of this reader interaction found that the PWWP domain has binding affinity orders of magnitude lower for a H3K36me peptide compared to H3K36me3 in a nucleosomal context. Interestingly, a similar observation was made for the Tudor domain of the H3K36me reader PHF1 [85]. Here, an isolated peptide model of the H3 tail showed decreased affinity as well. Due to the proximity of H3K36 to nucleosomal DNA, a role of DNA binding was hypothesized for both proteins. NMR studies showed for PSIP1 and PHF1 alike

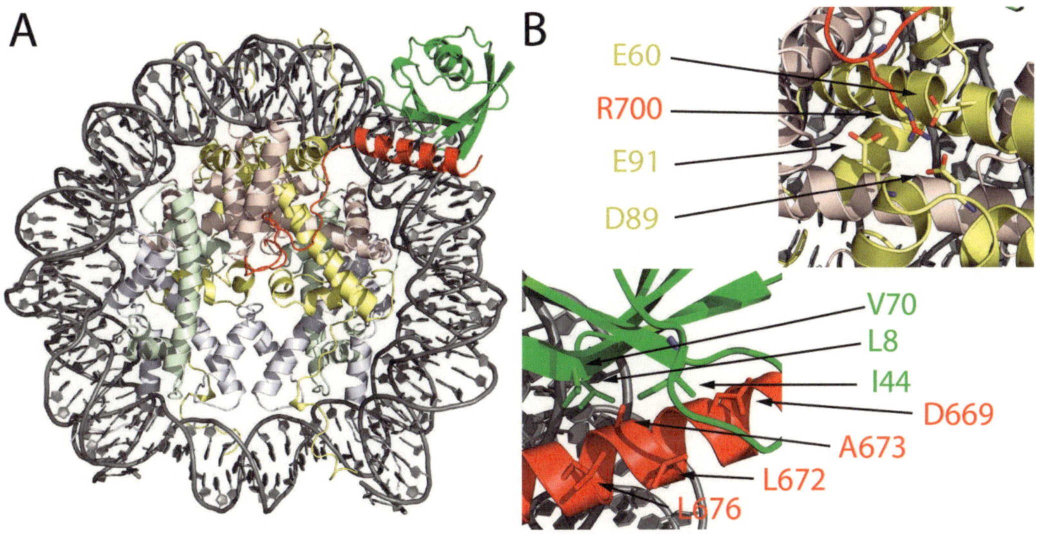

Figure 5. (A) Structural model of nucleosome-bound RNF169 (red) and ubiquitin (green). (B, top) The proposed main acidic patch anchoring residue R700 (conserved position throughout the docking solutions) is shown in the conserved arginine anchor position between the acidic triads (Glu 60, Asp 89, Glu 91). (B, bottom) Side chain interactions between RNF169 MIU2 (red) and ubiquitin (green). Figure generated using the author-provided PDB file [69].

a binding site for nucleosomal DNA, resulting in a simultaneous binding mechanism of both trimethyl lysine and nucleosomal DNA.

For PHF1-Tudor, a crystal structure bound to a trimethylated H3 tail peptide was already available to use. The additional importance of the nucleosomal context and synergetic binding mechanism can be understood from the corresponding nucleosome-bound structure (**Figure 6A**). In case of PSIP1-PWWP, the domain structure was solved by NMR and, together with NMR titration data, used to determine a structural model of nucleosome-bound protein (**Figure 6B**) [67, 68, 85]. The structural models of both highlighted the importance of the nucleosomal context in H3K36me3 recognition, emphasizing that complex formation critically depends on two synergetic binding processes. Firstly, the aromatic residues that form the aromatic cage bind to trimethylated lysine H3K36me3. This recognition of the PTM is crucial for the binding, but the readers reach their full binding affinity only when their positive surface residues interact with the nucleosomal DNA. This makes both studies outstanding examples of synergetic interplay of epitopes in nucleosome-binding proteins (**Figure 6C**, **D**).

The insights derived from these structural models were used to design experiments to validate the structural model and may offer possible tools for further research approaches. In case of PSIP1-PWWP, the structural model sparked current efforts in the design of nucleosome-mimicking peptides to modulate the PSIP1-chromatin interaction.

4.6. LANA goes solid state

The studies mentioned above illustrate the potential of data-driven modeling of nucleosome-protein complexes based on state-of the-art solution NMR. Recent advances in solid-state

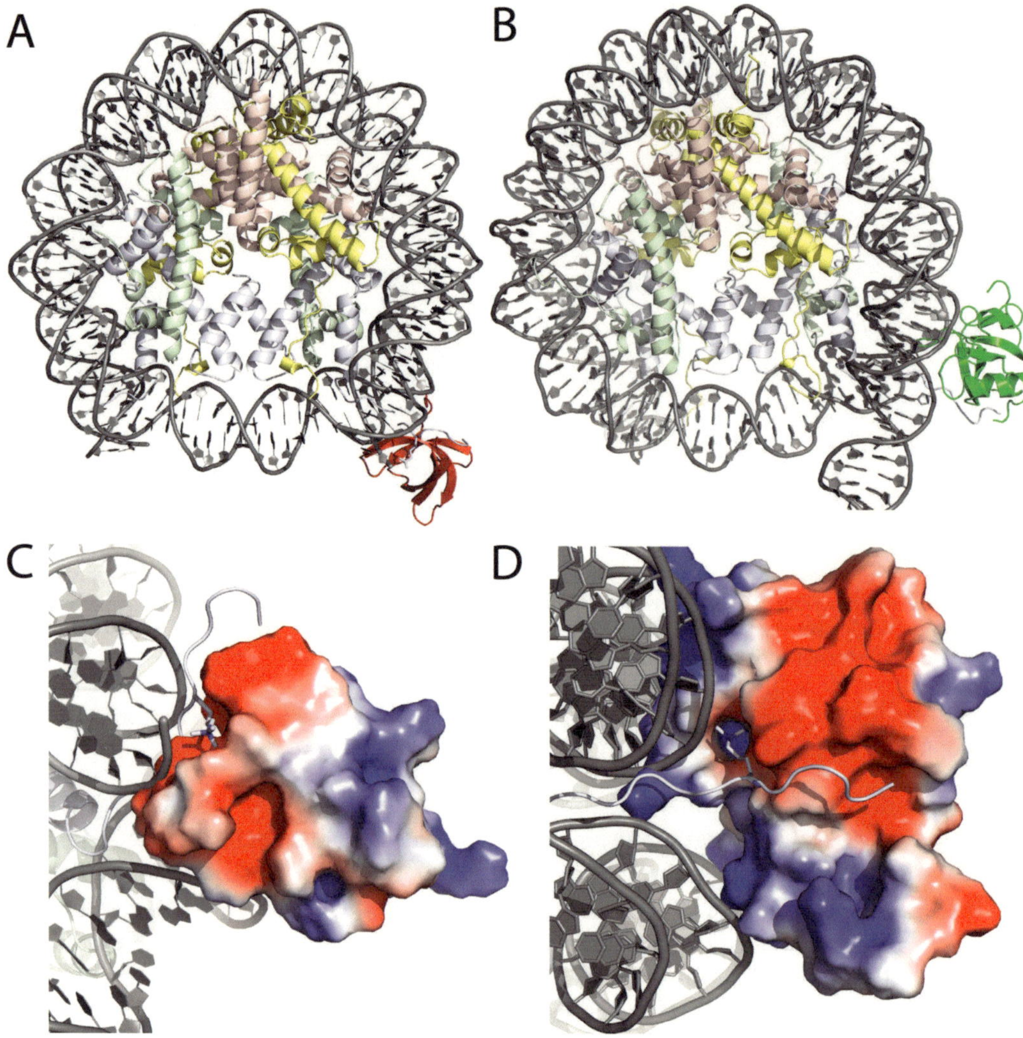

Figure 6. Structural model of nucleosome-bound PHF1 (red; A) and PSIP1-PWWP (green; B). The electrostatic potential of nucleosomal DNA and the surface of PHF1 (C) and PWWP (D), respectively, act in combination with H3K36me3 recognition by the aromatic cage motif (trimethyl lysine side chain shown as sticks). Figure generated using the author-provided PDB file [85].

NMR (ssNMR) have enabled the detailed investigation of large, soluble biomolecular complexes. Very recently, our lab capitalized on these advances and tailored them for application to nucleosome-protein complexes [87]. Unlike the methyl-TROSY methods, this approach allows observation of all residues, in principle allowing for a more complete mapping of binding interfaces. In this approach, NMR spectra are recorded on sediments, generated by ultracentrifugation, of nucleosomes or their complexes. After assignments of NMR signals of histone H2A in the unbound nucleosome, spectra were recorded on the nucleosome complex with the LANA peptide, analogous to the LANA crystal structure (**Figure 7A**) [61, 87]. Based on the chemical shift changes, the binding site of LANA could be mapped to the acidic patch and a structural model generated. The large agreement between the crystal structure and ssNMR-derived structural model (**Figure 7B**) illustrates the power of this approach. In our view, ssNMR, just as the solution NMR approach, is an attractive alternative for structure determination for nucleosome-protein complexes. While its application awaits to be extended to larger nucleosome-binding domains, we anticipate that it will be a valuable addition to the tool kit in chromatin structural biology.

4.7. Modeling nucleosome-bound Rad6-Bre1 based on cross-linking MS

Next to NMR, cross-linking mass spectrometry has found increasing application as a data source on nucleosome-protein interactions. With cross-linking, intermolecular contacts between the proteins of interest are captured and converted to covalent connections. These connections are introduced by small molecule linkers, specific for the fusion of well-defined side chains or less specific as radical-forming photo cross-linkers. Furthermore, cross-linkers possess a spacer between their terminal functional groups to define the range of cross-linking ability [99, 100]. Both characteristics can be tuned for the study of a specific system, resulting in a manifold of reported linker molecules. After cross-linking, the protein complex undergoes trypsin digestion resulting in peptide fragments of the complex. Here, covalently cross-linked fragments stay connected. An analysis of these fragments by liquid chromatography mass spectrometry (LC-MS) enables identification of the sequence positions. The cross-links can thus be converted to distance restraints between two residues, with the distance depending on the length of the cross-linker. These restraints can be used to guide structural modelling of the complex [80]. In one of the earliest examples for nucleosome complexes, XL-MS was used to map the binding sites of the various nucleosome-binding domains of the chromatin

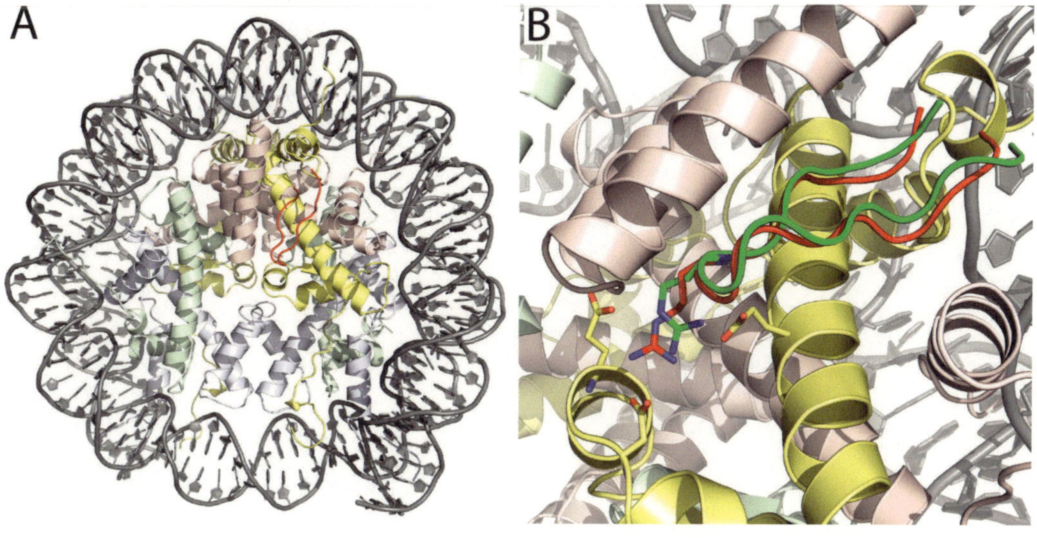

Figure 7. (A) Structural model for nucleosome-bound LANA peptide. ssNMR data derived from NMR titration experiments were used to direct the docking simulation. (B) Alignment of the ssNMR-derived model for LANA (red) and the crystal structure (green, pdb: 1zla) shows remarkable accuracy of the docking-derived solution. For both, the canonical arginine anchor is depicted as sticks in the typical central position between the acidic triad of H2A (yellow).

remodeling complex ISW2 onto the nucleosome surface [91]. These data were subsequently used to build a structural model of the ISW2-nucleosome complex. A recent case of cross-linking-based modeling in nucleosome research is the E2/E3 ubiquitin ligase complex Rad6-Bre1 (**Figure 8A**). Bre1 is known to act as a homodimer in a complex with Rad6 to specifically ubiquitinate H2B K123 [101, 102]. However, the molecular mechanism of specific ubiquitination remained unknown without any nucleosome-bound complex structure available. Gallego et al. addressed exactly this problem by using XL-MS data to identify the binding interface between the Bre1 RING domain and the nucleosome. Next to nucleosomal DNA binding, they observed binding of the homodimer to the acidic patch (**Figure 8B**), which was verified by LANA-induced inhibition of Bre1 RING nucleosome binding. As a first step in the modeling, the authors modeled the Rad6-Bre1 complex structures based on homology with known E2/E3 RING ligases. Importantly, the resulting model was supported by the observed cross-links. The Rad6-Bre1 model could then be docked onto the nucleosome guided by the observed cross-links. This provided the structural basis for the specificity of Bre1 towards H2B K123 ubiquitination [70].

4.8. Adding new perspective on binding modes

Data-driven structural models complement high-resolution structures in many ways. An interesting example is the RCC1-nucleosome interaction, which serves as binding platform for subsequent binding of Ran, a protein relevant during mitosis (see Section 3.2). Biochemical data have shown that Ran activity is increased in the nucleosome-bound complex. The crystal structure suggests no nucleosome-Ran interactions upon modeling Ran to the RCC1

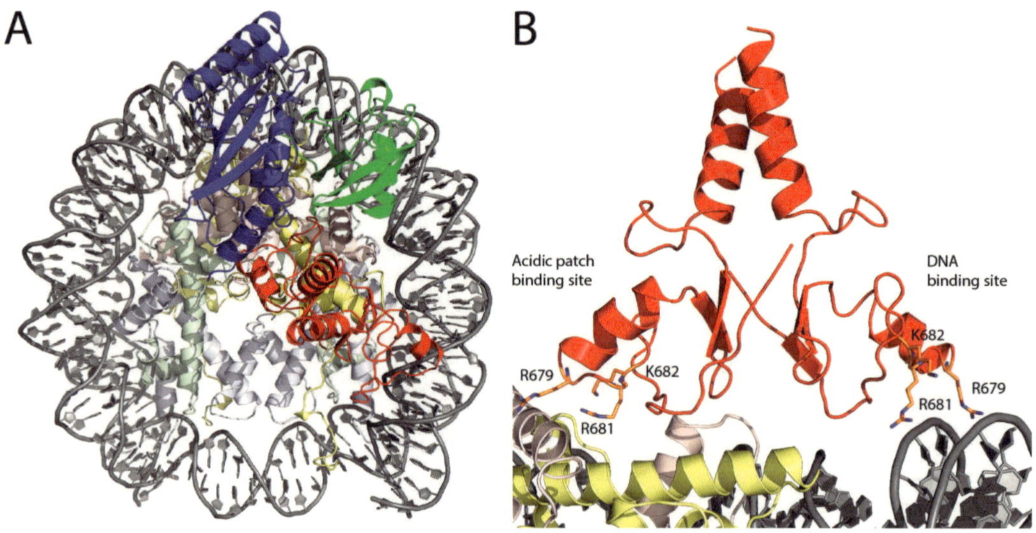

Figure 8. (A) Structural model of homodimeric Bre1 (red) bound to the nucleosome together with the E2 ligase Rad6 (blue) with attached ubiquitin (green). The study was conducted by identifying the interactions between positive Bre1 RING residues and the acidic patch. The docking was further facilitated due to the known target lysine residue. (B) Close view on Bre1 bound to both the acidic patch and nucleosomal DNA. The homodimeric nature allows the engagement of both epitopes in simultaneous binding. Figure generated using the author-provided PDB file [70].

Ran-binding interface. Before the crystal structure of nucleosome-bound RCC1 was solved, a data-driven model was reported, which does feature Ran-nucleosome interactions. [62]. The authors suggest that, upon Ran binding, the nucleosomal DNA contacts with RCC1 N-terminal tail observed in the crystal are broken in favor of Ran-nucleosome interactions as observed in model. Even though additional studies have to elucidate the exact mechanism of RCC1-Ran nucleosome binding, the use of crystal structure and data-driven model in combination outlines a possible mechanism to further investigate.

4.9. Debating H1

Another cardinal topic is the nucleosome-bound state of linker histone H1. To date, the structure of the chromatosome, consisting of the four canonical histones and 166bp of DNA in a complex with linker histones, is strongly debated. In this case as well, there are contradictions between structural models and a nucleosome-bound crystal structure of the chromatosome. The crystal structure reported by Zhou *et al.* displays the globular domain of linker histone H5 (chicken H5) with truncated tails in an on-dyad binding mode encountering both entering and leaving ends of linker DNA [75]. As for linker histone H1 (X. laevis H1.0b, human H1.5), a similar on-dyad binding mode was reported by cryo-EM and crystallography independently from absence or presence of H1 tails [45]. In fact, while not vital for linker histone positioning, the H1 C-terminal domain engages in binding of one of both linker DNAs preferably, introducing asymmetry into the nucleosome-bound complex.

In contrast to the proposed on-dyad complex, computational studies on linker histone binding suggest an alternative, off-dyad binding geometry of the complex in which the linker histone shows interactions with but one strand of linker DNA [103]. This binding mode was shown experimentally in the case of the globular domain of linker histone H1 (D. melanogaster). Here, NMR-based distance information, obtained through paramagnetic relaxation enhancement (PRE), was used to derive the nucleosome-binding mode of H1, showing an asymmetric, off-dyad binding [43]. Interestingly, it was shown by PRE as well that the mutation of a set of five crucial amino acids in H5 to its equivalents in H1 is sufficient to change the binding mode of H5 from on-dyad (crystal) to off-dyad [90]. This points out the importance of linker histone subtype sequence and the interacting residues in determining the binding mode towards the nucleosome [44].

5. Conclusions

Chromatin structural biology is an equally important as demanding field. This is not only clear from the tremendous efforts necessary for the first nucleosome structure but also from the limited number of structures for nucleosome-protein complexes. While crystallography and cryo-EM resulted in various high-resolution structures, not every interaction is accessible this way due to either of many experimental limitations, such as the need for crystallization, the fleeting nature of some complexes or the pervasive role of highly dynamic protein regions. Here, an increasing number of studies shift towards a combined approach utilizing various sources of interaction data to direct sophisticated data-driven docking. This way all knowledge on a nucleosome-interacting system can be integrated into a structural model that is otherwise inaccessible. These models strongly depend on the quality and quantity of data and contain an inherent ambiguity. However, as in the case of linker histone H1, structural models can point to alternative binding modes and thus result in new, testable hypotheses. Additionally, crucial residues for nucleosome binding can be identified, allowing design of, for example, loss of function or loss of binding mutants to silence specific pathways. It also offers the possibility to drive the design of competing small molecule or peptide structures as potential candidates for epigenetic drugs interfering with specific effector binding. Remarkably, these developments might be otherwise lost due to the lack of a structure. However, as for now, a database for such structural models, akin to the RCSB protein databank, remains to be established. This might however be essential to advance the study of chromatin effector proteins. Publicly available structures including their data-based restraints could be used for further refinements upon availability of new, additional datasets from an array of techniques. It also would offer the possibility of negative results, otherwise rarely reported, to contribute to drive or score the quality of already reported models. Data-driven modeling of nucleosome-protein complexes has the potential to yield unique fundamental insights into nucleosome-binding dynamics and enable advances in modulation of chromatin effector proteins, which would be otherwise inaccessible.

Acknowledgements

We thank all authors of the studies included in this work who kindly provided us with files of their structural models for review. This work is supported by the Netherlands Organization for Scientific Research (NWO) through a VIDI grant (723.013.010) to Hugo van Ingen.

Conflict of interest

The authors of this work declare no conflict of interest.

Author details

Velten Horn[1,2] and Hugo van Ingen[2*]

*Address all correspondence to: h.vaningen@uu.nl

1 Department of Macromolecular Biochemistry, Leiden Institute of Chemistry, Leiden University, Leiden, The Netherlands

2 NMR Group, Bijvoet Center for Biomolecular Research, Utrecht University, Utrecht, The Netherlands

http://dx.doi.org/10.5772/intechopen.81016

References

[1] Kornberg RD. Chromatin structure: A repeating unit of histones and DNA. Science. 1974;**184**(4139):868-871

[2] Thomas JO, Kornberg RD. An octamer of histones in chromatin and free in solution. Proceedings of the National Academy of Sciences. 1975;**72**(7):2626

[3] Luger K, Mäder AW, Richmond RK, Sargent DF, Richmond TJ. Crystal structure of the nucleosome core particle at 2.8 Å resolution. Nature. 1997;**389**:251

[4] Luger K, Dechassa ML, Tremethick DJ. New insights into nucleosome and chromatin structure: An ordered state or a disordered affair? Nature Reviews. Molecular Cell Biology. 2012;**13**:436

[5] Sarah H, Karolina L, Nicole S, Meghan L, Sarah R, Sibaji S. Use of epigenetic drugs in disease: An overview. Genetics & Epigenetics. 2014;**6**:GEG.S12270

[6] Gijsbers R, Vets S, De Rijck J, Ocwieja KE, Ronen K, Malani N, et al. Role of the PWWP domain of Lens epithelium-derived growth factor (LEDGF)/p75 cofactor in lentiviral integration targeting. Journal of Biological Chemistry. 2011;**286**(48):41812-41825

[7] Yokoyama A, Cleary ML. Menin critically links MLL proteins with LEDGF on cancer-associated target genes. Cancer Cell. 2008;**14**(1):36-46

[8] Egger G, Liang G, Aparicio A, Jones PA. Epigenetics in human disease and prospects for epigenetic therapy. Nature. 2004;**429**:457

[9] Yu X, Li Z, Shen J. BRD7: A novel tumor suppressor gene in different cancers. American Journal of Translational Research. 2016;**8**(2):742-748

[10] Pérez-Salvia M, Esteller M. Bromodomain inhibitors and cancer therapy: From structures to applications. Epigenetics. 2017;**12**(5):323-339

[11] Kouzarides T. Chromatin modifications and their function. Cell. 2007;**128**(4):693-705

[12] Bannister AJ, Kouzarides T. Regulation of chromatin by histone modifications. Cell Research. 2011;**21**(3):381-395

[13] Rothbart SB, Strahl BD. Interpreting the language of histone and DNA modifications. Biochimica et Biophysica Acta (BBA) - Gene Regulatory Mechanisms. 2014;**1839**(8):627-643

[14] Li Y, Sabari Benjamin R, Panchenko T, Wen H, Zhao D, Guan H, et al. Molecular coupling of histone crotonylation and active transcription by AF9 YEATS domain. Molecular Cell. 2016;**62**(2):181-193

[15] Andrews FH, Shinsky SA, Shanle EK, Bridgers JB, Gest A, Tsun IK, et al. The Taf14 YEATS domain is a reader of histone crotonylation. Nature Chemical Biology. 2016;**12**:396

[16] Zhao D, Guan H, Zhao S, Mi W, Wen H, Li Y, et al. YEATS2 is a selective histone crotonylation reader. Cell Research. 2016;**26**:629

[17] Filippakopoulos P, Knapp S. Structural genomics and drug discovery for chromatin-related protein complexes involved in histone tail recognition. In: Emili A, Greenblatt J, Wodak S, editors. Systems Analysis of Chromatin-Related Protein Complexes in Cancer. New York, NY: Springer New York; 2014. pp. 211-225

[18] Hughes RM, Wiggins KR, Khorasanizadeh S, Waters ML. Recognition of trimethyllysine by a chromodomain is not driven by the hydrophobic effect. Proceedings of the National Academy of Sciences. 2007;**104**(27):11184

[19] Chen C, Nott TJ, Jin J, Pawson T. Deciphering arginine methylation: Tudor tells the tale. Nature Reviews. Molecular Cell Biology. 2011;**12**:629

[20] Maurer-Stroh S, Dickens NJ, Hughes-Davies L, Kouzarides T, Eisenhaber F, Ponting CP. The Tudor domain 'Royal Family': Tudor, plant Agenet, chromo, PWWP and MBT domains. Trends in Biochemical Sciences. 2003;**28**(2):69-74

[21] Teske KA, Hadden MK. Methyllysine binding domains: Structural insight and small molecule probe development. European Journal of Medicinal Chemistry. 2017;**136**:14-35

[22] Yun M, Wu J, Workman JL, Li B. Readers of histone modifications. Cell Research. 2011; **21**(4):564-578

[23] Barbieri I, Cannizzaro E, Dawson MA. Bromodomains as therapeutic targets in cancer. Briefings in Functional Genomics. 2013;**12**(3):219-230

[24] Gupta A, Hunt CR, Chakraborty S, Pandita RK, Yordy J, Ramnarain DB, et al. Role of 53BP1 in the regulation of DNA double-Strand break repair pathway choice. Radiation Research. 2014;**181**(1):1-8

[25] Yang Y, Lu Y, Espejo A, Wu J, Xu W, Liang S, et al. TDRD3 is an effector molecule for arginine-methylated histone Marks. Molecular Cell. 2010;**40**(6):1016-1023

[26] Trojer P, Li G, Sims RJ, Vaquero A, Kalakonda N, Boccuni P, et al. L3MBTL1, a histone-methylation-dependent chromatin lock. Cell. 2007;**129**(5):915-928

[27] Boccuni P, MacGrogan D, Scandura JM, Nimer SD. The human L(3)MBT polycomb group protein is a transcriptional repressor and interacts physically and functionally with TEL (ETV6). Journal of Biological Chemistry. 2003;**278**(17):15412-15420

[28] Daugaard M, Baude A, Fugger K, Povlsen LK, Beck H, Sørensen CS, et al. LEDGF (p75) promotes DNA-end resection and homologous recombination. Nature Structural & Molecular Biology. 2012;**19**:803

[29] Ge H, Si Y, Roeder RG. Isolation of cDNAs encoding novel transcription coactivators p52 and p75 reveals an alternate regulatory mechanism of transcriptional activation. The EMBO Journal. 1998;**17**(22):6723-6729

[30] Flanagan JF, Mi L-Z, Chruszcz M, Cymborowski M, Clines KL, Kim Y, et al. Double chromodomains cooperate to recognize the methylated histone H3 tail. Nature. 2005;**438**:1181

[31] Sims RJ, Chen C-F, Santos-Rosa H, Kouzarides T, Patel SS, Reinberg D. Human but not yeast CHD1 binds directly and selectively to histone H3 methylated at lysine 4 via its tandem chromodomains. Journal of Biological Chemistry. 2005;**280**(51):41789-41792

[32] Eissenberg JC, Elgin SCR. The HP1 protein family: Getting a grip on chromatin. Current Opinion in Genetics & Development. 2000;**10**(2):204-210

[33] Luco RF, Pan Q, Tominaga K, Blencowe BJ, Pereira-Smith OM, Misteli T. Regulation of alternative splicing by histone modifications. Science. 2010;**327**(5968):996

[34] Alpatov R, Lesch Bluma J, Nakamoto-Kinoshita M, Blanco A, Chen S, Stützer A, et al. A chromatin-dependent role of the fragile X mental retardation protein FMRP in the DNA damage response. Cell. 2014;**157**(4):869-881

[35] LeRoy G, Rickards B, Flint SJ. The double bromodomain proteins Brd2 and Brd3 couple histone acetylation to transcription. Molecular Cell. 2008;**30**(1):51-60

[36] Winter S, Simboeck E, Fischle W, Zupkovitz G, Dohnal I, Mechtler K, et al. 14-3-3 proteins recognize a histone code at histone H3 and are required for transcriptional activation. The EMBO Journal. 2008;**27**(1):88-99

[37] Xu C, Bian C, Yang W, Galka M, Ouyang H, Chen C, et al. Binding of different histone marks differentially regulates the activity and specificity of polycomb repressive complex 2 (PRC2). Proceedings of the National Academy of Sciences. 2010;**107**(45):19266

[38] McGinty RK, Tan S. Recognition of the nucleosome by chromatin factors and enzymes. Current Opinion in Structural Biology. 2016;**37**:54-61

[39] Armache K-J, Garlick JD, Canzio D, Narlikar GJ, Kingston RE. Structural basis of silencing: Sir3 BAH domain in complex with a nucleosome at 3.0 Å resolution. Science (New York, NY). 2011;**334**(6058):977-982

[40] Wilson MD, Benlekbir S, Fradet-Turcotte A, Sherker A, Julien J-P, McEwan A, et al. The structural basis of modified nucleosome recognition by 53BP1. Nature. 2016;**536**:100

[41] Carroll CW, Milks KJ, Straight AF. Dual recognition of CENP-A nucleosomes is required for centromere assembly. The Journal of Cell Biology. 2010;**189**(7):1143

[42] Carroll CW, Silva MC, Godek KM, Jansen LE, Straight AF. Centromere assembly requires the direct recognition of CENP-A nucleosomes by CENP-N. Nature Cell Biology. 2009;**11**(7):896-902

[43] Zhou BR, Feng H, Kato H, Dai L, Yang Y, Zhou Y, et al. Structural insights into the histone H1-nucleosome complex. Proceedings of the National Academy of Sciences of the United States of America. 2013;**110**(48):19390-19395

[44] Fyodorov DV, Zhou B-R, Skoultchi AI, Bai Y. Emerging roles of linker histones in regulating chromatin structure and function. Nature Reviews. Molecular Cell Biology. 2017;**19**:192

[45] Bednar J, Garcia-Saez I, Boopathi R, Cutter AR, Papai G, Reymer A, et al. Structure and dynamics of a 197 bp nucleosome in complex with linker histone H1. Molecular Cell. 2017;**66**(3):384-397. e8

[46] Cirillo LA, Lin FR, Cuesta I, Friedman D, Jarnik M, Zaret KS. Opening of compacted chromatin by early developmental transcription factors HNF3 (FoxA) and GATA-4. Molecular Cell. 2002;**9**(2):279-289

[47] Soufi A, Donahue G, Zaret Kenneth S. Facilitators and impediments of the pluripotency reprogramming factors' initial engagement with the genome. Cell. 2012;**151**(5):994-1004

[48] Soufi A, Garcia Meilin F, Jaroszewicz A, Osman N, Pellegrini M, Zaret Kenneth S. Pioneer transcription factors target partial DNA motifs on nucleosomes to initiate reprogramming. Cell. 2015;**161**(3):555-568

[49] Harp JM, Uberbacher EC, Roberson AE, Palmer EL, Gewiess A, Bunick GJ. X-ray diffraction analysis of crystals containing twofold symmetric nucleosome core particles. Acta Crystallographica Section D. 1996;**52**(2):283-288

[50] Makde RD, England JR, Yennawar HP, Tan S. Structure of RCC1 chromatin factor bound to the nucleosome core particle. Nature. 2010;**467**(7315):562-566

[51] Wang F, Li G, Altaf M, Lu C, Currie MA, Johnson A, et al. Heterochromatin protein Sir3 induces contacts between the amino terminus of histone H4 and nucleosomal DNA. Proceedings of the National Academy of Sciences of the United States of America. 2013;**110**(21):8495-8500

[52] Arnaudo N, Fernandez IS, McLaughlin SH, Peak-Chew SY, Rhodes D, Martino F. The N-terminal acetylation of Sir3 stabilizes its binding to the nucleosome core particle. Nature Structural & Molecular Biology. 2013;**20**(9):1119-1121

[53] Yang D, Fang Q, Wang M, Ren R, Wang H, He M, et al. Nalpha-acetylated Sir3 stabilizes the conformation of a nucleosome-binding loop in the BAH domain. Nature Structural & Molecular Biology. 2013;**20**(9):1116-1118

[54] Kato H, Jiang J, Zhou B-R, Rozendaal M, Feng H, Ghirlando R, et al. A conserved mechanism for centromeric nucleosome recognition by centromere protein CENP-C. Science (New York, N.Y.). 2013;**340**(6136):1110-1113

[55] McGinty RK, Henrici RC, Tan S. Crystal structure of the PRC1 ubiquitylation module bound to the nucleosome. Nature. 2014;**514**(7524):591-596

[56] Morgan MT, Haj-Yahya M, Ringel AE, Bandi P, Brik A, Wolberger C. Structural basis for histone H2B deubiquitination by the SAGA DUB module. Science (New York, N.Y.). 2016;**351**(6274):725-728

[57] Girish TS, McGinty RK, Tan S. Multivalent interactions by the Set8 histone methyltransferase with its nucleosome substrate. Journal of Molecular Biology. 2016;**428**(8):1531-1543

[58] Pinotsis N, Waksman G. Crystal structure of the legionella pneumophila Lpg2936 in complex with the cofactor S-adenosyl-L-methionine reveals novel insights into the mechanism of RsmE family methyltransferases. Protein Science: A Publication of the Protein Society. 2017;**26**(12):2381-2391

[59] Eustermann S, Schall K, Kostrewa D, Lakomek K, Strauss M, Moldt M, et al. Structural basis for ATP-dependent chromatin remodelling by the INO80 complex. Nature. 2018; **556**(7701):386-390

[60] Ayala R, Willhoft O, Aramayo RJ, Wilkinson M, McCormack EA, Ocloo L, et al. Structure and regulation of the human INO80–nucleosome complex. Nature. 2018;**556**(7701):391-395

[61] Barbera AJ, Chodaparambil JV, Kelley-Clarke B, Joukov V, Walter JC, Luger K, et al. The nucleosomal surface as a docking station for Kaposi's sarcoma herpesvirus LANA. Science. 2006;**311**(5762):856

[62] England JR, Huang J, Jennings MJ, Makde RD, Tan S. RCC1 uses a conformationally diverse loop region to interact with the nucleosome: A model for the RCC1-nucleosome complex. Journal of Molecular Biology. 2010;**398**(4):518-529

[63] Kan P-Y, Caterino TL, Hayes JJ. The H4 tail domain participates in intra- and internucleosome interactions with protein and DNA during folding and oligomerization of nucleosome arrays. Molecular and Cellular Biology. 2009;**29**(2):538-546

[64] Mattiroli F, Uckelmann M, Sahtoe DD, van Dijk WJ, Sixma TK. The nucleosome acidic patch plays a critical role in RNF168-dependent ubiquitination of histone H2A. Nature Communications. 2014;**5**:3291

[65] Clarke PR, Zhang C. Spatial and temporal coordination of mitosis by ran GTPase. Nature Reviews. Molecular Cell Biology. 2008;**9**:464

[66] Nishitani H, Ohtsubo M, Yamashita K, Iida H, Pines J, Yasudo H, et al. Loss of RCC1, a nuclear DNA-binding protein, uncouples the completion of DNA replication from the activation of cdc2 protein kinase and mitosis. The EMBO Journal. 1991;**10**(6):1555-1564

[67] Eidahl JO, Crowe BL, North JA, McKee CJ, Shkriabai N, Feng L, et al. Structural basis for high-affinity binding of LEDGF PWWP to mononucleosomes. Nucleic Acids Research. 2013;**41**(6):3924-3936

[68] van Nuland R, van Schaik FM, Simonis M, van Heesch S, Cuppen E, Boelens R, et al. Nucleosomal DNA binding drives the recognition of H3K36-methylated nucleosomes by the PSIP1-PWWP domain. Epigenetics & Chromatin. 2013;**6**(1):12

[69] Kitevski-LeBlanc J, Fradet-Turcotte A, Kukic P, Wilson MD, Portella G, Yuwen T, et al. The RNF168 paralog RNF169 defines a new class of ubiquitylated histone reader involved in the response to DNA damage. eLife. 2017;**6**

[70] Gallego LD, Ghodgaonkar Steger M, Polyansky AA, Schubert T, Zagrovic B, Zheng N, et al. Structural mechanism for the recognition and ubiquitination of a single nucleosome

residue by Rad6–Bre1. Proceedings of the National Academy of Sciences of the United States of America. 2016;**113**(38):10553-10558

[71] Liu X, Li M, Xia X, Li X, Chen Z. Mechanism of chromatin remodelling revealed by the Snf2-nucleosome structure. Nature. 2017;**544**(7651):440-445

[72] Chittori S, Hong J, Saunders H, Feng H, Ghirlando R, Kelly AE, et al. Structural mechanisms of centromeric nucleosome recognition by the kinetochore protein CENP-N. Science. 2018;**359**(6373):339

[73] Pentakota S, Zhou K, Smith C, Maffini S, Petrovic A, Morgan GP, et al. Decoding the centromeric nucleosome through CENP-N. eLife. 2017;**6**

[74] Farnung L, Vos SM, Wigge C, Cramer P. Nucleosome-Chd1 structure and implications for chromatin remodelling. Nature. 2017;**550**(7677):539-542

[75] Zhou B-R, Jiang J, Feng H, Ghirlando R, Xiao TS, Bai Y. Structural mechanisms of nucleosome recognition by linker histones. Molecular Cell. 2015;**59**(4):628-638

[76] Amamoto Y, Aoi Y, Nagashima N, Suto H, Yoshidome D, Arimura Y, et al. Synthetic post-translational modifications: Chemical catalyst-driven regioselective histone acylation of native chromatin. Journal of the American Chemical Society. 2017;**139**(22):7568-7576

[77] Fang Q, Chen P, Wang M, Fang J, Yang N, Li G, et al. Human cytomegalovirus IE1 protein alters the higher-order chromatin structure by targeting the acidic patch of the nucleosome. eLife. 2016;**5**:e11911

[78] Lesbats P, Serrao E, Maskell DP, Pye VE, O'Reilly N, Lindemann D, et al. Structural basis for spumavirus GAG tethering to chromatin. Proceedings of the National Academy of Sciences. 2017;**114**(21):5509

[79] van Ingen H, Bonvin AMJJ. Information-driven modeling of large macromolecular assemblies using NMR data. Journal of Magnetic Resonance. 2014;**241**:103-114

[80] Rappsilber J. The beginning of a beautiful friendship: Cross-linking/mass spectrometry and modelling of proteins and multi-protein complexes. Journal of Structural Biology. 2011;**173**(3):530-540

[81] Xue LC, Dobbs D, Bonvin AMJJ, Honavar V. Computational prediction of protein interfaces: A review of data driven methods. FEBS Letters. 2015;**589**(23):3516-3526

[82] Clore GM, Schwieters CD. Docking of protein–protein complexes on the basis of highly ambiguous intermolecular distance restraints derived from 1HN/15N chemical shift mapping and backbone 15N–1H residual dipolar couplings using conjoined rigid body/torsion angle dynamics. Journal of the American Chemical Society. 2003;**125**(10):2902-2912

[83] Dominguez C, Boelens R, Bonvin AMJJ. HADDOCK: A protein–protein docking approach based on biochemical or biophysical information. Journal of the American Chemical Society. 2003;**125**(7):1731-1737

[84] van Dijk Aalt DJ, Boelens R, Bonvin Alexandre MJJ. Data-driven docking for the study of biomolecular complexes. The FEBS Journal. 2004;**272**(2):293-312

[85] Musselman CA, Gibson MD, Hartwick EW, North JA, Gatchalian J, Poirier MG, et al. Binding of PHF1 Tudor to H3K36me3 enhances nucleosome accessibility. Nature Communications. 2013;**4**:2969

[86] Yang M, Gocke CB, Luo X, Borek D, Tomchick DR, Machius M, et al. Structural basis for CoREST-dependent demethylation of nucleosomes by the human LSD1 histone demethylase. Molecular Cell. 2006;**23**(3):377-387

[87] Xiang S, Le Paige UB, Horn V, Houben K, Baldus M, van Ingen H. Site-specific studies of nucleosome interactions by solid-state NMR. Angewandte Chemie International Edition. 2018;**57**(17):4571-4575

[88] Qiao Q, Li Y, Chen Z, Wang M, Reinberg D, Xu RM. The structure of NSD1 reveals an autoregulatory mechanism underlying histone H3K36 methylation. The Journal of Biological Chemistry. 2011;**286**(10):8361-8368

[89] Hu Q, Botuyan MV, Cui G, Zhao D, Mer G. Mechanisms of ubiquitin-nucleosome recognition and regulation of 53BP1 chromatin recruitment by RNF168/169 and RAD18. Molecular Cell. 2017;**66**(4):473-487. e9

[90] Zhou BR, Feng H, Ghirlando R, Li S, Schwieters CD, Bai Y. A small number of residues can determine if linker histones are bound on or off dyad in the chromatosome. Journal of Molecular Biology. 2016;**428**(20):3948-3959

[91] Dang W, Bartholomew B. Domain architecture of the catalytic subunit in the ISW2-nucleosome complex. Molecular and Cellular Biology. 2007;**27**(23):8306-8317

[92] Musselman CA, Avvakumov N, Watanabe R, Abraham CG, Lalonde M-E, Hong Z, *et al.* Molecular basis for H3K36me3 recognition by the Tudor domain of PHF1. Nature Structural & Molecular Biology. 2012;**19**:1266

[93] Kato H, van Ingen H, Zhou BR, Feng H, Bustin M, Kay LE, et al. Architecture of the high mobility group nucleosomal protein 2-nucleosome complex as revealed by methyl-based NMR. Proceedings of the National Academy of Sciences of the United States of America. 2011;**108**(30):12283-12288

[94] Forneris F, Binda C, Vanoni MA, Battaglioli E, Mattevi A. Human histone demethylase LSD1 reads the histone code. Journal of Biological Chemistry. 2005;**280**(50):41360-41365

[95] Ollerenshaw JE, Tugarinov V, Kay LE. Methyl TROSY: Explanation and experimental verification. Magnetic Resonance in Chemistry. 2003;**41**(10):843-852

[96] Miller TCR, Simon B, Rybin V, Grötsch H, Curtet S, Khochbin S, et al. A bromodomain–DNA interaction facilitates acetylation-dependent bivalent nucleosome recognition by the BET protein BRDT. Nature Communications. 2016;**7**:13855

[97] Hass MAS, Ubbink M. Structure determination of protein–protein complexes with long-range anisotropic paramagnetic NMR restraints. Current Opinion in Structural Biology. 2014;**24**:45-53

[98] Schilder J, Liu WM, Kumar P, Overhand M, Huber M, Ubbink M. Protein docking using an ensemble of spin labels optimized by intra-molecular paramagnetic relaxation enhancement. Physical Chemistry Chemical Physics. 2016;**18**(8):5729-5742

[99] Back JW, de Jong L, Muijsers AO, de Koster CG. Chemical cross-linking and mass spectrometry for protein structural modeling. Journal of Molecular Biology. 2003;**331**(2): 303-313

[100] Sinz A. Chemical cross-linking and mass spectrometry for mapping three-dimensional structures of proteins and protein complexes. Journal of Mass Spectrometry. 2003;**38**(12):1225-1237

[101] Turco E, Gallego LD, Schneider M, Köhler A. Monoubiquitination of histone H2B is intrinsic to the bre1 RING domain-Rad6 interaction and augmented by a second Rad6-binding site on Bre1. Journal of Biological Chemistry. 2015;**290**(9):5298-5310

[102] Kumar P, Wolberger C. Structure of the yeast Bre1 RING domain. Proteins. 2015;**83**(6): 1185-1190

[103] Pachov GV, Gabdoulline RR, Wade RC. On the structure and dynamics of the complex of the nucleosome and the linker histone. Nucleic Acids Research. 2011;**39**(12):5255-5263

Chapter 3

Chromatin Dynamics upon DNA Damage

Judith Miné-Hattab and Xavier Darzacq

Additional information is available at the end of the chapter

http://dx.doi.org/10.5772/intechopen.83559

Abstract

The dynamics organization of the nuclear genome is essential for many biological processes and is often altered in cells from diseased tissue. In the presence of double-strand break (DSBs) in *S. cerevisiae* and some mammalian cell lines, DNA mobility is dramatically altered. These changes in DNA mobility act as a double-edged sword since they promote homologous pairing in diploid yeast for example, but in some cases, they lead to potentially mutagenic DNA repair event and are the source of chromosomal translocations. In this chapter, we will present the state of the art in the field of chromosomes mobility in response to DNA damage. After introducing the importance of genome organization and dynamics, we will present in a clear and accessible manner several methods used in the literature to measure and quantify chromatin mobility inside living cells. We will then give an overview of the important findings in the field, both in yeast and in mammalian cells.

Keywords: chromatin dynamics, DNA repair, homologous recombination, global and local increased mobility, tracking

1. Introduction

1.1. Levels of chromatin organization

The eukaryotic genome is highly packaged into chromatin, a nucleoprotein complex composed of DNA wrapped into nucleosomes. Chromatin displays several levels of organization ranging from the 2-nanometers diameter of the DNA double helix to a few micrometers of chromosome territories in the nucleus.

The primary chromatin structure consists in nucleosomes distributed along a DNA fiber similar to a "beads on a string" structure. For many years, it has been proposed that this structure spontaneously folds into a thicker 30-nm fiber observed *in vitro* [1, 2]. However, several studies failed to observe such a structure inside living cells, and this classical view has been revised [3, 4]. More recently, a study of chromatin structure at super resolution in mammalian cells proposed that nucleosomes associate along the chromatin fiber in heterogeneous groups of varying sized named "nucleosomes clutches" [5]. These clutches are spaced by nucleosomes-depleted regions, and the number of nucleosomes per clutch is very heterogeneous in a given nucleus, arguing against the existence of a well-ordered chromatin fiber. The size of these nucleosomes clutches is dependent on the differentiation state with differentiated cells containing on average larger and denser clutches than stem cells [5]. The existence of nucleosomes clutches remains to be confirmed, and future microscopy studies at super resolution will probably clarify the precise chromatin structure at this scale.

The secondary level of chromatin organization consists of "chromatin loops" formed by long-distance interactions along the chromatin fiber [6]. These loops frequently link promoters and enhancers, correlate with gene activation, and show conservation across cell types and species [6]. The size of chromatin loops is around 100 kilobases long; however, their precise distribution along the genome and their dynamics remain unknown. At a larger scale, these chromatin loops group together to form a larger level of chromatin organization named "Topological Associated Domains" or TADs. TADs are continuous regions of enriched contact frequency, corresponding to about 1 megabases of DNA, in which physical interactions occur relatively frequently [7–9]. It has been proposed that TADs are formed through a dynamics process of loop extrusion [10], and it has been proposed that cohesin and CTCF proteins are associated with the dynamics formation of TADs and loops [11]. Chromatin organization in TADs is not found in all organisms: for example, TADs appear to be absent in *A. thaliana* (genome size 135 Mb) [12, 13] but present in other plants [14]; in *M. pneumonia*, bacterial TAD-like domains of 15–33 kb (named chromosomal interaction domains, CIDs) have been described [15]. In *Saccharomyces cerevisiae* yeast, the primary level of organization appears to be shorter than TADs, with domains of 1–5 genes forming compact gene crumples, or globules, rather than loops [16], although the conclusion that TADs and loops are absent in budding yeast remains contentious [17, 18].

The third level of chromatin organization corresponds to chromosomes compartments [19]. These compartments are constituted of several DNA megabases and regroup several TADs sharing similar characteristics such as chromatin compaction, genes density, etc. The chromosome positioning inside the nucleus is not random and can be altered in cells from diseased tissue [20].

Within these complex levels of organization, chromatin is generally constrained in its motion, but large movements can occur in specific situations [21]. Indeed, chromatin movements up to 1 μm have frequently been reported both in yeast and in mammalian cells [22–24]. Chromatin mobility plays an essential role in many biological processes such as transcription, DNA repair [23, 25, 26], or differentiation [27]. Here, we will focus on the changes in chromatin mobility in response to DNA damage in both mammalian and *Saccharomyces cerevisiae* yeast cells.

1.2. Studying chromatin dynamics in the context of DNA damage

Our genome is constantly damaged by a variety of exogenous and endogenous agents. These DNA damages can result in missing or altered bases, bubbles due to deletion or insertion of a nucleotide, linked pyrimidines, single or double strand breaks, or cross-linked strands [28]. Each human cell undergoes tens of thousands lesions per day, among them, 50 endogenous double-strand breaks (DSBs) per cell cycle [29]. In contrast with their small numbers, DSBs are the most cytotoxic and genotoxic, since both strands of the DNA double helix are simultaneously cut [30]. Failure to repair such lesions leads to genomic instability and/or cell death. In higher eukaryotes, mutations in DNA repair genes lead to diseases such as Werner, Bloom, and other cancer predisposition syndromes [31].

DNA repair is an essential process for genome integrity preservation. Chromatin dynamics have mainly been studied following DSBs, the most deleterious form of DNA damage. Eukaryotic organisms use two major mechanisms to repair DSBs: non-homologous end-joining (NHEJ) and homologous recombination (HR). NHEJ consists in directly ligating the two broken ends with no or minimal end processing [32]. NHEJ can occur throughout the cell cycle and is the preferred pathway in mammalian cells.

In contrast to NHEJ, HR occurs primarily in S/G2 phase cells and uses the undamaged homologous sister chromatid DNA sequence as a template for copying the missing information. The genetics and biochemistry of DSB repair by homologous recombination have been extensively investigated *in vitro* and *in vivo* [33–37]. In eukaryotes, HR is orchestrated by mega-Dalton multiprotein complexes of 500–2000 proteins that co-localize with the DSB [38]. These protein centers can be visualized in living cells as fluorescent foci using fluorescently tagged HR proteins [38, 39]. Among the proteins occupying these centers are the enzymes of the highly conserved Rad52 epistasis group, including Rad51, Rad52, and Rad54 [40, 41]. When a DSB forms and the HR recombination pathway is chosen, the 5′ ends of the DNA break undergo resection by nucleases to yield 3′ single-stranded DNA (ssDNA) tails on which the repair proteins polymerize [34, 42]. This repair protein-ssDNA complex, called a nucleoprotein filament, then searches for homologous sequences among neighboring double-stranded DNA (dsDNA) molecules. A common source for an intact duplex DNA donor is the undamaged sister chromatid; however, homologous sequences on either the homolog or on a different chromosome can be captured by the presynaptic nucleofilament to perform inter-homolog recombination or ectopic recombination, respectively. Once homology is found, the invading strand primes DNA synthesis on the homologous template, ultimately restoring genetic information disrupted by the DSB. The search for a homologous dsDNA across the genome is a key step of HR; however, the pairing of homologous sequences remains the most enigmatic stage of HR with implications reaching beyond the range of DNA repair alone [43].

Considerable progress has been recently made to understand the molecular basis of repair pathway choice, pointing toward cell cycle stage and chromatin landscape as key determinants for the choice of the repair pathway. Although chromatin packing may protect the genome against DNA damage [44], multiple studies suggest that DNA repair processes are less efficient in densely packed heterochromatin [45], leading to an accumulation of mutations in these regions [46]. Recent findings also suggest that DSB occurring in transcriptionally active genes displays dedicated repair mechanisms. Indeed, in contrast to the rest of the euchromatic genome (intergenic and inactive genes), damaged active genes are preferentially repaired by HR in G2 cells, thanks to a chromatin modification (H3K36me3) dependent pathway [47], while in G1, they exhibit delayed repair and enhanced clustering [48]. Such results thus highlight the major impact of the chromatin landscape on the DNA repair processes.

Investigating the nature of DNA diffusion in the context of DNA repair is particularly relevant to understand how cells maintain genome integrity. When a DSB occurs, the two broken ends first need to stay in close proximity, both for NHEJ and HR. Following this first step, chromatin mobility probably differs depending on the repair pathway used. Since HR requires the search for a homologous sequence, many studies investigated chromatin mobility in response to DSB repaired by HR. In the following section, we will present several techniques that have been used in the literature to investigate chromatin mobility.

2. Methods to quantify chromatin dynamics

2.1. Microscopy techniques to visualize chromatin mobility

During the last 15 years, powerful microscopy techniques have allowed the visualization of chromatin mobility inside living cells. One method to image chromatin dynamics consists in uniformly labeling chromatin, using for example fluorescently tagged histones, or DNA intercalant. Local chromatin movements can then be investigated by FRAP (Fluorescence Recovery After Photo-bleaching) or image correlation methods for example [49]. However, this approach is limited in resolution. In the FRAP approach, the size of the laser spot used to photo-bleach or photo-convert the tagged chromatin or labeled DNA probably encompasses several megabases of DNA wrapped around thousands of nucleosomes. Another common labeling approach uses repeated bacterial sequences (*lac* or *tet* operators) integrated into the genome [50]. These *lacO*/*tetO* arrays are bound by LacI/TetR proteins, which are fused to fluorescent proteins (**Figure 1A**). These arrays are visible by microscopy as distinct spots that can be tracked through time to measure the chromatin dynamics. Importantly, the *lacO*/LacI and *tetO*-TetR systems offer the possibility to fluorescently tag genomic loci at a defined genomic locus. Another tagging method, consisting of the ParB-INT DNA labeling system, has also been developed to fluorescently mark genomic loci [51]. To increase the resolution and have access the position and the dynamics of individual histones, a uniform chromatin labeling can be performed using photo-activable fluorophors. Such approach allows

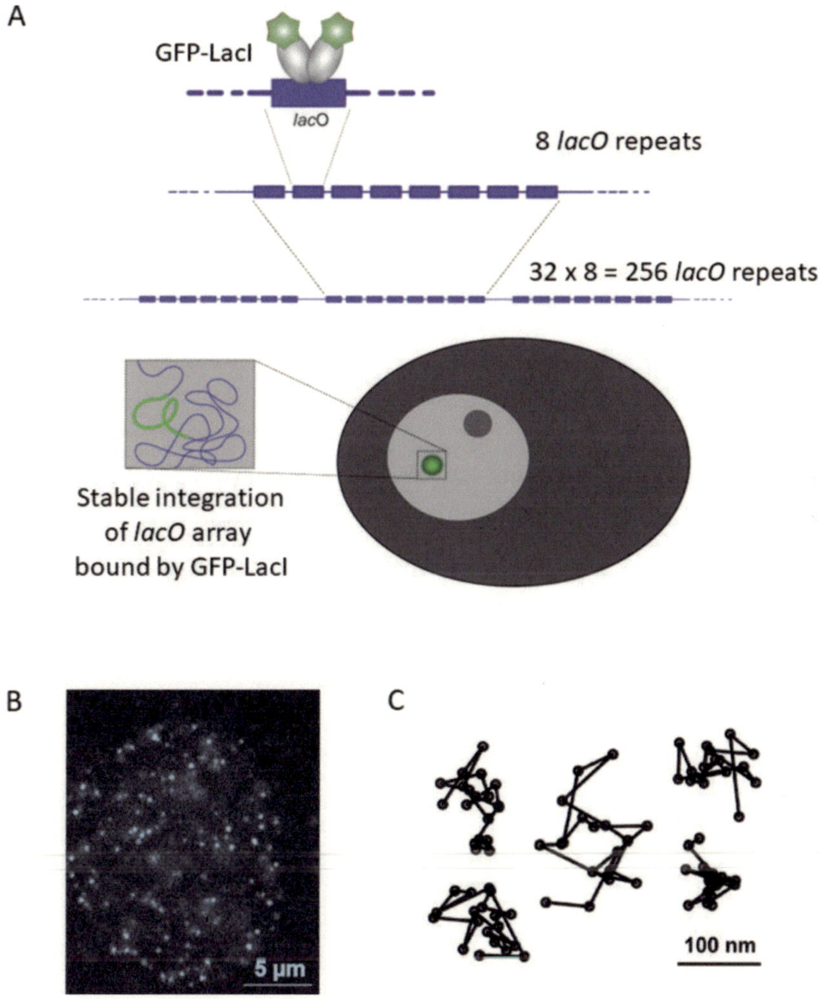

Figure 1. (A) Illustration of the *lacO*/LacI-GFP system to fluorescently mark a specific genomic locus. (B) Single-nucleosome image of a human HeLa cell nucleus expressing H2B-PA-mCherry. Each dot represents single nucleosome [54]. (C) Representative trajectories of fluorescently labeled single nucleosome (50 ms per frame), with permission from [54].

the visualization of chromatin organization at 20 nm resolution [52] or the measurement of histones dynamics [53, 54] (**Figure 1B**). However, in contract with the previous approaches (*lacO*, *tetO* arrays or ParB-INT DNA labeling system), a uniform chromatin labeling does not give access to the DNA sequence to which a specific histone is bound to. To combine the visualization of a single region of the genome at super resolution, a recent technique has been developed, named Oligopaint FISH [55]. Overall, these different approaches allow us to access different scales of chromatin organization and dynamics, from several megabases to a single nucleosome.

2.2. Quantification of chromatin loci mobility

Several theoretical studies have shown that the mode of diffusion of a moving object drastically changes the way it explores the available space. The time to reach a specific target can dramatically change depending on the way a particle samples its surrounding environment [56]. To quantify the mobility of a chromatin locus marked using a lacO/LacI system, the most common method consists in measuring its position (*x, y, z*) over time and calculating its mean square displacement (MSD) (Eq. (1)) [57]. The MSD curve represents the amount of space a locus has explored in the nucleus (**Figure 2A**).

$$MSD(n \cdot \Delta t) = \frac{1}{N-n} \sum_{i=0}^{N-1-n} \left[(x_{i+n} - x_i)^2 + (y_{i+n} - y_i)^2 + (z_{i+n} - z_i)^2 \right] \tag{1}$$

The shape of MSD curves then reveals the nature of DNA motion. Four main types of motion have been described in the literature (**Figure 2B**): confined motion, anomalous sub-diffusion, Brownian motion, and directive motion.

The simplest type of motion is Brownian diffusion: when a particle freely diffuses, its MSD curve is linear with time and its motion is called "Brownian." However, in living cells, DNA motion is often slower than Brownian diffusion and is called "sub-diffusive" [58]. Several

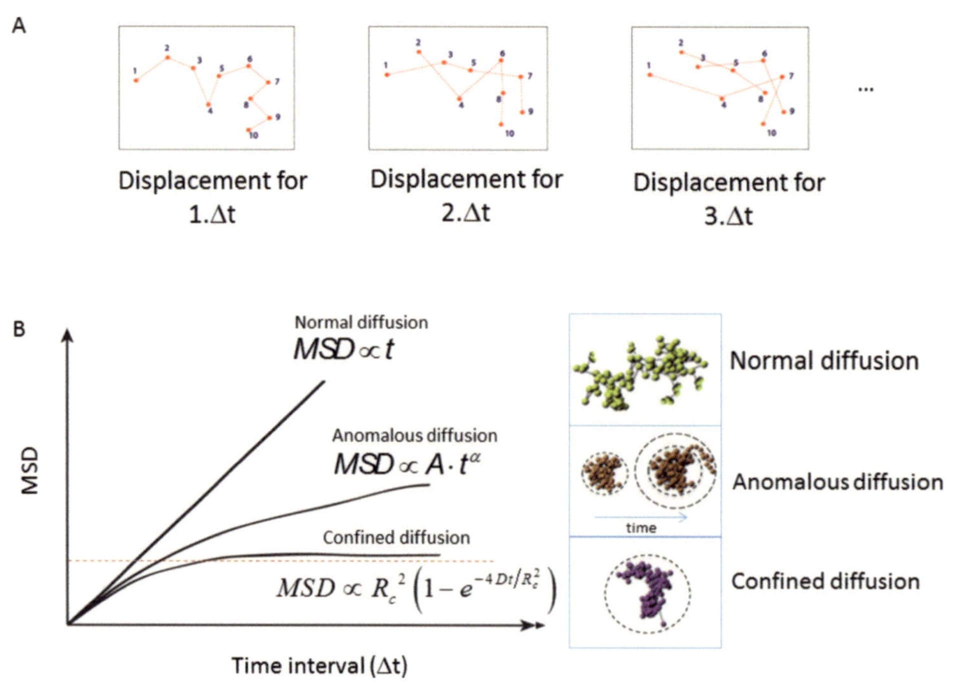

Figure 2. (A) Illustration of trajectories: from the left to the right, points are represented spaced by 1.Δt, 2.Δt and 3.Δt and (B) mean square displacement for normal anomalous and confined diffusion, with a representation of the corresponding trajectories.

types of sub-diffusive motion have been observed. When a chromosomal locus is confined inside a sub-volume of the nucleus, the motion is called *confined sub-diffusion* and the MSD exhibits a plateau. Confined motion has been observed and quantified in living cells, when chromatin motion is examined during several minutes [23, 59, 60]. In that case, the MSD curves can be fitted by $MSD \propto L^2(1 - e^{-4Dt/L^2})$, where L is the plateau of the MSD curve (proportional to the radius of confinement) and D is the diffusion coefficient.

When the force or structure that restricts the motion is not a simple confinement but is modulated in time and space with scaling properties, the motion is called *anomalous sub-diffusion* [58, 61]. In this case, sub-diffusive loci are constrained, but unlike confined loci, they can diffuse without boundary and thus reach further targets if given enough time. For sub-diffusive motion, the MSD exhibits a power law ($MSD \sim At^{\alpha}$), where α, the anomalous exponent, is smaller than 1. The anomalous exponent α is linked to the degree of recurrence of DNA exploration, that is, the number of times a DNA locus reiteratively scans neighboring regions before reaching a distant position [62]. When α is small, the locus explores recurrently the same environment for a long time, while a large α indicates that the locus is able to explore new environments often. The anomalous diffusion coefficient A represents the amplitude of DNA motion; it is proportional to the diffusion coefficient only in the case of normal diffusion (when $\alpha = 1$), which is rarely observed in biological systems [58]. Finally, a moving particle moves in a directive manner toward a target, and the motion is called directive.

The MSD is a standard statistical tool that describes a set of trajectories of similar objects. However, numerous artifacts perturb this statistic. The localization accuracy can have a strong impact on the MSD curve, even computed on simple Brownian motion [63]. Considering the movement of a single photon emitter, the localization accuracy can be divided into:

i. The error in the determination of the accurate particle position due to convolution with the point spread function (PSF) and the finite number of photons. This error is more important for short acquisition times since the number of photons collected is small.

ii. The error due to the movement of the particle during the camera acquisition. This error is more important with higher exposure times and is sometimes referred to as "motion blur."

For 2D Brownian motion with a diffusion coefficient D, Michalet computed the formula of the converged MSD including the corrections for localization accuracy (see Eq. (2)) [64]:

$$MSD(t) = 4Dt + \sigma_0^2\left(1 + \frac{Dt_E}{s_0^2}\right) - \frac{4}{3}Dt_E \quad (2)$$

where σ_0^2 is the localization accuracy of an immobile particle; s_0^2 is the variance of the PSF; t_E is the exposure time of the camera.

The term $4Dt$ is the theoretical MSD for simple Brownian motion. The term $\sigma_0^2\left(1 + \frac{Dt_E}{s_0^2}\right)$ accounts for the *motion blur* of the particle along its path during acquisition. Since the localization accuracy σ_0^2 is inversely proportional to the number of collected photons, we have $\sigma_0^2 \propto \frac{1}{t_E}$. The motion blur term therefore converges to a fixed value as the exposure time increases. The term $\frac{4}{3}Dt_E$ accounts for the correlation between successive displacements due to the exposure overlap.

For anomalous motion with a diffusion coefficient D, the MSD formula including the corrections for localization accuracy is described in [65].

3. Increased mobility in response to DNA damage

3.1. Evidence of increased mobility in response to DSB

Most of the studies on chromatin mobility are based on the analysis of MSD curves calculated from the trajectories of fluorescently labeled chromatin loci. Using this approach, the diffusion coefficients reported in the literature varies from 5×10^{-5} to 10^{-3} $\mu m^2/s$ depending on the organisms, the loci studied, and the type of damage [66, 67]. In several studies, chromatin undergoes confined diffusion [22, 59, 68–71], while others have reported anomalous diffusion [69, 70, 72–75]. So far, no consensus has been reached to describe the nature of DNA motion probably because these studies have been performed using different microscopy techniques and illumination settings. Indeed, multi time-scales observation of chromatin motion revealed that chromatin is driven by different types of diffusion at each time scale [76–78]. As a consequence, the type of diffusion depends on the time scale of observation. While comparing studies on chromatin dynamics, it is thus important to compare studies performed at similar time scales.

Several studies have investigated DNA motion in the context of DNA damage. Since DSB is the most deleterious type of damage in the cell, chromatin dynamics in the context of DNA repair has been mainly investigated in response to DSBs. The changes in chromatin architecture and dynamics following DSB have been studied mostly in yeast, Drosophila and mammalian nuclei. In budding yeast, chromatin mobility has been investigated during the process of HR, when a Rad52 focus is already formed at the damaged locus [23, 25]. In diploid, where a homologous template is available, chromatin mobility is dramatically increased at the damaged site, allowing the damaged locus to explore a nuclear volume 10 times larger [23]. Increased mobility may facilitate homology search; however, haploid yeast cells, where no homologous template is present, also exhibit increased mobility in response to DSBs [25]. Since the two broken ends stay in contact during the process of HR repair [79], the current view is that the two broken ends explore the nuclear space together.

Importantly, only induced DSB associated to Rad52 foci display increased mobility. Indeed, during resection, the early stage of HR, a strong inhibition of chromatin mobility has been reported in yeast, highlighting the importance of the stage of DNA repair in chromatin mobility changes [51]. Finally, different types of damages have very different consequences on DNA mobility. For example, in yeast, spontaneous DSBs occurring during DNA replication exhibit decreased mobility [80]; DSBs induced by a protein-DNA adduct display no change in motion and camptothecin (CPT)-induced Rad52 foci display no increased mobility [25].

In mammalian cells, while several studies have reported increased chromatin mobility upon DNA damage, others fail to observe significant changes. In HeLa cells, after α-particle-induced DSBs, γH2AX foci are more mobile [81, 82]. Similarly, uncapped telomeres exhibit increased mobility in mouse cells, and this movement is dependent on the 53BP1 repair protein [83]. Movement of heterochromatic DSBs toward euchromatin was observed in mouse embryo fibroblasts (MEFs), HeLa cells [84], and Drosophila cells [45]. It was proposed that re-localization of heterochromatic DSBs close to euchromatin regions prevents rearrangements between repetitive DNA sequences present in heterochromatin. Taken together, these studies suggest that chromosome mobility increases significantly in the presence of DSBs. By contrast, in other studies using MEFs [85], HeLa, or U2OS cells [86], DSBs generated by UV laser or γ-irradiation did not significantly alter chromosome mobility. Only energy-dependent local expansion of chromatin was observed around the initial damaged zone immediately after DNA damage [85]. These contradictory observations in mammalian cells probably result from variation between cell lines, the regions of chromatin damaged, and the type of damage induced. Recent studies suggest that DSBs induced in active genes, naturally enriched in the trimethyl form of histone H3 lysine 36 (H3K36me3), are repaired by HR [87]. These DSBs are susceptible to exhibit increased mobility, while DSBs repaired by Non-Homologous End Joining are rather immobile. Further systematic studies will be necessary to confirm these observations.

3.2. Local versus global increased mobility

Changes in chromatin conformation have been extensively described around the site of damage but an important question is whether these changes in chromatin mobility also affect the rest of the genome. Interestingly, in budding yeast, increased chromatin mobility is not an intrinsic property of the damaged locus. Indeed, in diploid yeast, after induction of four random DSBs per nucleus by γ-irradiation, undamaged loci explore a 2.4 times larger nuclear volume than in the absence of irradiation [23]. Moreover, the global increased mobility is dose-dependent since upon induction of approximately 20 DSBs, the chromosomes explore almost the entire yeast nucleus [23]. **Figure 3** illustrates the mobility observed for different levels of γ-irradiation in a diploid yeast cell. Global increased mobility is observed in haploid yeast, although it required higher doses of damages and it has been tested with a different type of DSBs (zeocin-induced DSBs) [88].

In mammalian cells, changes in mobility far from a damaged locus are not reported in the literature. Since mammalian nuclei are much larger than the yeast nuclei, but chromatin motion exhibits very similar constrained (~0.5 μm of confined radius), it is likely that global mobility is specific to organisms with small nuclei and is therefore not present in mammalian cells.

Importantly, most of these studies investigated chromatin mobility at one specific time scale. However, when studying the diffusion of a specific locus, the time scale at which data are collected reflects a specific spatial scale of the exploration studied. From nucleosomes to fiber, the different scales of chromatin organization might exhibit different diffusion behaviors. Using fast microscopy, a recent study investigated DNA mobility at several time scales, up to 1000 times faster than previously observed [65]. These experiments revealed that DNA motion following DNA damage is more complex than what had been previously described. Chromatin dynamics therefore appears to be scale-dependent: in response to DNA damage,

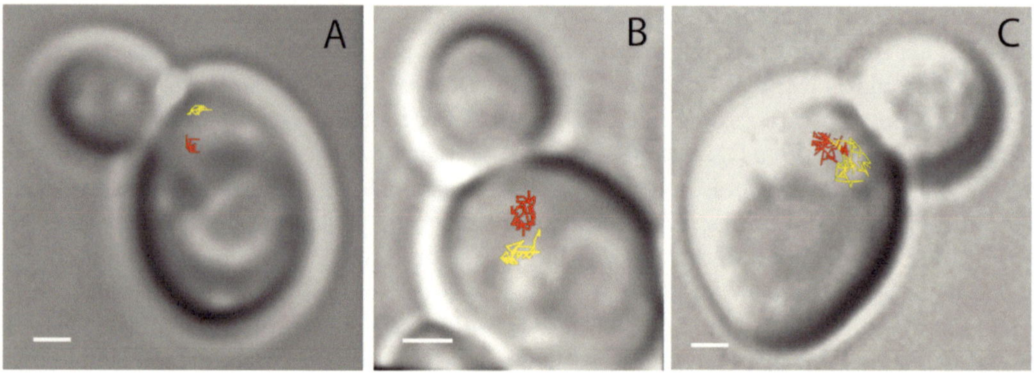

Figure 3. Examples of the dynamics of *URA3* loci (chromosome V) in budding yeast as a function of the number of double-strand breaks (DSBs) in the nucleus. The lines indicate a 2D projection of the trajectories of the two *URA3* loci taken at between 10- and 30-s time intervals for approximately 15 min. (A) In the absence of DSBs, the two homologous loci are distant and explore only 3% of the nuclear volume. (B) In the presence of one to four DSBs induced on chromosomes III, the mobility of the *URA3* loci increases and each locus can explore 11% of the nuclear volume. (C) After about 20 random γ-irradiation-induced DSBs per nucleus (200 Gy), *URA3* loci explore almost the entire nuclear volume and their trajectories overlap. The scale is 1 μm. These three examples illustrate that more DSBs in the nucleus induce greater DNA mobility, thereby increasing the probability of collisions between loci (with permission from [66]).

chromatin is more mobile at large time scales but, surprisingly, its mobility is reduced at short time scales, this effect being stronger at the damaged site. Such a pattern of dynamics is consistent with a global chromatin stiffness that has been proposed to arise in response to DNA damage [65, 89, 90]. These results underline the importance of performing multiscale tracking to fully understand the complex dynamics of chromatin at each scale.

3.3. DSBs clustering

In addition to the increased chromatin mobility, several studies indicate that multiple DSBs cluster together into a single "repair center." Clustering of unrelated DSBs might be a consequence of increased mobility, although such clusters of DSBs might promote translocations and their functions is not clear. In yeast, several DNA lesions collapse together into the same repair focus, suggesting that these multiple DSBs are driven to a shared location, in so-called "repair centers" or "repair factories" [91]. In mammalian cells, clusters of radiation-induced foci have been observed in many independent studies [45, 81, 85, 86]. DSB clusters could be formed by collisions and fusion of several DSBs. One interesting hypothesis is that repair foci have a higher viscosity than the rest of the nucleus, due to their high concentration of repair proteins around the damaged site. It has been proposed that several repair foci act as dynamic liquid droplets that are able to fuse together when they collide [92].

3.4. Factors controlling local and global increased mobility

To investigate the mechanism of increased chromatin mobility in response to DSBs, several studies have tested chromatin mobility in mutant cells [23, 25, 65, 81, 83, 88, 90, 93–95].

These studies revealed that several genes involved in DSBs repair and chromatin remodeling are involved in chromatin mobility changes. In diploid yeast, Rad51, the central protein of HR, is required for chromate in mobility [23], as well as for increased chromatin rigidity [65]. Several checkpoint proteins acting earlier than Rad51 nucleoprotein filament formation are also essential for increasing the mobility of the damaged locus. For example, Rad9, a protein containing a BRCT domain roughly equivalent to human MDC1, BRCA1, and 53BP1, is also required for increased mobility at the damaged site in haploid yeast [25]. MEC1 and SML1, but not RAD53 and Tel1, are essential for increasing the mobility of the damaged locus [23, 25, 66]. Interestingly, the activation of the checkpoint protein Mec1 at a specific locus is sufficient to promote increased mobility, even in the absence of physical DSB [88]. Finally, in mouse cells, increased chromosome movements are associated with uncapped telomeres, and this movement is dependent on the 53BP1 repair protein [83].

Another study proposed that centromere and telomere release following DSBs are at the origin of chromatin changes in mobility. Strecker et al. found that a combined disruption of telomeres and centromeres can reproduce chromatin mobility observed after a DSB [96]; they identified the Mec1-dependent phosphorylation of Cep3, a kinetochore component, as an essential player in global increased chromatin mobility on DSBs.

More recently, it has been shown that molecular motors play an important role in DSBs mobility [93–95]. For example, in fly in human genomes, heterochromatin constitutes about 30% of the genome [97], and "safe" repair of heterochromatic DSBs by homologous recombination relies on the relocalization of repair foci to the nuclear periphery. During this process, nuclear actin filaments form at repair sites to drive heterochromatin DSBs at the periphery and disassemble after relocalization [93]. Actin filaments act in concert with Smc5/6, Arp2/3, Arp2/3 activators Scar and Wash, nuclear myosins Myo1A, Myo1B, and MyoV. Interestingly, in U2OS cells, ARP2/3-mediated actin polymerization enhances DSBs motion during homologous recombination, increasing the clustering of repair foci [94]. In budding yeast, DNA-damaged induced nuclear microtubule filaments (DIMs) form in response to endogenous or exogenous DNA damage [95]. These DIM filaments, formed at repair sites, reach the nuclear periphery to dive irreparable DSBs and disassemble after relocalization. Such DSBs motion is mediated by the Rad9 DNA damage response mediator and the Kar3 kinesin motor. Another model implicating microtubules has been proposed by Lawrimore et al. to explain the global increased mobility observed in yeast in response to DNA damage: in their model, microtubules would be responsible for a global chromatin shake-up that would be essential for global increase mobility on DSBs [98].

In these different examples, DSBs mobility is promoted by molecular motors, and DSBs exhibit a complex motion including mixture of directive and Brownian motions [93] or nonlinear directive motion [95]. Overall, these recent studies revealed the essential role of molecular motors in DSBs mobility to drive heterochromatic or irreparable DSBs to the nuclear periphery or in clustering of multiple DSBs. Importantly, these mechanisms are conserved through several organisms (human, *Drosophila melanogaster*, *Xenopus laevis*, and budding yeast).

3.5. Modifications of chromatin compaction in response to DSB and perspectives

In addition to chromatin mobility, many studies investigated the modulation of the chromatin compaction state both at a specific damaged site and throughout the genome. Several studies showed a chromatin decondensation visible at the micrometer scale accessible by conventional light microscopy [85, 99]. In the recent studies, super resolution imaging of a *lacO* array before and after damage allows the visualization of chromatin decompaction at the damaged site in haploid yeast [90, 100]. In mammalian cells, most of the studies report chromatin decondensation at the damaged site. However, it has been shown that following this initial fast decondensation, the damaged chromatin area slowly recondenses to reach higher compaction levels than before damage induction [101]. In addition, both chromatin expansion and compaction occur at the same time but in different regions of the chromatin near the DSBs [49]. Overall, chromatin changes in compaction are tuned in space and time upon DSBs, but the precise role of each step remains to be elucidated.

An interesting way to interpret chromatin changes in dynamics and compaction upon DSBs is to think in terms of mechanical properties of chromatin, such as chromatin stiffness (or persistence length). As illustrated in **Figure 4**, the persistence length of a polymer is a mechanical property that quantifies its stiffness. The persistence length is the length over which correlations in the direction of the tangent are lost.

Changes in chromatin persistence length following DNA damage have been discussed, however, with contradictories interpretations. While some studies suggest that chromatin is more flexible following DSBs [100, 102, 103], other results indicate that chromatin is globally stiffer upon DSBs [65, 90]. One proposed explanation for chromatin stiffening upon DSBs could be the presence of negative charges due to H2A S129 phosphorylation [90]. Further studies will be required to solve to this open question and more generally to understand the physical mechanisms underlying the modifications of chromatin dynamics in response to DNA damage.

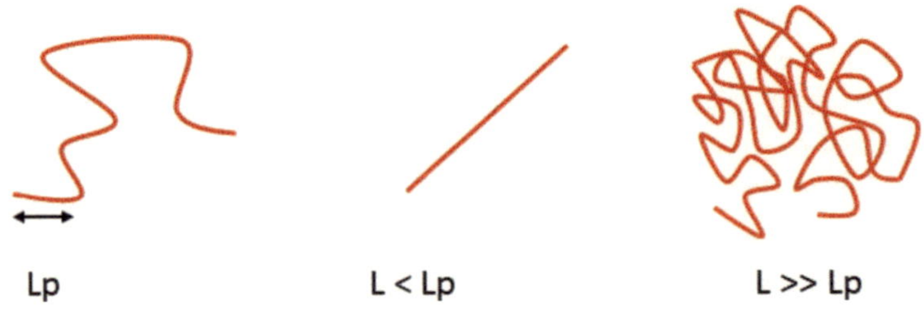

Figure 4. Illustration of the persistence of a polymer, where Lp is the polymer persistence length and L is the polymer length.

4. Conclusion

Thanks to recent advanced in fast and high-resolution microscopy, it became possible to quantify chromatin mobility with unprecedented precision and to understand how chromatin explores the nuclear space. Following DSBs, chromatin mobility is dramatically altered in budding yeast both at the site of DNA damage and genome wide. In mammalian cells, DSBs mobility is strongly influenced by cell cycle stage, chromatin state, and repair pathway choice. For example, while DSBs induced in pericentric heterochromatin during the S and G2 phases of the cell cycle are more mobile and relocate to the nuclear periphery, DSBs generated during the G1 phase remain stable. Changes of chromatin mobility upon DNA damage is an intriguing phenomenon, and over the last five years, several views to explain it have been proposed in the literature. They can be grouped into two classes: (1) increase in chromatin motion is due to intrinsic chromatin modifications which require chromatin remodelers, kinases, and repair proteins involved in the DNA response machinery; (2) increase in chromatin motion due to changes of external mechanical constraints that maintain chromatin and to the action of molecular motors. It is likely that both mechanisms act in concert to drive differently damaged chromatin depending on the type of damages, the chromatin state, or the cell cycle.

Several questions remain open: how changes in chromatin dynamics alter its organization at the scale of TADs? Is there a global change in mechanical properties of the chromatin upon DSBs, such as an increase in chromatin rigidity? In the future, it will be an exciting challenge to investigate changes in chromatin organization and dynamics upon DNA damage combining different approaches from live cell microscopy, super resolution imaging, and Hi-C.

Acknowledgements

This work was funded by the ANR-12-PDOC-0035-01.

Author details

Judith Miné-Hattab[1,2]* and Xavier Darzacq[3]

*Address all correspondence to: judith.mine@curie.fr

1 Institut Curie, PSL Research University, CNRS, UMR3664, Paris, France

2 Institut Curie, Sorbonne Université, CNRS, UMR3664, Paris, France

3 Department of Molecular and Cell Biology, Li Ka Shing Center for Biomedical and Health Sciences, CIRM Center of Excellence, University of California, Berkeley, United States

References

[1] Robinson PJ et al. EM measurements define the dimensions of the "30-nm" chromatin fiber: Evidence for a compact, interdigitated structure. Proceedings of the National Academy of Sciences of the United States of America. 2006;**103**(17):6506-6511

[2] Schalch T et al. X-ray structure of a tetranucleosome and its implications for the chromatin fibre. Nature. 2005;**436**(7047):138-141

[3] Joti Y et al. Chromosomes without a 30-nm chromatin fiber. Nucleus. 2012;**3**(5):404-410

[4] Maeshima K, Hihara S, Eltsov M. Chromatin structure: Does the 30-nm fibre exist in vivo? Current Opinion in Cell Biology. 2010;**22**(3):291-297

[5] Ricci MA et al. Chromatin fibers are formed by heterogeneous groups of nucleosomes in vivo. Cell. 2015;**160**(6):1145-1158

[6] Rao SS et al. A 3D map of the human genome at kilobase resolution reveals principles of chromatin looping. Cell. 2014;**159**(7):1665-1680

[7] Nora EP et al. Spatial partitioning of the regulatory landscape of the X-inactivation centre. Nature. 2012;**485**(7398):381-385

[8] Dixon JR et al. Topological domains in mammalian genomes identified by analysis of chromatin interactions. Nature. 2012;**485**(7398):376-380

[9] Sexton T et al. Three-dimensional folding and functional organization principles of the Drosophila genome. Cell. 2012;**148**(3):458-472

[10] Fudenberg G et al. Formation of chromosomal domains by loop extrusion. Cell Reports. 2016;**15**(9):2038-2049

[11] Hansen AS et al. Recent evidence that TADs and chromatin loops are dynamic structures. Nucleus. 2018;**9**(1):20-32

[12] Feng S et al. Genome-wide Hi-C analyses in wild-type and mutants reveal high-resolution chromatin interactions in Arabidopsis. Molecular Cell. 2014;**55**(5):694-707

[13] Wang C et al. Genome-wide analysis of local chromatin packing in *Arabidopsis thaliana*. Genome Research. 2015;**25**(2):246-256

[14] Wang M et al. Evolutionary dynamics of 3D genome architecture following polyploidization in cotton. Nature Plants. 2018;**4**(2):90-97

[15] Trussart M et al. Defined chromosome structure in the genome-reduced bacterium *Mycoplasma pneumoniae*. Nature Communications. 2017;**8**:14665

[16] Hsieh TH et al. Mapping nucleosome resolution chromosome folding in yeast by micro-C. Cell. 2015;**162**(1):108-119

[17] Eser U et al. Form and function of topologically associating genomic domains in budding yeast. Proceedings of the National Academy of Sciences of the United States of America. 2017;**114**(15):E3061-E3070

[18] Lazar-Stefanita L et al. Cohesins and condensins orchestrate the 4D dynamics of yeast chromosomes during the cell cycle. The EMBO Journal. 2017;**36**(18):2684-2697

[19] Lieberman-Aiden E et al. Comprehensive mapping of long-range interactions reveals folding principles of the human genome. Science. 2009;**326**(5950):289-293

[20] Misteli T. Higher-order genome organization in human disease. Cold Spring Harbor Perspectives in Biology. 2010;**2**(8):a000794

[21] Chubb JR, Bickmore WA. Considering nuclear compartmentalization in the light of nuclear dynamics. Cell. 2003;**112**(4):403-406

[22] Heun P et al. Chromosome dynamics in the yeast interphase nucleus. Science. 2001; **294**(5549):2181-2186

[23] Miné-Hattab J, Rothstein R. Increased chromosome mobility facilitates homology search during recombination. Nature Cell Biology. 2012;**14**(5):510-517

[24] Levi V et al. Chromatin dynamics in interphase cells revealed by tracking in a two-photon excitation microscope. Biophysical Journal. 2005;**89**(6):4275-4285

[25] Dion V et al. Increased mobility of double-strand breaks requires Mec1, Rad9 and the homologous recombination machinery. Nature Cell Biology. 2012;**14**(5):502-509

[26] Roukos V et al. Spatial dynamics of chromosome translocations in living cells. Science. 2013;**341**(6146):660-664

[27] Meshorer E et al. Hyperdynamic plasticity of chromatin proteins in pluripotent embryonic stem cells. Developmental Cell. 2006;**10**(1):105-116

[28] Lodish H, Berk A, Zipursky S. DNA damaged and repair and their role in carcinogenesis. In: Molecular Cell Biology. 4th ed. New York: W. H. Freeman; 2000

[29] Vilenchik MM, Knudson AG. Endogenous DNA double-strand breaks: Production, fidelity of repair, and induction of cancer. Proceedings of the National Academy of Sciences of the United States of America. 2003;**100**(22):12871-12876

[30] Wyman C, Kanaar R. DNA double-strand break repair: All's well that ends well. Annual Review of Genetics. 2006;**40**:363-383

[31] Rassool FV. DNA double strand breaks (DSB) and non-homologous end joining (NHEJ) pathways in human leukemia. Cancer Letters. 2003;**193**(1):1-9

[32] Symington LS, Rothstein R, Lisby M. Mechanisms and regulation of mitotic recombination in *Saccharomyces cerevisiae*. Genetics. 2014;**198**(3):795-835

[33] Lisby M et al. Choreography of the DNA damage response: Spatiotemporal relationships among checkpoint and repair proteins. Cell. 2004;**118**(6):699-713

[34] Finkelstein IJ, Greene EC. Single molecule studies of homologous recombination. Molecular BioSystems. 2008;**4**(11):1094-1104

[35] Sugawara N, Haber JE. Repair of DNA double strand breaks: In vivo biochemistry. Methods in Enzymology. 2006;**408**:416-429

[36] Sugawara N, Wang X, Haber JE. In vivo roles of Rad52, Rad54, and Rad55 proteins in Rad51-mediated recombination. Molecular Cell. 2003;**12**(1):209-219

[37] Lisby M, Rothstein R. Cell biology of mitotic recombination. Cold Spring Harbor Perspectives in Biology. 2015;**7**(3):a016535

[38] Lisby M, Rothstein R, Mortensen UH. Rad52 forms DNA repair and recombination centers during S phase. Proceedings of the National Academy of Sciences of the United States of America. 2001;**98**(15):8276-8282

[39] Nagy Z, Soutoglou E. DNA repair: Easy to visualize, difficult to elucidate. Trends in Cell Biology. 2009;**19**(11):617-629

[40] Krogh BO, Symington LS. Recombination proteins in yeast. Annual Review of Genetics. 2004;**38**:233-271

[41] Lisby M, Rothstein R. Choreography of recombination proteins during the DNA damage response. DNA Repair (Amst). 2009;**8**(9):1068-1076

[42] Miné J et al. Real-time measurements of the nucleation, growth and dissociation of single Rad51-DNA nucleoprotein filaments. Nucleic Acids Research. 2007;**35**(21):7171-7187

[43] Barzel A, Kupiec M. Finding a match: How do homologous sequences get together for recombination? Nature Reviews. Genetics. 2008;**9**(1):27-37

[44] Takata H et al. Chromatin compaction protects genomic DNA from radiation damage. PLoS One. 2013;**8**(10):e75622

[45] Chiolo I et al. Double-strand breaks in heterochromatin move outside of a dynamic HP1a domain to complete recombinational repair. Cell. 2011;**144**(5):732-744

[46] Schuster-Bockler B, Lehner B. Chromatin organization is a major influence on regional mutation rates in human cancer cells. Nature. 2012;**488**(7412):504-507

[47] Aymard F et al. Transcriptionally active chromatin recruits homologous recombination at DNA double-strand breaks. Nature Structural & Molecular Biology. 2014;**21**(4):366-374

[48] Aymard F et al. Genome-wide mapping of long-range contacts unveils clustering of DNA double-strand breaks at damaged active genes. Nature Structural & Molecular Biology. 2017;**24**(4):353-361

[49] Hinde E et al. Chromatin dynamics during DNA repair revealed by pair correlation analysis of molecular flow in the nucleus. Biophysical Journal. 2014;**107**(1):55-65

[50] Robinett CC et al. In vivo localization of DNA sequences and visualization of large-scale chromatin organization using lac operator/repressor recognition. The Journal of Cell Biology. 1996;**135**(6 Pt 2):1685-1700

[51] Saad H et al. DNA dynamics during early double-strand break processing revealed by non-intrusive imaging of living cells. PLoS Genetics. 2014;**10**(3):e1004187

[52] Recamier V et al. Single cell correlation fractal dimension of chromatin: A framework to interpret 3D single molecule super-resolution. Nucleus. 2014;**5**(1):75-84

[53] Nozaki T et al. Dynamic organization of chromatin domains revealed by super-resolution live-cell imaging. Molecular Cell. 2017;**67**(2):282-293 e7

[54] Shinkai S et al. Dynamic nucleosome movement provides structural information of topological chromatin domains in living human cells. PLoS Computational Biology. 2016;**12**(10):e1005136

[55] Beliveau BJ et al. Single-molecule super-resolution imaging of chromosomes and in situ haplotype visualization using Oligopaint FISH probes. Nature Communications. 2015;**6**:7147

[56] Guerin T, Benichou O, Voituriez R. Non-Markovian polymer reaction kinetics. Nature Chemistry. 2012;**4**(7):568-573

[57] Meister P et al. Visualizing yeast chromosomes and nuclear architecture. Methods in Enzymology. 2010;**470**:535-567

[58] Barkai E, Garini Y, Metzler R. Strange kinetics of single molecules in living cells. Physics Today. 2012;**65**(8):29-35

[59] Marshall WF et al. Interphase chromosomes undergo constrained diffusional motion in living cells. Current Biology. 1997;**7**(12):930-939

[60] English BP et al. Single-molecule investigations of the stringent response machinery in living bacterial cells. Proceedings of the National Academy of Sciences of the United States of America. 2011;**108**(31):E365-E373

[61] Metzler R et al. Anomalous diffusion models and their properties: Non-stationarity, non-ergodicity, and ageing at the centenary of single particle tracking. Physical Chemistry Chemical Physics. 2014;**16**(44):24128-24164

[62] Ben-Avraham D, Havlin S. Diffusion and Reactions in Fractals and Disordered Systems. Cambridge United Kingdom: Cambridge University Press; 2000

[63] Saxton MJ, Jacobson K. Single-particle tracking: Applications to membrane dynamics. Annual Review of Biophysics and Biomolecular Structure. 1997;**26**:373-399

[64] Michalet X. Mean square displacement analysis of single-particle trajectories with localization error: Brownian motion in an isotropic medium. Physical Review. E, Statistical, Nonlinear, and Soft Matter Physics. 2010;**82**(4 Pt 1):041914

[65] Mine-Hattab J et al. Multi-scale tracking reveals scale-dependent chromatin dynamics after DNA damage. Molecular Biology of the Cell. 2017

[66] Miné-Hattab J, Rothstein R. DNA in motion during double-strand break repair. Trends in Cell Biology. 2013;**23**(11):529-536

[67] Lebeaupin T et al. Chromatin dynamics at DNA breaks: What, how and why? AIMS Biophysics. 2015;**2**(4):458-475

[68] Masui O et al. Live-cell chromosome dynamics and outcome of X chromosome pairing events during ES cell differentiation. Cell. 2011;**145**(3):447-458

[69] Backlund MP, Joyner R, Moerner WE. Chromosomal locus tracking with proper accounting of static and dynamic errors. Physical Review. E, Statistical, Nonlinear, and Soft Matter Physics. 2015;**91**(6-1). DOI: 062716

[70] Cabal GG et al. SAGA interacting factors confine sub-diffusion of transcribed genes to the nuclear envelope. Nature. 2006;**441**(7094):770-773

[71] Taddei A et al. Nuclear pore association confers optimal expression levels for an inducible yeast gene. Nature. 2006;**441**(7094):774-778

[72] Weber SC, Spakowitz AJ, Theriot JA. Bacterial chromosomal loci move subdiffusively through a viscoelastic cytoplasm. Physical Review Letters. 2010;**104**(23):238102

[73] Hajjoul H et al. High-throughput chromatin motion tracking in living yeast reveals the flexibility of the fiber throughout the genome. Genome Research. 2013;**23**(11): 1829-1838

[74] Lucas JS et al. 3D trajectories adopted by coding and regulatory DNA elements: First-passage times for genomic interactions. Cell. 2014;**158**(2):339-352

[75] Burnecki K et al. Universal algorithm for identification of fractional Brownian motion. A case of telomere subdiffusion. Biophysical Journal. 2012;**103**(9):1839-1847

[76] Bronstein I et al. Transient anomalous diffusion of telomeres in the nucleus of mammalian cells. Physical Review Letters. 2009;**103**(1):018102

[77] Miné-Hattab J et al. Fast imaging of DNA motion reveals distinct sub-diffusion regimes at the site of DNA damage. BioRXiv. 2016

[78] Vazquez J, Belmont AS, Sedat JW. Multiple regimes of constrained chromosome motion are regulated in the interphase Drosophila nucleus. Current Biology. 2001;**11**(16): 1227-1239

[79] Lisby M, Rothstein R. DNA repair: Keeping it together. Current Biology. 2004;**14**(23): R994-R996

[80] Dion V et al. Cohesin and the nucleolus constrain the mobility of spontaneous repair foci. EMBO Reports. 2013;**14**(11):984-991

[81] Aten JA et al. Dynamics of DNA double-strand breaks revealed by clustering of damaged chromosome domains. Science. 2004;**303**(5654):92-95

[82] Krawczyk PM et al. Chromatin mobility is increased at sites of DNA double-strand breaks. Journal of Cell Science. 2012;**125**(Pt 9):2127-2133

[83] Dimitrova N et al. 53BP1 promotes non-homologous end joining of telomeres by increasing chromatin mobility. Nature. 2008;**456**(7221):524-528

[84] Jakob B et al. DNA double-strand breaks in heterochromatin elicit fast repair protein recruitment, histone H2AX phosphorylation and relocation to euchromatin. Nucleic Acids Research. 2011;**39**(15):6489-6499

[85] Kruhlak MJ et al. Changes in chromatin structure and mobility in living cells at sites of DNA double-strand breaks. The Journal of Cell Biology. 2006;**172**(6):823-834

[86] Jakob B et al. Live cell microscopy analysis of radiation-induced DNA double-strand break motion. Proceedings of the National Academy of Sciences of the United States of America. 2009;**106**(9):3172-3177

[87] Clouaire T, Legube G. DNA double strand break repair pathway choice: A chromatin based decision? Nucleus. 2015;**6**(2):107-113

[88] Seeber A, Dion V, Gasser SM. Checkpoint kinases and the INO80 nucleosome remodeling complex enhance global chromatin mobility in response to DNA damage. Genes & Development. 2013;**27**(18):1999-2008

[89] Faller R, Müller-Plathe F. Chain stiffness intensifies the reptation characteristics of polymer dynamics in the melt. Chemphyschem. 2008;**2**(3):180-184

[90] Herbert S et al. Chromatin stiffening underlies enhanced locus mobility after DNA damage in budding yeast. The EMBO Journal. 2017;**36**(17):2595-2608

[91] Lisby M, Mortensen UH, Rothstein R. Colocalization of multiple DNA double-strand breaks at a single Rad52 repair centre. Nature Cell Biology. 2003;**5**(6):572-577

[92] Altmeyer M et al. Liquid demixing of intrinsically disordered proteins is seeded by poly(ADP-ribose). Nature Communications. 2015;**6**:8088

[93] Caridi CP et al. Nuclear F-actin and myosins drive relocalization of heterochromatic breaks. Nature. 2018;**559**(7712):54-60

[94] Schrank BR et al. Nuclear ARP2/3 drives DNA break clustering for homology-directed repair. Nature. 2018;**559**(7712):61-66

[95] Oshidari R et al. Nuclear microtubule filaments mediate non-linear directional motion of chromatin and promote DNA repair. Nature Communications. 2018;**9**(1):2567

[96] Strecker J et al. DNA damage signalling targets the kinetochore to promote chromatin mobility. Nature Cell Biology. 2016;**18**(3):281-290

[97] Ho JW et al. Comparative analysis of metazoan chromatin organization. Nature. 2014; **512**(7515):449-452

[98] Lawrimore J et al. Microtubule dynamics drive enhanced chromatin motion and mobilize telomeres in response to DNA damage. Molecular Biology of the Cell. 2017;**28**(12): 1701-1711

[99] Ziv Y et al. Chromatin relaxation in response to DNA double-strand breaks is modulated by a novel ATM- and KAP-1 dependent pathway. Nature Cell Biology. 2006;**8**(8): 870-876

[100] Amitai A et al. Visualization of chromatin decompaction and break site extrusion as predicted by statistical polymer modeling of single-locus trajectories. Cell Reports. 2017;**18**(5):1200-1214

[101] Burgess RC et al. Activation of DNA damage response signaling by condensed chromatin. Cell Reports. 2014;**9**(5):1703-1717

[102] Seeber A, Dion V, Gasser SM. Remodelers move chromatin in response to DNA damage. Cell Cycle. 2014;**13**(6):877-878

[103] Hauer MH et al. Histone degradation in response to DNA damage enhances chromatin dynamics and recombination rates. Nature Structural & Molecular Biology. 2017;**24**(2): 99-107

Chapter 4

A Consistent Systems Mechanics Model of the 3D Architecture and Dynamics of Genomes

Tobias A. Knoch

Additional information is available at the end of the chapter

http://dx.doi.org/10.5772/intechopen.89836

Abstract

Already for thousands of years mankind is aware of inheritance and its manipulation by mating and breeding. The discovery of the cell nucleus by A. van Leeuwenhook in the 17th century marks a start to elucidate the epically discussed evolutionary transfer of information in detail. Now after more than 170 years of research on the 3D architecture and dynamics of genomes and the co-evolved interaction networks of regulatory elements creating genome function - i.e. the storage, replication, and expression of genetic information—a consistent systems statistical mechanics genomics framework emerges for the first time. Obviously the structure and function of genomes co-evolved as an inseparable system allowing the physical storage, expression, and replication of genetic information. The DNA double helix and the nucleosome had been determined structurally at the very highest level already, including genome sequences and epigenetic histone modifications. That chromosomes form territories with functional relevant positioning within the cell nucleus and that chromosomal subdomains exist has been also determined to a fair degree of detail. Only recently, however, we were finally able to fill the much debated gap in-between by establishing that nucleosomes compact into a quasi-fibre folded into stable loops which form stable multi-loop aggregates/rosettes connected by linkers and hence creating chromosome arms and entire chromosomes. Interestingly, this has lead immediately to a consistent and cross-proven systems statistical mechanics genomic framework which is balancing stability/flexibility ensuring genome integrity, enabling expression/regulation of genetic information, as well as genome replication - all this in evolutionary perspectives as the natural outcome of Darwinian natural selection and Lamarkian self-referenced manipulation. Thus, genotype and phenotype are multilisticly entangled and beyond are embedded in genome ecology i(!)n- and environments. This not only opens the door to a true universal sequencing of genetic information, but also is the key for a general understanding of genomes, their function and evolution, as well as for applied diagnostics and treatment of disease, for future genome manipulation and engineering efforts, as far as the creation of artificial or extra-terrestrial live contexts.

Keywords: chromatin structure, genome organization, systems genomics, genomic statistical mechanics, genotype phenotype entanglement, evolution, genome ecology

1. Introduction to the History and State of the Art

Inheritance has always played a central part in the quest for elucidating the origin of nature, life and mankind. Beyond the epic mythical assumptions, it also has been obvious for millennia that the evolutionary transfer of information plays a key role during the manipulation of inheritance by mating and breeding. Already in antique times many a "theory" was devoted to the apparent, as well as especially to the obvious fact that nature seemed to be composed of small, similar, and consistent subcomponents—so called atoms. With the description of the tissue of plants (including its substructures of vesicles and bubbles) by Robert Hooke or in the case of the cell nucleus by Anton van Leeuwenhook, in the 17th century new momentum entered the field. Nevertheless, it took until 1830 when Robert Brown defined the cell nucleus as such and until 1939 when Theodor Schwann established the cell as the fundamental unit of all plant and animal tissues while linking to the assumed fundamental design principle of life as well as nature in general. Despite fast growing microscopic resolutions there were huge challenges: not only staining and visualization methods were lacking, but also huge preparatory issues were faced especially concerning the "notorious" hard to stain cell nucleus. With the development of the natural sciences many a discovery was made culminating in the structural description of the DNA double helix [1] and the discovery of the nucleosome [2–4] at the atomic level, full genome sequences and finally histone modifications defining epigenetic landscapes. It also became obvious that the structure and function of genomes co-evolved as an inseparable system allowing the physical storage, replication, and expression of genetic information [5–7].

However, the immense size and structural complexity of genomes spanning many orders of magnitude has always imposed huge experimental challenges. Thus, the higher-order architecture has been and still is widely discussed with many interesting details yet to be described. Already how nucleosomes are spaced, positioned, remodelled, and whether and how nucleosome chains fold into fibres at physiological salt concentrations have been matters of continuing debate: e.g. Finch and Klug [8] proposed a relatively regular solenoid and *in vivo* neutron scattering experiments revealed a fibre diameter of 30 ± 5 nm as a dominant nuclear feature [9–12]. In contrast more recent work suggested no compaction at all (rev. [13, 14]), and highly polymorphic, nucleosome position- [15] and dynamic function-dependent structures [16, 17], which are essential to explain nucleosome concentration distributions [18–20], or dynamic and functional properties such as the nuclear diffusion of macromolecules. Moreover, the fine-structured multi-scaling long-range correlation behaviour of the DNA sequence also predicts a compacted chromatin fibre [21–24]. With a novel chromatin interaction technique—*T2C*—we were, however and indeed, able to show that nucleosomes form in general a quasi-fibre with a differential compaction of $\sim 5 \pm 1$ nucleosomes/11 nm [25, 26], which is in agreement with a novel *in vivo* fluorescence correlation spectroscopy (FCS) approach measuring the dynamics of chromatin [27].

The higher-order chromatin architecture has been a matter of even greater debate: Pioneering light microscopy studies by Rabl [28] and Boveri [29] hinted towards a hierarchical self-similar, territorial organization. Electron microscopy suggested a more random interphase organization as in the models of Comings [30, 31] or Vogel and Schroeder [32]. In the radial-loop-scaffold model of Paulson and Laemmli [33] ~60 kbp-sized chromatin loops attached to a nuclear matrix/scaffold explained the condensation degree of metaphase chromosomes. According to Pienta and Coffey [34], these loops persisted in interphase and formed stacked rosettes in metaphase. Micro-irradiation studies by C. Cremer and T. Cremer [35, 36] and fluorescence *in situ* hybridization (FISH) by Lichter [37] as well as C. Cremer and T. Cremer [38] and publications thereafter [39, 40], confirmed a territorial organization of chromosomes, their arms, and stable sub-chromosomal domains during interphase, including their structural persistence during metaphase (de-)condensation. The assumption since then has been that the ~850 G, Q, R, and C ideogram bands [41, 42] split into and thus also consist actually of ~2500 subchromosomal interphase domains. Chromatin rosettes explaining a (sub-)territorial folding were first visualized using electron microscopy by Jekatrina Erenpreisa [43] and others [44] but remained unappreciated, until Belmont and Bruce proposed the EM-based helical hierarchy chromonema fibre (CF) model [45]. Spatial distance measurements between small FISH-labelled genetic regions, led to the Random-Walk/Giant-Loop (RW/GL) model with the first analytical looped polymer description by Sachs [46–48]. Here, 1 to 5 Mbp loops are attached to a non-protein backbone, following the line of Pienta and Coffey [34]. Later, a combination of distance measurements using structure-preserving FISH protocols, high-resolution microscopy, and huge parallel polymer simulations of chromosomes and entire cell nuclei, only were compatible with the rosette-like Multi-Loop-Subcompartment (MLS) model in which around 60 to 120 kbp loops form rosettes connected by similar sized linkers [7, 21–24, 49, 50]. Thereafter, the RW/GL model has then been discussed in terms of methodological "demolition" of the architecture [21, 22, 51, 52]. This is also in agreement with studies on replication (see [39] and thereafter). Again *in vivo* FCS measurements of nucleosome concentration distributions and dynamic and functional properties such as the diffusion of macromolecules are only in agreement with a small multi-loop aggregate/rosette-like chromatin folding [18–20, 22, 53, 54]. The fine-structured multi-scaling long-range correlations of the DNA sequence once again also predict this [22–24, 55].

To further distinguish between the different architecture proposals, proximity crosslinking techniques (developed and used already in the last century) were further developed into a family of interaction capture techniques such as 3C [56, 57], 3C-qPCR [58], 4C [59], 3C-seq/4C-seq [60], 5C [61], and Hi-C [62]. They once more confirmed the existence of looping and sub-chromosomal domains, now inconsistenly referred to as topologically associated domains (TAD; [63]) with a somewhat higher localization accuracy when compared to FISH. These approaches also led to a number of - although by the underlying (raw) data basically unsupported - conjectures (Imam et al., in preparation), e.g. the fractal globule model [62], the loop array architecture of mitotic chromosomes [64], and the highly dynamic loop formation based on single-cell experiments [65] or in a genome wide assay [66]. In contrast, with the introduction of targeted chromatin capture *T2C* [25, 67–69], we were able to show that the chromatin quasi-fibre forms small stable loops of ~30-100 kbp which form stable multi-loop aggregates/rosettes connected by linkers of similar sizes as the loops [25, 26]. The development of our novel *in vivo* FCS approach came to the same conclusion [27].

2. Finalizing the 3D Genome Architecture & Dynamics

Heuristically, it is very instructive how the central part of the 3D genome architecture and dynamics could now be determined by us in detail, and how out of this process immediately an also evolutionary consistent model (**Figure 1**) arises in agreement with the entire history and heuristics of the field. This has been achieved by a highly integrated systems approach linking holistically: i) a novel high-quality selective high-throughput high-resolution chromosome interaction capture (T2C) technique [25, 26, 67–69] (elucidating the structure with unprecedented resolution of some base pairs), ii) a novel *in vivo* FCS approach [27] exploring the structure and dynamics by measuring chromatin movement, and iii) a novel analytical approach [27] and improvement of super-computer simulations of individual chromosomes and entire cell nuclei [7, 21–24, 26, 49–52, 70] to predict, analyse, and interpret the 3D architecture and dynamics from a theoretical standpoint, and combining all these with iv) scaling analysis of the 3D-architecture [21, 22, 26] and the DNA sequence itself [22, 24, 26] since the architecture and its dynamics leaves sequence "footprints" due to the co-evolutionary entanglement of structure and sequence. The combination of these resulted not only in a consistent

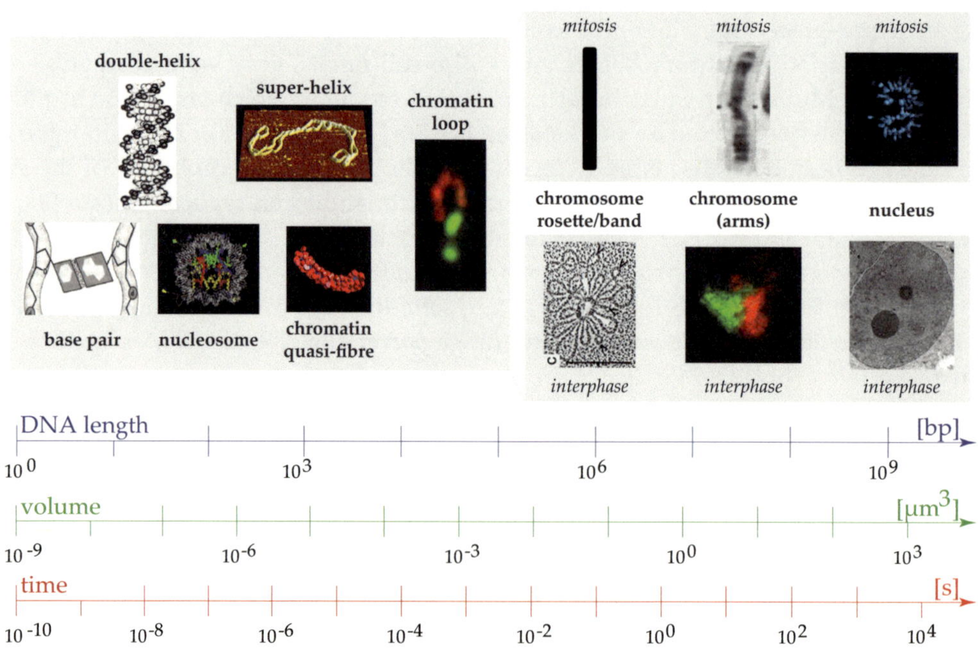

Figure 1. Overview on the size and time scaling of genome organization: The scaling and the levels of organization range over 9, 12, and 14 orders of magnitude! Initially base pairs are formed composing the DNA double helix (image see [22]), forming with a histone core complex the nucleosome (image from [22]), which condense into a chromatin quasi-fibre (simulation image; courtesy G. Wedemann). The DNA double helix forms also superhelices (AFM image of plasmid DNA; courtesy K. Rippe). The next compaction step consists of stable chromatin loops (FISH image; courtesy P. Fransz) forming stable loop aggregates/rosettes connected by a linker (EM image from [44]), which make up interphase chromosome arms and territories (FISH image; courtesy S. Dietzel) and the metaphase ideogram bands (image see [22]). 46 chromosomes compose the human nucleus and are decondensed in interphase (EM image; courtesy K. Richter) and condensed for separation during mitosis (image from [22]).

model for genome organization, but re-evaluation of the development of the entire field in the last ~170 years fostered this conclusion also tremendously and directly resulted in an evolutionary consistent model of genome organization in general.

2.1. Detailed Structure Determination by T2C

To finally determine and structurally sequence with highest resolution, signal-to-noise ratio, interaction frequency range, and statistical significance the 3D genome architecture we developed targeted chromatin capture (T2C) - a chromatin interaction technique though with far-better quality specifically addressing the needs for genome architectural "sequencing" [25, 26 67–69]. Briefly: i) after chromatin crosslinking, ii) cell permeabilization for intra-nuclear enzymatic DNA restriction, iii) the extracted and largely diluted cross-linked DNA is re-ligated primarily within the crosslinked complexes. After iv) de-crosslinking, purification, and final shortening to <500 bp of the chimeric DNA ligates, v) a purified region-specific DNA interaction fragment library is selected by using DNA capture arrays, before finally vi) high-throughput sequencing, mapping to the reference genome, interaction partner determination and visual/quantitative analysis is conducted (**Figure 2**). Notably, we use only uniquely mapped sequences without applying any other corrections

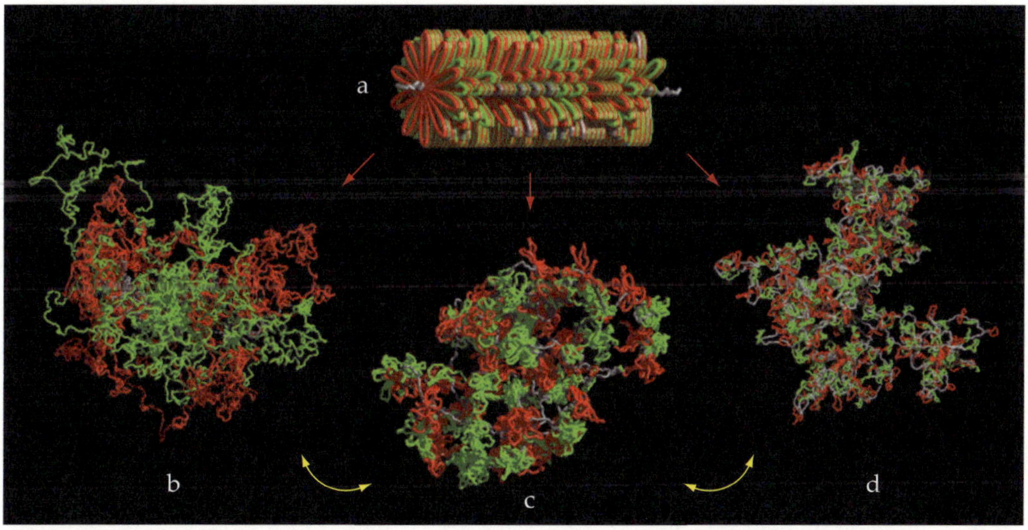

Figure 2. Simulated chromosome models [7, 21–23, 26, 49–52]: Volume rendered images of simulated Random-Walk/Giant-Loop (RW/GL) and Multi-Loop-Subcompartment (MLS) models. As a starting conformation with metaphase chromosome form and size (top), rosettes were stacked (a). Thereof, interphase chromosomes in thermodynamic equilibrium, were decondensed by Monte-Carlo and relaxing Brownian Dynamics. The simulated RW/GL model containing here large 5 Mbp loops notably shows that the large loops do not form distinct structures but intermingle freely (b). In contrast, in the MLS model with 126 kbp loops and linkers, the rosettes form distinct subchromosomal domains and chromatin territories in which the loops do not intermingle freely (c). In an RW/GL model with 126 kbp loops and 63 kbp linkers, again distinct chromatin territories are formed but in contrast to the MLS model without subchromosomal domains (d). It is obvious that the MLS model not only balances stability and flexibility considerations in storage and transcriptional respects, but also is optimal for replication due to its in essence two-dimensional topology allowing controlled duplication and separation during mitosis.

bearing information loss due to the very nature of T2C. This specific setup is not only far superior due to its improvement of 3 to 4 orders of magnitude compared to other interaction approaches (see Introduction), but also allows nearly unlimited opportunities e.g. such as multiplexing for complex research and diagnostics.

Most importantly, however, T2C allows reaching fundamental resolution limits where "genomic" statistical mechanics and uncertainty principles apply [26]: With fragment length and thus resolutions of a couple of base pairs, a high interaction frequency range, and high signal-to-noise ratio, not only molecular resolution is reached and thus the fundamental limits of cross-linking techniques, but also the mechanism of observation is now on the same scale as the observables (in analogy to classic and quantum mechanics). Actually due to the stochastics following the bias of the system behaviour, the observables, the observation, and thus the measured values are constrained by what we call “genomic” statistical mechanics with corresponding uncertainty principles. This originates from the individual complexity of each highly resolved interaction with a unique but coupled individual probabilistic fragment setting in each cell at a given time. Hence, the actual conditions and components can be determined only partially with high accuracy while with low accuracy otherwise and are eventually even entirely destroyed by the measurement. Thus, the central limit theorem applies with an overlap of system inherent and real noise stochastics, and hence in the end only probabilistic analyses and statements can be drawn as hitherto is well known from classical mechanics, and more so from quantum (mesoscopic) systems. Consequently, population based or multiple single-cell experiments have to be interpreted and understood in a “genome” statistical mechanics manner with uncertainty principles due to the inseparability of factors/parameters also seen there. Thus, in practical terms, valid results are obtained when the statistical limit is reached, i.e. when scaling up the experiment does not narrow down the distribution any further and does not lead to fundamental (overall) changes anymore in observables. Nevertheless, if the statistical limit is reached and if the quality parameters like resolution, frequency range, and signal-to-noise ratio are sound, conclusions could be drawn as in the many cases of classic mechanics, and more so of quantum (mesoscopic) systems.

Consequently, due to this sensitivity of T2C, we [26] were able to determine finally the missing parts of the 3D architecture on scales where a "genomic" statistical mechanics applies with stable reproducibility as one can already see visually in colour coded interaction maps (**Figure 2**): Not only are rare interactions stably detected within an unprecedented frequency range spanning 5-6 orders of magnitude, but also the maps are reproducibly mostly empty (<10% of possible signals are taken). Both interactions and non-interactions show clearly dedicated interaction patterns on all spatial scales within and between domains, including their re-emergence as attenuated repetition on other scales since obviously genomes are scale-bridging systems [22, 23]—all of which can be immediately identified as structural features - briefly (**Figure 2**):

i. On the largest genomic and thus spatial scale, subchromosomal domains are visible as square-like interaction domains (often unfortunately called TADs; [63]) featuring in general a higher average uniform interaction degree compared to interactions between domains, with a sharp drop at the edge of domains, as well as a clear linker region between the domains that connects them. The borders of the domains can be determined down to the single fragment level and thus a very high resolution (see below). The interaction of domains with each other and a closer inspection of the interactions in the vicinity of the linker interacting often more frequently compared to other domain parts are mainly due to the breaking of spatial isotropy.

ii. At intermediate scales within the subchromosomal domains, the interaction pattern shows clearly distinct gaps and a quantifiable grid-like arrangement of interactions, which also continues outside and "crosses" with the linear pattern originating from sequentially subsequent domain(s). These interactions on scales of tens of kilo base pairs are doubt-free originating from stable chromatin loops, forming a stable loop aggregate/rosette like architecture, due to several consecutive loops coinciding.

iii. On the smallest scale, a dense and high interaction frequency pattern is observed in the region from 3 to 10 kbp (i.e. < ~5-15, and ~50 nucleosomes, respectively) along the diagonal. It varies independently of the local fragment size with distinct interactions and non-interacting "gaps". This suggests, that there are defined stable interactions on the nucleosome scale forming an irregular yet locally defined and compacted structure, i.e. a quasi-fibre with average properties (e.g. an average linear mass density).

A detailed quantification [26, 27] of several regions leads to a quasi-fibre compaction of 5 ± 1 nucleosomes per 11 nm, with an average chromatin quasi-fibre persistence length of ~80 to 120 nm, loops and linkers of ~30 to 100 kbp, forming multi-loop aggregates/rosettes with typically 300 kbp to 1.5 Mbp subchromosomal domain sizes. Different cell types, species, or functional conditions showed only a relatively small variation of this theme [26, 27].

All this is consistent with a variety of previous observations and predictions such as compacted fibre structures described throughout the literature (see e.g. [16, 17]), the internal structure of subchromosomal domains [7, 21, 22, 24, 38–40, 43, 49, 50] agreeing on all structural levels with the absolute nucleosome concentration distributions [18, 19], the dynamic and functional properties such as the architectural stability and movement of chromosomes [7, 22, 54, 71, 72], chromatin dynamics [73], as well as the diffusion of molecules inside nuclei (e.g. [22, 54, 72]), and recent genome wide *in vivo* FCS measurements of the chromatin quasi-fibre dynamics [27] also suggesting such a chromatin quasi-fibre with variable, function-dependent properties. Beyond, other hypothesis (see Introduction; [26, 27]) about the 3D genome organization on these scales can clearly be ruled out: e.g. no-compaction or a highly-regular chromatin fibre, unstable/dynamic loops or unstable/dynamic loop aggregates/rosettes can clearly be ruled out, because they simple would lead to other interaction patterns and the intrinsic chromatin fibre dynamics with movements on the milli-second scale (Movies 1, 2 [26]) would lead to immediate structural dissolution. Most importantly no other model leads to a consistent functional framework bridging consistently the here described scales as can also be shown by the agreement with scaling analysis of the 3D-architecture [21, 22, 26] and the DNA sequence itself [22, 24, 26]. Beyond, not only functional aspects as the easy (de-)condensation during mitosis can be easily explained, but we were also able to find this organization in the data of others across species and even across specie-kingdoms (Imam et al., in preparation).

2.2. Dynamics and Structure Revealed by FCS

To investigate the 3D genome architecture and dynamics also by an orthogonal genome wide and *in vivo* approach, a novel *in vivo* FCS technique exploring the structure and dynamics by measuring chromatin movement combined with a novel analytical approach was introduced [27]. It is based on the fact that a specific chromatin quasi-fibre and its higher-order architecture directly influences its intrinsic dynamics. Thus, the concept dissects intra-molecular polymer dynamics from fluorescence intensity fluctuations measured with FCS to investigate meso-scale chromatin dynamics in living cells and connects this to the underlying three-dimensional organization. Besides, the classical analytical polymer models where extended to include dynamics, physical properties, and accessibility. As primary tracer protein for chromatin movement a linker histone H1.0-EGFP construct was chosen [18, 19, 22]. On the one hand, H1.0 decorates chromatin globally and reflects its density. On the other hand, it binds only transiently such that photobleached molecules are constantly replaced by fluorescent ones, and thus chromatin dynamics becomes amenable to FCS analysis (see also [20, 54]): Here, topologically and dynamically independent chromatin domains of 500 kbp to 1.5 Mbp in size were identified that are best described by a compacted chromatin fibre and a loop-cluster polymer model under theta-solvent conditions. In more detail again the formation of stable loops and stable multi-loop aggregates/rosettes from a chromatin fibre with certain density and flexibility properties emerged as prominent structural feature of dynamically independent domains - and this throughout the cell nucleus in living cells! The detailed quantitative values for the involved parameters again lead in essence to the same values as found already in the T2C data: a quasi-fibre compaction of 5 ± 1 nucleosomes per 11 nm, with an average persistence length of ~80 to 120 nm, and loops and linkers of ~30 to 100 kbp [27]. Notably, it cannot be stressed enough that the loops and multi-loop aggregates/rosettes form *stable* entities on the time scales which were approachable by FCS (between 10 µs and 10 to 20 s) and do neither open, close, or in any other way reform (longer timescale up to hours are historically known). This not only moves many an assumption currently proposed (see Introduction) into the realm of fairy tales—conceptually and by hard experimental facts in agreement with the research of the last ~30 years (e.g. [18–20, 22, 54, 71]). Visualization of simulated structures illustrates this clearly (Movies 1, 2 [26]): structures described consistently throughout the literature would dissolve immediately - what has never been observed (though attempted to be measured) - and also in consistent agreement with the T2C results measured at the limit of resolution. Beyond, also characteristic variations were found between eu- and heterochromatin: Hydrodynamic relaxation times and gyration radii of independent chromatin domains are larger for open (161 ± 15 ms, 297 ± 9 nm) than for dense chromatin (88 ± 7 ms, 243 ± 6 nm) and increase globally upon chromatin hyperacetylation or ATP depletion. Thus, functional changes are a variation of a basic theme, e.g. more compact heterochromatic domains have a larger inaccessible volume fraction than more open euchromatic ones. Nevertheless, molecular diffusion is fast enough to roam a complete domain within few microseconds, during which the domain itself appears static. Relaxation of domains in the 100 ms range affects genome access in a protein concentration-dependent manner: highly abundant molecules at several 100 nM concentrations 'fill' the fluctuating domain so that a larger volume fraction than for a static TAD becomes adiabatically accessible. In contrast,

http://dx.doi.org/10.5772/intechopen.89836

for low-abundance molecules encounters with specific loci within a domain are diffusion-limited. They sense a higher inaccessible volume fraction. Thus, domain dynamics result in a concentration-dependent differential accessibility that is more pronounced in heterochromatin than in euchromatin due to its shorter relaxation times [20, 22, 27, 54]. In this manner the FCS approach can be extended to acquire complete nuclear maps and thus to "sequence" the dynamic organization of nuclei in living cells.

2.3. Analytical and Computer Simulations Theoretic Evaluation

To better understand the 3D genome organisation suggested e.g. by the above results, to evaluate hypotheses, and to plan future experiments, we were the first who have - since 1996 - developed polymer models with pre-set conditions for *in silico* super-computer simulations (i.e. without attempting to fit data; [7, 21–23, 26, 49–52, 70]) and later also an analytical mathematics framework [27]. The simulations use a stretchable, bendable, and volume excluded polymer (hydrodynamic) approximation of the 30 nm chromatin fibre consisting of individual homogenous segments with a resolution of ~1.0 to 2.5 kbp while combining Monte Carlo and Brownian Dynamics approaches (**Figures 2–4**). The analytical polymer approach extends and applies for the first time Gaussian chain and Kratky-Porod model descriptions in combination with the Rouse and Zimm models for polymer dynamics to complex star and rosette topologies under real excluded volume conditions as well as dilute and semi-dilute solvent conditions [27]. Whereas the analytical model is exact, the simulations explore emerging effects not explicitly introduced into the analytical model.

Simulations (**Figure 2**) of the Random-Walk/Giant-Loop model in which large individual loops (0.5–5.0 Mbp) are connected by a linker resembling a flexible backbone, as well as the Multi-Loop Subcompartment (MLS) model with rosette-like aggregates (0.5–2 Mbp) with smaller loops (60–250 kbp) connected by linkers (60–250 kbp), have already predicted that only an MLS model, i.e. a compacted quasi-fibre forming stable loops and stable loop aggregates/rosettes connected by a linker, can properly explain the formation of chromosome arms and territories [22], the spatial distances measured both using fluorescence *in situ* hybridization (FISH) experiments [7, 21–23, 26, 49–52, 70], and beyond even the general morphology of nuclei *in vivo* using histone fluorescence fusion proteins [22, 51], nucleosome concentration distributions, as well as dynamic and functional properties such as the diffusion of macromolecules [18, 19, 22, 53, 54]. These models also contained already enough information/aspects to cover other architectures such as free random-walks, random or fractal globules as well as their stability and dynamics. Additionally, the visualization (**Figures 2–4**, Movies 1, 2 [26]) creates an immediate feeling for the behaviour of genomes in 3D - a fact which already by pure visual inspection rules out many of the introduction mentioned obscure suggestions immediately.

With the unprecedented quality of both the interaction mapping by T2C and the FCS dynamic measurements (see above) the introduction of simulation and analytical models complex enough to approximate the 3D genome organization adequately showed even more clearly that only a quasi-fibre, stable loop, stable loop aggregate/rosette-like architecture is compatible with the measurements: In essence the simulations and analytical models describe even the slightest details of the T2C and FCS measurements correctly including many at first sight

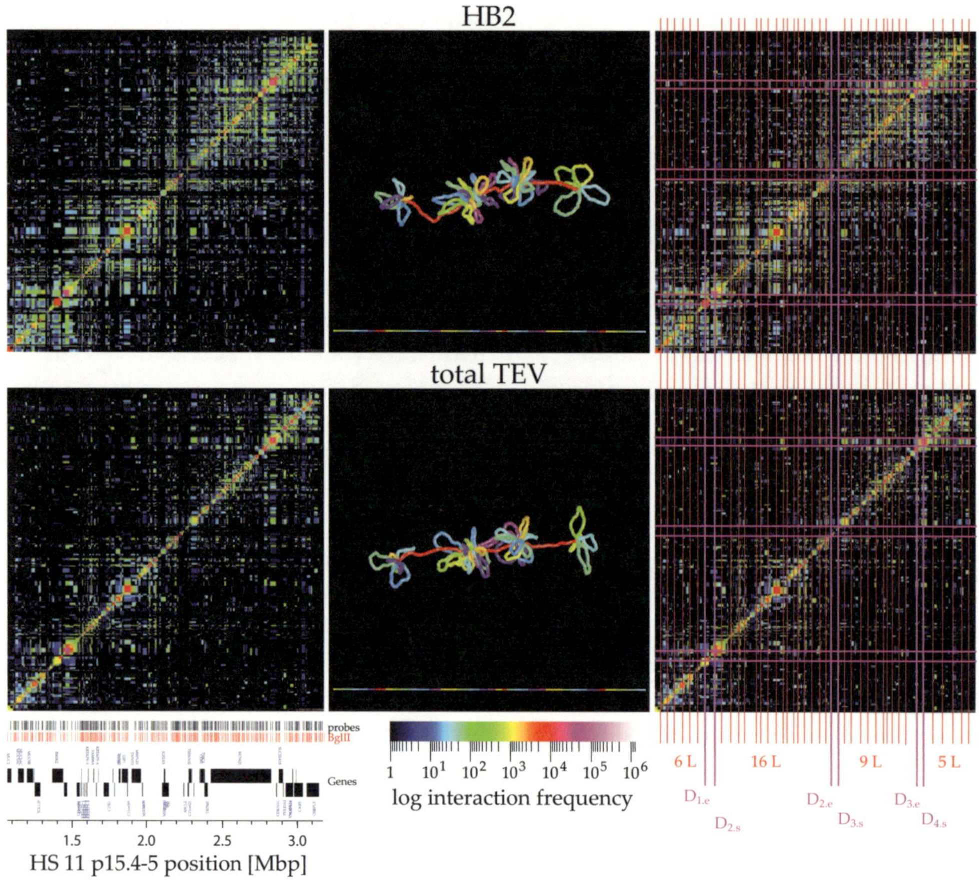

Figure 3. Determination of the 3D architecture in the IGF/H19 11p 15.5-15.4 region by T2C interaction mapping and computer simulations: Interaction matrices (logarithmic and colour coded scale; left & right) in HB2 and HEK293T TEV cells [26] show in unprecedented clarity the formation of a quasi-chromatin fibre, folding into stable loops (red lines; xL: number of loops), forming due to the grid-like pattern stable multi-loop aggregates/rosettes, i.e. subchromosomal domains separated by a linker (borders: pink lines, right; D1s, D1e: start and end of domains). A grid-like pattern is also visible in the interactions between the domains and corresponds to trans-domain loop interactions. The aggregation into a chromatin quasi-fibre is visible near the diagonal and loop internal structures are also detectable. Between different cell types or functional states only some local differences are visible resulting in a consensus architecture and allowing simulation of the 3D architecture (middle; resolution < ~1 kbp). Note that the simulation is driven by the dominant consensus architecture.

paradoxical results as e.g. i) that high numbers of especially small loops in a rosette result due the high density in steric exclusion and thus stretched loops eventually even "shielding" inner-rosette parts, ii) that inter-domain interactions are influenced by the connecting linker, loop size and numbers, and how non-equilibrium effects would appear, as well as iii) the isotropy breaking of consecutive subchromosomal domains as seen in the interactions at the border of domains and the domain-domain interactions. On a more general level the simulations support also the large and at first sight remarkable emptiness of interaction matrices and its link to the existence of a dedicated chromatin quasi-fibre. Additionally, the simulations hint

http://dx.doi.org/10.5772/intechopen.89836

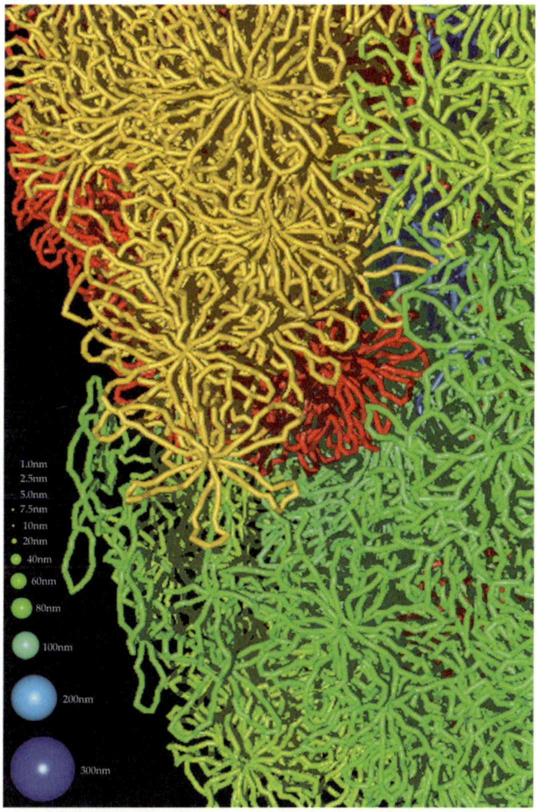

Figure 4. Insight into the spatial and dynamic/diffusional properties and morphology of the 3D organization of entire nuclei: The detailed view from the outside into a simulation for an MLS model with 126 kbp loops and linkers [22] shows the structure and low overlap of chromosome territories, the rosette like subchromosomal domains, and that the mean spacing between quasi-fibres ranges at least from 50 to 100 nm. Hence, the obstruction of diffusing particles (see spherical legend) is proportional to their size. Thus, small molecules as nucleotides and most (subunits of) proteins or gene transcripts reach every location of the nucleus by moderately obstructed diffusion. Consequently, active transport of molecules should be restricted to few exceptions and a channel like network for transportation (proposed by the Inter Chromosomal Domain model) is not necessary. Nevertheless, obviously the rosette core is denser leading to a general diffusion limited access. Thus, the interplay between accessibility and obstruction while considering the fast Brownian dynamics of the quasi-fibre and the entire system is of functional relevance.

to a relatively low crosslink probability, radius, and frequency in experiments comparing the clearly visible fine-structure (such as the (anti-)parallel neighbouring of the chromatin quasi-fibre at loop bases [26]. Also both the simulation and analytical approach describe in detail every aspect of the experimentally found multi-scaling behaviour with a fine-structure not only of the architecture and dynamics, but also of the DNA sequence (see below) to a degree of detail even we are still astonished about. The stability of the architecture with respect to the intrinsic chromatin fibre dynamics can also be illustrated by e.g. the decondensation from a mitotic chromosome into interphase (Movie 1 [26]) or just in a normal interphase state (Movie 2 [26]). This also shows that any 3D architecture would dissolve within seconds if it would not be stabilised. Consequently, both theoretic approaches came with old and new data consistently to the same conclusion whatever orthogonal high-quality method is used and thus are a theoretical framework for the understanding, test, and engineering of genomes.

2.4. DNA-Sequence Fine-Structured Multi-Scaling

Since what is near in physical space should also be near (i.e. in terms of similarity) in DNA sequence space and this presumably genome-wide [22–24, 55], and because evolutionary surviving mutations of all sorts will be biased by the genome architecture itself and vice versa, the correlation and thus scaling behaviour of the DNA sequence [22–24, 26, 55] and its connection to the 3D genome architecture scaling - either from T2C interaction mapping [26] or from simulations [21–23] - allows for comprehensive investigation of genome organization in a unified scale-bridging manner from a few to the mega base pair level. Using to this end, the perhaps simplest correlation analysis possible (to avoid information loss or biases), we calculated the mean square deviation of the base pair composition (purines/pyrimidines) within windows of different sizes and calculating the function $C(l)$ and its local slope $\delta(l)$, which measures the correlation degree, or in more practical lay-men terms, is similar to a spectral measure [22–24, 26]: in relation to mammalian genome organization for each of two different human and mouse strains i) long-range power-law correlations were found on almost the entire observable scale, ii) with the local correlation coefficients showing a species specific multi-scaling behaviour with close to random correlations on the scale of a few base pairs, a first maximum from 40 bp to 3.6 kbp, and a second maximum from 8×10^4 to 3×10^5 bp, and iii) an additional fine-structure is present in the first and second maxima. The correlation degree and behaviour within the species are nearly identical comparing different chromosomes (with larger differences for the X and Y chromosomes). The behaviour on all scales is equivalent concerning the different measures used to investigate the long-range multi-scaling of the genome architecture with the transitions of behaviours even at similar scaling positions [26] and can be associated with a single base pair resolution i) the nucleosome, ii) the compaction into a quasi-fibre, iii) the chromatin fibre regime, iv) the formation of loops, v) subchromosomal domains, and vi) their connection by linkers. Additionally, the already previously proven association to nucleosomal binding on the fine-structural level [22–24] is not only found again, but also is in agreement with the fine-structure found in the interaction scaling. Since the correlation analysis is genome-wide (in contrast to the T2C analysed regions so far) and since individual chromosomes show a highly similar scaling this clearly shows the genome-wide validity of the 3D organization. Moreover, the existence and details of this behaviour show the stability and persistence of the architecture since sequence reshuffling or other destructive measures would result in a loss of this pattern. This would also be the case for an unstable architecture, which would not leave a defined footprint within the sequence. This is again in agreement with our simulations of the dynamics or the genome wide *in vivo* FCS measurements [27]. Consequently, this shows not only by two analysis of completely independent "targets" (the T2C interaction experiments and the analysis of the DNA sequence) the compaction into a chromatin quasi-fibre and a stable multi-loop aggregate/rosette genome architecture again, but proved here also the long discussed notion that what is near in physical space is also near, i.e. more similar, in sequence space. Hence, the 3D architecture and DNA sequence organization are co-evolutionarily tightly entangled (review of previous notions in [22, 24]). Thus, in the future from the DNA sequence and other higher-order codes (e.g. the epigenetic code) most architectural genome features can be potentially determined, since most structural/architectural features left a footprint on the DNA sequence and other code levels and vice versa as one would expect from a stable scale bridging systems genomic entity.

3. Systems Consistency of the 3D Genome Organization

The above described holistic combination of several new orthogonal approaches [26, 27] including the heuristics of the field leads interestingly undoubtedly to a consistent picture of genome architecture, dynamics, and in general organization, by establishing that nucleosomes compact into a quasi-fibre folded into stable loops, forming stable multi-loop aggregates/rosettes connected by linkers creating chromosome arms and entire chromosomes. Nevertheless, the heuristics of the field immediately questions whether i) we really now have an evolutionary consistent picture of genome organization, ii) whether this is the unavoidable outcome of Darwinian natural selection and Lamarkian self-referenced manipulation (what we introduce here), and iii) finally whether we can understand now genome organization in its systems context within cells, organs, and the entire organism? This in essence already relates back to the fundamental question of how life emerged from the primordial soup [5, 6, 22]; see details in following sections) but in the context discussed here can be addressed by first reflecting on the existing major functions of genomes, thus setting the stage: i) genomes need to stably store genetic information, ii) the information needs to be differentially read out to give rise to and regulate the molecular machinery, and iii) genomes need to replicate and mutate to spread and evolve:

i. Obviously the by far most important function is to stably store over long periods of time genetic information though with enough flexibility including mutations - or in short: without proper storage neither information retrieval, nor replication, nor evolutionary development exist. This involves obviously being resistant against physical/chemical and/or in- or external mechanical destruction. Whereas, the first act mainly as from the bottom up involving one or a group of chemical bonds in proximity by direct interactions in the molecular soup, the latter depends on the large-scale structure of the basic molecular components and thus acts indirectly top-down on chemical bonds, i.e. that in- or external global stress is transferred and eventually accumulated via the global structure down to molecular levels while leading to mechanical failure. Both this physico-chemical and structural conformation-based destruction paradigms, influence genome architecture on all its levels under evolutionary pressure. They can be formulated such that a) mechanical failure rates are minimized regarding very long time spans, and b) in- or external mechanical failure rates reach an optimum due to the right balance between internal stability increasing with scale (for sensible ranges) and external stress decreasing the stability with increasing scale. From the well known average DNA breaking length of ~300–500 bp after already relatively severe sonication, this translates right away to the nucleosome and chromatin quasi-fibre level assuming that internal nucleosomal attachment increases the stability and elongating it by a factor 146 bp to 200 bp (repeat length), i.e. the average breakage length of an uncompacted chromatin fibre is 44 kbp or in the extreme 100 kbp balancing the quasi-fibre internal stability increase by further compaction counteracted by the bigger mechanical susceptibility due to local compaction clusters. Thus, the found loops size of 30-100 kbp as well as its chromatin quasi-fibre persistence length of 80-120 nm is just what one would theoretically expect as the evolutionary outcome. The same holds for the formation of stable multi-loop aggregates/rosettes where the major player is internal stability, which is a function of quasi-fibre compaction, loops sizes, and loop numbers [51, 52], giving rise to the natural found size distribution between ~0.3-1.5 Mbp [21–24, 26, 27, 40–42]. Also on the entire chromosome level again in- and external stability criteria have reached an optimum during evolution concerning the number of subchromosomal domains as well as their total size and number within a genome which again would just fit what one would theoretically expect: subchromosomal domain linkers are in the ballpark of loop sizes, the number of subchromosomal domains is <200-300 which just is the optimum size where mechanical stress does not too much destruct mitotic chromosomes under normal conditions. Consequently, the stability criteria are clearly satisfied while obviously still allowing enough flexibility by variation of this theme within the relatively broad boundary limits and various levels compensating individual stretching of limits (e.g. bigger loops might be stabilised by higher quasi-fibre compaction). Beyond, destruction of a complete structural element (e.g. nucleosome, loop) in relation to the characteristic scale seems never really to exceed 1-5% - an important criterion for overall system resilience.

ii. Access to and obstruction of genetic information, i.e. genetic information retrieval in a regulated fashion is, of course, next to pure storage the major task for a genome, although without a stable information storage retrieval gets arbitrarily complicated whether replication takes place or not. Since the information is readout with similar means as the storage itself, i.e. in a molecular way in contrast e.g. to an optical readout, this relies in principle on two major conditions: a) the physical space for the regulation of the 3D architecture needed that a readout takes place, and b) accessibility/obstruction to the genetic information for the readout-machinery as well as post-processing and transport of the transcribed information. For the first the DNA, nucleosomes, chromatin quasi-fibre, loops and loop aggregates/rosettes, need to have the space to be modified and get rearranged, i.e. that a volume several times bigger than the actual structure exists for ease of change. This involves, naturally a certain compaction, since a homogenous soup would not allow this. Since the regulation and readout is done by molecular mechanisms, it is also obvious that a low spatial occupancy allows moderately obstructed diffusional access of both the regulation and readout machinery only for DNA with a certain compaction degree. For such a scenario the volume occupancy of the architecture in aqueous solution should be well (!) below the limit of ~50% (model depending) as known from percolation studies [74], i.e. in terms of the performance expected for genomes, volume occupancy should be <10% since both the genomic architecture as well as the machinery should be able to access it for regulation by modification as well as readout. For chromatin, experimental values are between 2.5% to ~8% with a homogenous mesh spacing ranging from 115 to 65 nm ([22] and literature cited therein). Together with other factors and molecules in the cell nucleus like proteins and RNA, which all have a similar density, the volume occupancy is still <25%. These percolation assumptions hold, of course, also for the dynamics of the structure itself as pointed out above. At first sight this seems to be a dense system but the architecture is moving constantly by Brownian motion like in a spaghetti soup with additional floating components [18–20, 22, 27, 53, 54]. For chemical reactions this is well known for diffusion limited aggregation processes [75] as well as for percolating systems [75]. Due to the described consistent multi-layered 3D organizations showing also a multi-scaling of its volume occupancy as well as the space in-between this creates now even more and especially a scale dependent accessibility and obstruction to enhance the theoretic predictions of homogeneous though compacted systems with percolating space. Thus, under such conditions the necessary machinery for transcription as well as transcript transport is based mainly on moderately obstructed diffusion and despite of its high overall concentrations acts as an adequate multi-scale space [22, 53]. Consequently, similar to diffusion limited (catalytic) processes modification of the intrinsic architecture and dynamics of the entire genome organization is used for locally or globally fine-tuning of processes and thus functional regulation. Concerning, the stability of the 3D architecture only a quasi-fibre with stable loop aggregates/rosettes allows in terms of stability and flexibility local containment of large-scale interactions during the initiation of transcription e.g. by enhancer promoter interactions. For knot-free replication of the genome these (spatial) arguments also apply: whereas accessibility allows access of the machinery and space for the duplication, spatial obstruction protects the structural integrity. Interestingly, none of the described alternative architectures and dynamics hypothesis (see Introduction) agree to even a sufficient degree with these fundamental necessities to guaranty genome function.

iii. Replication and extinction of genetic information is the most crucial intervention into genome organization, since in contrast to the readout and regulation of genetic information by transcription, the entire structure and dynamics are affected by copying every single component of the organization. Here, an exact copy within a constrained space not only sequence wise, but also of its 3D architecture and dynamics as well as its disentanglement are the crucial parameters while still allowing structural stability/flexibility and even the access/obstruction of genetic information. From protein folding it is well known, that already during the amino-acid chain synthesis in the ribosome folding takes place, leading to a different 3D folding compared to the relaxation of finished and stretched out amino-acid chains. Obviously, also chromosome replication is such an adiabatic process (also chromosomes never fold from scratch, i.e. *de novo*, and always go continuously from one state to another), which takes place in parallel in the entire cell nucleus. And here again, genome architecture and dynamics are enabling replication to take place easily in principle only compatible with a chromatin quasi-fibre arranged in stable multi-loop aggregates/rosettes. This is due to the fact this architecture on the level of stable multi-loop aggregates/rosettes follows a knot-free two-dimensional topology. Of course, genome architecture is not a simple two-dimensional object in space considering the DNA-double helix and nucleosomal twist and writhe but nevertheless in terms of replication disentanglement it is. Consequently, replication origins can be situated and start replication everywhere in each chromatin loop with replication forks leading towards both directions until they hit a loop base (which is the reason for the bidirectional CTCF sites functioning as linear DNA markers for the directional oriented replication machinery). During this procedure even the twist and writhe are copied and need to be untangled as in the case of transcription. While hitting the loop bases then the two forks coming from two loops have to be joined and untangled, but no complex network of knots as they would appear even in a Random-Walk/Giant-Loop or even more so in a fractal globule like replication scenario would have to be cut and re-joined. Again here theoretical predictions for loop size and loop numbers are just fitting the experimental findings (see e.g. [39] and thereafter). Due to the two-dimensional topology of the multi-loop aggregates/rosettes, they can just be separated very easily in 3D space (this idea was proposed and illustrated to the author by his at the time 6 year old son Leander Aurelius!). And again the compaction and volume occupancy in the cell nucleus play an important role: the compaction into a chromatin fibre reduces not only the formation of DNA knots largely (perhaps almost to zero), but also provides with the volume occupancy in the cell nucleus the room for undisturbed replication, with the right flexibility provided by the intrinsic dynamics, allowing the disentanglement of replicated structures with minimal e.g. topoisomerase/decatenase driven active processes.

In summary, the above proves even further and especially in a holistic combination with the presented new orthogonal approaches [26, 27] and including the heuristics of the field, that indeed the described 3D genome organization - DNA forming nucleosomes compacted into a quasi-fibre folded into stable loops, forming stable multi-loop aggregates/rosettes connected by linkers creating chromosome arms and entire chromosomes (**Figure 1**) - presents without doubt a consistent scale bridging systems statistical mechanics genomics fulfilling the functional conditions necessary for storage, transcription, and replication. Additionally, the actual values found for the various parameters involved are just found in those "regions" one would expect as the unavoidable outcome of Darwinian natural selection and Lamarkian self-referenced manipulation (see below).

4. A Systems Genomics Statistical Mechanics

The heuristics leading to the here described consistent 3D genome organization has also resulted in another fundamental breakthrough besides merely clarifying the missing gap(s): the emergence of a multilistic systems statistical mechanics with uncertainty principles by reaching the fundamental resolution limits (see Section 2.1 above; [26]. Hence, this allows directly not only i) to extend the atomic theory based on ancient Greek philosophy and the notion of Theodor Schwann of cells being the fundamental atomic unit of tissues to the mesoscopic scale of genome architecture/dynamics, but also ii) to analyse and to describe how from the collective behaviour of these elements a holistic meta level, i.e. a phenotype, emerges. Thus, by reaching fundamental resolution limits now the statistical and uncertainty properties of each architectural/dynamic level can be determined both by experimental measurements as well as theoretical descriptions. Hence, from each of these "atomistic" basic units/elements their collective behaviour can be derived by a statistical mechanics on each individual level as wells as a complex interwoven scale-bridging, i.e. a hierarchic back referencing networked systems statistical mechanics - which obviously exists - can now be established in detail. This exceeds and is much more complex than establishing the statistical mechanics at the turn of the 20th century where from the individual components e.g. gas molecules a statistical mechanics established the collective properties of the entire system, e.g. the entire gas, because genome organization is not only a simple dualistic system of e.g. two levels but a complex multilistic network system with back references: In detail this means determining experimentally the behaviour of a genome structural/dynamic level precisely with its entire statistics and then doing the same on the level emerging from the underlying level. In principle this is what we have started already by setting up an experimental and theoretic framework over the past 20 years to elucidate genome organization [7, 18–24, 26, 27, 49, 50], although only now with the complete description of the general 3D genome architecture/ dynamics it is possible to fill the existing lack of knowledge in detail, determine the values for parameters with high precision, and in constant cycles of refinement adjust the description to an ever higher degree of approximation. Thus, the difference to the development of statistical mechanics in classical and later quantum physics at the turn to the 20th century is that in biology many and also much higher levels still are determined by and also act back even on the very first level to a much higher degree. This also immediately unites the at first sight contradicting theoretic descriptions of living systems of Ilia Prigogine [75], stating that living systems are far away from thermodynamic equilibrium, with those proposed by Georgi Gladyshev [76] stating that hierarchic substance stability is locally in thermodynamic equilibrium. Actually, these descriptions are even extended due to the multilistic statistical systems mechanics, i.e. manifold recursive hierarchically back-referencing, which are until now not described but e.g. envisioned in efforts to extend quantum mechanics to higher order complexities [77]. Consequently, a genomic multilistic statistical systems mechanics allows not only to describe and test basic properties of life, but also to answer perhaps the most fundamental questions of life as e.g. whether life time-wise can be extended beyond the currently obvious or thought of limits by manipulated engineering in one of its most central parts - the genome - a quest of epic dimensions appearing already at least between the lines in "What Is Life ?" by Erwin Schrödinger [78].

5. Genotype-Phenotype-Entanglement and Genome Ecology

The most important implication from the findings described above is most likely the multilistic entanglement between genotype and phenotype being the natural outcome of Darwinian natural selection and Lamarkian self-referenced manipulation in a genome ecology framework, which is connected directly to the origin of genomes and life itself: While entropy grows like an inexorable river, local disturbances lead to ever more ordered self-organizing and self-sustaining resistors, more complex structures, and finally life. In the 1970s Manfred Eigen [5, 6] showed how from the primordial soup autocatalytic chemical reaction-networks emerged and how they form ever more complex cooperatively organized networks and systems of so called hypercycles. With environmental separation by the emergence of units as cells and specialization of subunits, then genomes have developed as specialized keepers of the blueprint needed to maintain, regulate, and develop this syntropic machinery. Since genetic information is physically stored in molecular structures with dedicated architecture and dynamics, it is thus also obvious that the material carrier for the storage, usage, and replication of genetic information co-evolved inseparably. Yet another inevitable consequence of our results leading to the consistent statistical systems genome mechanic framework is indeed our proof [26, 27] that architecture, dynamics, and DNA sequence are co-evolutionary unseparably entangled (in a quantum mechanical sense): All architecture/dynamics levels have not only left a footprint on the DNA sequence level but beyond also all levels have left a footprint on all other levels with an astonishing degree of detail (see Section 2.4). Consequently, the co-evolution of all levels has also co-evolved not only to a higher degree than previously thought, but also indeed as an entire system where all levels are (equally ?) determinant (**Figure 5**).

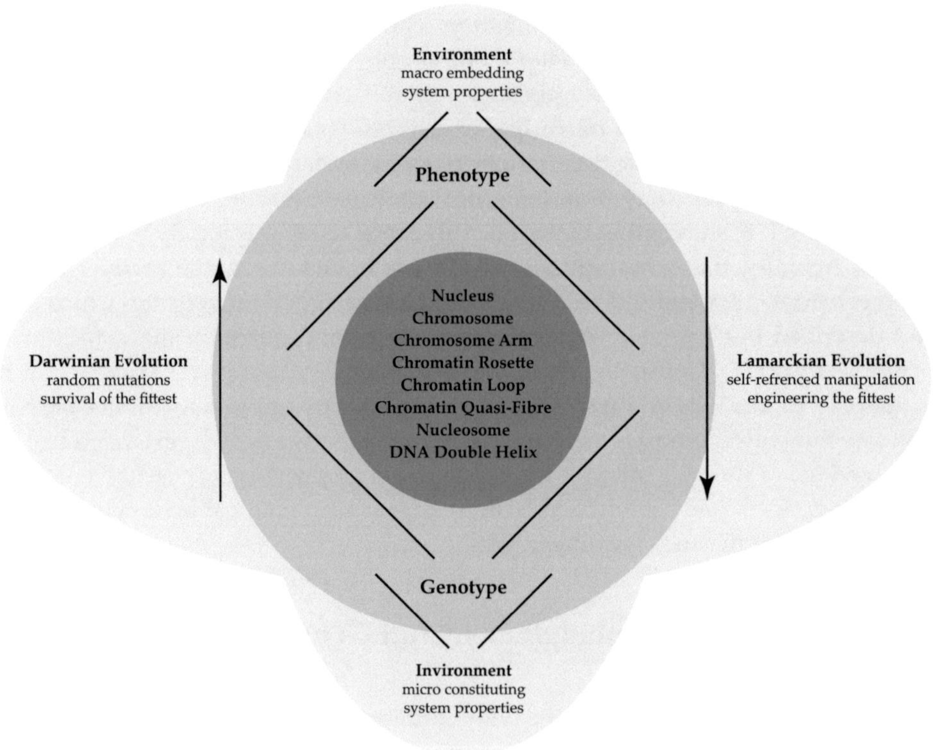

Figure 5. Genome ecology emerging from the system mechanics of genomes in relation to the genotype-phenotype entanglement and its embedding in- and environment: Genomes are interwoven holistic multi-scale hierarchic systems entities in which all organizational levels are also manifest, i.e. fingerprinting, on all other levels. Thus, immediately each level is a phenotype of its underlying genotype immediately conditioning back on it recursively. Thus, both genotype and phenotype are entangled inseparably in a (due to the involvement and entanglement of all levels) multilistic manner. In consequence this not only unites Darwinian and Lamarckian evolutionary paradigms, but also embeds and relates genomes with their in- and environment, and thus giving rise to a general genome ecology.

In evolutionary terminology the genotype (i.e. the double helix) creates a phenotype (the nucleosome) and this phenotype recursively conditions the genotype (i.e. again the double helix). The nucleosome is also a genotype conditioning the quasi-fibre phenotype, recursively conditioning the nucleosome and DNA, etc. Since this is happening with all levels simultaneously this inseparable dualism extends in the present genome organisation to a multilism, shaping evolutionary development in hierarchical terms from bottom to top by Darwinian natural selection as well as from top to bottom by Lamarkian self-referenced manipulation. Thus, our finding that indeed all genome architecture/dynamic levels are tightly entangled with each other also immediately resolves the falsely assumed paradoxes between Darwinian and Lamarckian evolution by uniting them at least on the genome level. This is remarkable not only in historic terms considering the even politically and religiously extremely hot debates/fights about "man evolving from apes" as well as the "intentionally planed long neck of giraffes", but also heuristically, since the in principle relatively simple final completion of the 3D genome architecture/dynamics at the limit of the resolution leads not only to a consistent 3D genome organization and statistical systems genome mechanics, but beyond reveals in one go some and perhaps the most important fundamentals of life (**Figure 5**).

Beyond, this strong entanglement over several orders of magnitude (**Figures 1**, **2**) within the genome, the described genotype-phenotype-entanglement can be driven conceptually even further considering the influence of both the a) hierarchically constituting elements giving rise to the system, i.e. chemical molecular base, atomic, and subatomic units, which will be called here i(!)nvironment, and b) the hierarchical higher levels, i.e. tissues, organs, animal etc., which are the environment. Although this may seem far fetched, but influences from both "directions" are well known (see e.g. Section 3), although due to their complexity this is often hard to track down in a reductionistic manner, thus hence their degree of influence is just emerging. In this respect the found entanglements bridging so many multi-scale levels and orders of magnitude in space and time, are on the one hand already astonishing in terms of the obviously wrong assumption that such influences would die-off very fast, while on the other hand this has general implications for all hierarchic systems showing that complex inter-, cross-, and even multi-cross-level influences are much more frequent and far reaching. Actually, the here shown multilistic genotype-phenotype entanglement shows a highly interwoven, networked, and recursive structure: instead of more or less separate hierarchic layers where only first or at the most secondary neighbour layers are connected, there are also influential connections to more distant layers at least locally if not in every part of the layer space. Thus, the genotype-phenotype entanglement embedded within an i(!)n- and environment actually results in a genome ecology in direct analogy to e.g. human ecology, autopoieses of social systems, or just any kind of systems theoretic entity [77–82].

6. Conclusion

Nature has created ever more complex forms of life by creating structural and dynamical islands of systems with specialized organelles such as genomes being responsible for storage, access, and replication of the information for their persistence and development. Despite the epic quest to determine the details and origin of inheritance, only recently we were finally able to fill the debated gaps of the central part of genome architecture and dynamics - despite the pioneering works of the last 170 years - by establishing that nucleosomes compact into a quasi-fibre which is folded into stable loops, forming stable multi-loop aggregates/rosettes connected by linkers creating chromosome arms and entire chromosomes [26, 27]. Although the heuristics of the field leads already to a sound basis, this could only be achieved - as we summarized here - by a highly integrated systems approach linking holistically i) a by far superior selective chromosome interaction (T2C) technique, ii) a novel *in vivo* FCS dynamic method, iii) a novel analytical approach and improved super-computer simulations, and iv) finally scaling analysis of the 3D-architecture and the DNA sequence itself. Including the heuristics of the field this leads to a consistent picture of genome organization, which match all the criteria necessary for storage, transcription, and replication as one would expect them as the outcome of Darwinian natural selection and Lamarkian self-referenced manipulation as shown here. In parallel, a multilistic systems statistical mechanics with uncertainty principles has emerged while reaching the fundamental resolution limits in the above holistic approach, which represents a theoretical framework which also reunites the overall far from thermodynamic equilibrium notion with local hierarchic substance stability. Beyond, the tight entanglement of genome levels having left footprints on all levels, has not only shown that genomes have evolved as an entire system, but also the multilistic entanglement between genotype and phenotype. Hence, the natural outcome of Darwinian natural selection and Lamarkian self-referenced manipulation is united in a genome ecology framework, which we consider a major step in the systems theory of life. Thus, this not only leads to a solid basis for sequencing genetic information holistically and thus for applied diagnostics and treatment of disease, as well as future genome manipulation and engineering efforts, but more importantly paves the path to a true understanding of genomes, their function and evolution, and thus of life in general - earthbound, extra-terrestrial, or artificial.

Acknowledgements

For supporting and influencing this long lasting work of T.A.K thanks go to: M. Wachsmuth, T. Weidemann, K. Fejes-Toth, M. Göker, R. Lohner, M. Stör, E. Spiess, K. Rippe, W. Waldeck, C. Cremer, T. Cremer, K. Erenpreisa, A. Ollins, D. Ollins, K. Sullivan, C. C. Murre, J. Skok, A. M. A. Imam, F. G. Grosveld, K. Egger, O. Zimina, and last but not least L. A. Knoch, as well as the German and International Societies for Human Ecology. T2C was invented by T.A.K. and F. G. Grosveld, with many thanks to M. Lesnussa, N. Kepper, A. Abuseiris, P. Kolovos, Jessica Zuin, R. W. W. Brouwer, H. J. G. van de Werken, W. F. J. van IJken, and Kerstin S. Wendt. This work was also part of the EpiGenSys consortium setup and coordinated by T.A.K., funded by ERASysBio+/FP7 and the national funding organizations (the Dutch Ministry for Science and Education, the Netherlands Science Organization, the UK Biotechnology and Biological Sciences Research Council, and the Bundesministerium für Bildung und Forschung (BMBF)). Further support came from the BMBF under grants # 01 KW 9602/2 (Heidelberg 3D Human Genome Study Group, German Human Genome Project), #01AK803A (German MediGRID), #01IG07015G (Services@MediGRID), as well as the Erasmus Medical Centre and the Hogeschool Rotterdam. The High-Performance Computing Center Stuttgart (HLRS; grant HumNuc), the Supercomputing Center Karlsruhe (SCC; grant ChromDyn), and the Computing Facility of the German Cancer Research Center (DKFZ) are thanked for access to their CRAY T3E and IBM SP2s in the initial part of this work. Thanks also go to all those institutions, universities, and companies providing us computational grid resources: the German D-Grid, the European Grid Initiative EGEE, as well as the Erasmus Computing Grid the Almere Grid, and all the unnamed computing grids there is access through via these. Very special thanks go also to all the world-wide distributed and unnamed donors of desktop computer power of our world-wide Correlizer@home BOINC grid!

Legends Movie 1 and 2

Video files available at: https://bit.ly/34KCuxe

Movie 1. Brownian Dynamics simulated decondensation from a metaphase starting configuration of a simulated Multi-Loop-Subcompartment model with 126 kbp loops and linkers with segment length of 50 nm (~5.2 kb) [26]. The whole 750 ms long movie shows how abruptly the metaphase chromosome expands due to its high density while opening the linker, which is constrained/condensed/pulled into a loop in metaphase. Nevertheless, the rosettes form distinct chromatin territories in which the loops do not intermingle freely in contrast to other models (see Introduction) such as the RW/GL model (**Figure 2**). The final shape and form in a whole nucleus would be determined by the limitations the other adjacent chromosomes provide (for more details see [22]). The different densities during decondensation also resemble nicely the conditions of shorter linkers, general genome regions with higher densities, or also the variation of nuclear volumes. Notably, the intrinsic movement of the chromatin fibre is clearly taking place on the millisecond scale, and hence, obviously a topological preformed architecture would dissolve within seconds if it would not be stable [26, 27].

Movie 2. Brownian Dynamics simulation of the consensus architecture (i.e. with the real measured loop and linker sizes) of the of the IGF/H19 region at HS11p15.5–15.4 (**Figure 3**), with a segment length of 20 nm (~2.0 kbp; colours of loops like in **Figure 3** middle, with additional linkers at the beginning and end of the region in red; for details see [26]). The whole movie encompasses 146 ms and shows the high intrinsic dynamics of the loops and the loop aggregate/rosette. Obviously, the single subchromosomal domains are constrained by the subsequent subchromosomal domains. Hence, and also obviously a topological preformed architecture would dissolve within seconds if it would not be stable [26, 27]. Nevertheless, the loop aggregates/rosettes form distinct subchromosomal domains in which the loops do not intermingle freely in contrast to other models (see Introduction) such as the RW/GL model (**Figure 2**). The final shape and form in a whole nucleus would be determined by the limitations adjacent chromosomes provide (for more details see [22]).

Author details

Tobias A. Knoch

Address all correspondence to: ta.knoch@taknoch.org

Biophysical Genomics, Department of Cell Biology and Genetics, Erasmus MC, Rotterdam, The Netherlands

References

[1] Watson JD, Crick FH. Molecular structure of nucleic acids; a structure for deoxyribose nucleic acid. Nature. 1953;**171**(4356):737-738

[2] Kornberg RD, Klug A. The nucleosome. Scientific American. 1981;**2**:28-44

[3] Olins AL, Olins DE. Spheroid chromatin units (v bodies). Science. 1974;**183**:330-332

[4] Luger C, Mäder AW, Richmond RK, Sargent DF, Richmond TJ. Crystal structure of the nucleosome core particle at 2.8 Å resolution. Science. 1997;**389**:251-260

[5] Eigen M, Winkler-Oswatitsch R. Transfer-RNA, an early gene? Naturwissenschaften. 1981;**68**:217-228

[6] Eigen M, Winkler-Oswatitsch R. Transfer-RNA: The early adapter. Naturwissenschaften. 1981;**68**:217-228

[7] Knoch TA, Münkel C, Langowski J. Three-dimensional organization of chromosome territories and the human cell nucleus—About the structure of a self replicating nano fabrication site. In: Foresight Institute—Article Archive. Palo Alto, CA, USA: Foresight Institute; 1998. pp. 1-6. Available at: http://www.foresight.org

[8] Finch JT, Klug A. Solenoidal model for the superstructure in chromatin. Proceedings of the National Academy of Sciences of the United States of America. 1976;**73**:1897-1901

[9] Baudy P, Bram S. Chromatin fiber dimensions and nucleosome orientation: A neutron scattering investigation. Nucleic Acids Research. 1978;**5**(10):3697-3714

[10] Baudy P, Bram S. Neutron scattering on nuclei. Nucleic Acids Research. 1979;**6**(4): 1721-1729

[11] Ibel K. Neutron diffraction of interphase nuclei. Journal of Molecular Biology. 1982;**160**(1):77-85

[12] Notbohm H. Small angle scattering of cell nuclei. European Biophysics Journal. 1986;**13**(6):367-372

[13] Dubochet J. Cryo-EM—The first thirty years. Journal of Microscopy. 2012;**245**(3):221-224

[14] Eltsov M, Maclellan KM, Maeshima K, Frangakis AS, Dubochet J. Analysis of cyro-electron microscopy images does not support the existence of 30-nm chromatin fibers in mitotic chromosomes in situ. PNAS. 2008;**105**(50):19732-19737

[15] Müller O, Kepper N, Schöpflin R, Ettig R, Rippe K, Wedemann G. Changing chromatin fiber conformation by nucleosome repositioning. Biophysical Journal. 2014;**107**(9): 2141-2150

[16] Kepper N, Foethke D, Stehr R, Wedemann G, Rippe K. Nucleosome geometry and inter-nucleosomal interactions control the chromatin fiber conformation. Biophysical Journal. 2008;**95**(8):3677-3691

[17] Stehr R, Schöpfling R, Ettig R, Kepper N, Rippe K, Wedemann G. Exploring the conformational space of chromatin fibers and their stability by numerical dynamic phase diagrams. Biophysical Journal. 2010;**98**(6):1028-1037

[18] Weidemann T, Wachsmuth M, Knoch TA, Müller G, Waldeck W, Langowski J. Counting nucleosomes in living cells with a combination of fluorescence correlation spectroscopy and cofocal imaging. Journal of Molecular Biology. 2003;**334**(2):229-240

[19] Wachsmuth M, Weidemann T, Müller G, Hoffmann-Rohrer UW, Knoch TA, Waldeck W, et al. Analyzing intracellular binding and diffusion with continuous fluorescence photobleaching. Biophysical Journal. 2003;**84**(5):3353-3563

[20] Capoulade J, Wachsmuth M, Hufnagel L, Knop M. Quantitative fluorescence imaging of protein diffusion and interaction in living cells. Nature Biotechnology. 2011;**29**(9):835-839

[21] Knoch TA. Dreidimensionale Organisation von Chromosomen-Domänen in Simulation und Experiment. (Three-dimensional organization of chromosome domains in simulation and experiment.). Mannheim, Germany: TAK Press; 1998. ISBN 3-00-010685-5

[22] Knoch TA. Approaching the Three-Dimensional Organization of the Human Genome: Structural-, Scaling- and Dynamic Properties in the Simulation of Interphase Chromosomes and Cell Nuclei, Long-Range Correlations in Complete Genomes, In Vivo Quantification of the Chromatin Distribution, Construct Conversions in Simultaneous Co-transfections. Mannheim, Germany: TAK Press; 2002. ISBN 3-00-009959-X

[23] Knoch TA. Towards a holistic understanding of the human genome by determination and integration of its sequential and three-dimensional organization. In: Krause E, Jäger W, Resch M, editors. High Performance Computing in Science and Engineering, High-Performance Computing Center (HLRS). Stuttgart, Berlin-Heidelberg-New York: University of Stuttgart, Springer; 2003. pp. 421-440. ISBN 3-540-40850-9

[24] Knoch TA, Göcker M, Lohner R, Abuseiris A, Grosveld FG. Fine structured multi-scaling long-range correlations in completely sequenced genomes—Features, origin and classification. European Biophysics Journal. 2009;**38**(6):757-779

[25] Knoch TA, Grosveld FG. Method for analysing the interaction of nucleotide sequences in a three-dimensional DNA structure. GB Patent Application GB20130020351 and International Patent Application WO2014IB02485 20141118

[26] Knoch TA, Wachsmuth M, Kepper N, Lesnussa M, Abuseiris A, Imam AMA, et al. The detailed 3D multi-loop aggregate/rosette chromatin architecture and functional dynamic organization of he human and mouse genomes. Epigenetics & Chromatin. 2016;**9**(58):1-22

[27] Wachsmuth M, Knoch TA, Rippe K. Mapping properties of dynamic chromatin domains by fluorescence correlation spectroscopy. Epigenetics & Chromatin. 2016;**9**(57):1-22

[28] Rabl C. Über Zellteilung. Morphologisches Jahrbuch. 1885;**10**:214-330

[29] Boveri T. Die Blastomerenkerne von Ascaris meglocephala und die Theorie der Chromosomenindiviualität. Archiv für Zellforschung. 1909;**3**:181-268

[30] Comings DE. The rationale for an ordered arrangement of chromatin in the interphase nucleus. American Journal of Human Genetics. 1968;**20**:440-460

[31] Comings DE. Mechanisms of chromosome banding and implications for chromosome structure. Annual Review of Genetics. 1978;**20**:440-460

[32] Vogel F, Schroeder TM. The internal order of the interphase nucleus. Humangenetik. 1974;**25**(4):265-297

[33] Paulson JR, Laemmli UK. The structure of histone-depleted metaphase chromosomes. Cell. 1980;**12**:817-828

[34] Pienta KJ, Coffey DS. A structural analysis of the role of the nuclear matrix and DNA loops in the organization of the nucleus and chromosome. Journal of Cell Science. Supplement. 1984;**1**:123-135

[35] Cremer C, Zorn C, Cremer T. An ultraviolet microbeam for 257 nm. Microskopy Acta. 1974;**75**:331-337

[36] Cremer T, Cremer C, Baumann H, Luedtke EK, Sperling K, Teubner V, et al. Rabl's model of the interphase chromosome arrangement, tested in Chinese hamster cells by premature chromosome condensation and laser-UV-microbeam experiments. Human Genetics. 1982;**60**:46-46

[37] Lichter P, Cremer T, Borden J, Manuelidis L, Ward DC. Delineation of individual human chromosomes in metaphase and interphase cells by in situ suppression hybridization using recombinant DNA libraries. Human Genetics. 1988;**80**:224-234

[38] Cremer T, Cremer C. Chromosome territories, nuclear architecture and gene regulation in mammalian cells. Nature Reviews. Genetics. 2001;**2**:292-301

[39] Berezney R, Dubey DD, Huberman JA. Heterogeneity of eukaryotic replicons, replicon clusters, and replication foci. Chromosoma. 2000;**108**(8):471-484

[40] Cremer T, Cremer M. Chromosome territories. Cold Spring Harbor Perspectives in Biology. 2010

[41] Yunis JJ. Mid-prophase human chromosomes. The attainment of 2000 bands. Human Genetics. 1981;**56**:293-298

[42] Francke U. Digitized and differentially shaded human chromosome ideograms for genomic applications. Cytogenetics and Cell Genetics. 1994;**65**:206-219

[43] Erenpreisa J. Large rossettes—The element of the suprachromonemal organisation of interphase cell nucleus. Proceedings of the Latvian Academy of Sciences Series. 1989; **B7**:68-71 (Russ)

[44] Reznik NA, Yampol GP, Kiseleva EV, Khristolyubova NB, Gruzdev AD. In: Harris JR, Zbarsky IB, editors. Possible Functional Structures in the Chromomere, Nuclear Structure and Function. New York-London: Plenum Press; 1990. pp. 27-29

[45] Belmont AS, Bruce K. Visualization of G1 chromosomes: A folded, twisted, supercoiled chromonema model of interphase chromatid structure. The Journal of Cell Biology. 1994;**127**(2):287-302

[46] Sachs RK, van den Engh G, Trask B, Yokota H, Hearst JE. A random-walk/giant-loop model for interphase chromosomes. Proceedings of the National Academy of Sciences of the United States of America. 1995;**92**:2710-2714

[47] Yokota H, Singer MJ, van den Engh GJ, Trask BJ. Regional differences in the compaction of chromatin in human G0/G1 interphase nuclei. Chromosome Research. 1997;**5**(3):157-166

[48] Yokota H, van den Engh G, Hearst J, Sachs RK, Trask BJ. Evidence for the organization of chromatin in megabase pair-sized loops arranged along a random walk path in the human G0/G1 interphase nucleus. The Journal of Cell Biology. 1995;**130**(6):1239-1249

[49] Jhunjhunwala S, van Zelm MC, Peak MM, Cutchin S, Riblet R, van Dongen JJM, et al. The 3D-structure of the immunoglobulin heavy chain locus: Implications for long-range genomic interactions. Cell. 2008;**133**(2):265-279

[50] Rauch J, Knoch TA, Solovei I, Teller K, Stein S, Buiting K, et al. Lightoptical precision measurements of the Prader-Willi/Angelman Syndrome imprinting locus in human cell nuclei indicate maximum condensation changes in the few hundred nanometer range. Differentiation. 2008;**76**(1):66-82

[51] Knoch TA. Simulation of different three-dimensional polymer models of interphase chromosomes compared to experiments—An evaluation and review framework of the 3D genome organization. Seminars in Cell & Developmental Biology. 2019;**90**:19-42

[52] Knoch TA. Simulation of different three-dimensional polymer models of whole interphase nuclei compared to experiments. submitted

[53] Knoch TA, Waldeck W, Müller G, Alonso A, Langowski J. DNA-Sequenz und Verfahren zur in vivo Markierung und Analyse von DNA/Chromatin in Zellen. German Patent Application 10013204.9-44 and International Patent Application PCT/DE01/01044. 2000

[54] Baum M, Erdel F, Wachsmuth M, Rippe K. Retrieving the intracellular topology from multi-scale protein mapping in living cells. Nature Communications. 2014;**5**: 4494-4506

[55] Knoch TA, Göcker M, Lohner R. Methods for the analysis, classification and/or tree construction of sequences using correlation analysis. US Patent Application 60/436.056 and International Patent Application PCT/EP03/14854

[56] Dekker J, Rippe K, Dekker M, Kleckner N. Capturing chromosome conformation. Science. 2002;**295**:1306-1311

[57] Tolhuis B, Palstra RJ, Splinter E, Grosveld F, de Laat W. Looping and interaction between hypersensitive sites in the active beta-globin locus. Molecular Cell. 2002;**10**:1453-1465

[58] Hagège H, Klous P, Braem C, Splinter E, Dekker J, Cathala G, et al. Quantitative analysis of chromosome conformation capture assays (3C-qPCR). Nature Protocols. 2007;**2**: 1722-1733

[59] Simonis M, Klous P, Splinter E, Moshkin Y, Willemsen R, de Wit E, et al. Nuclear organization of active and inactive chromatin domains uncovered by chromosome conformation capture-on-chip (4C). Nature Genetics. 2006;**38**:1348-1354

[60] Stadhouders R, Kolovos P, Brouwer R, Zuin J, van den Heuvel A, Kockx C, et al. Multiplexed chromosome conformation capture sequencing for rapid genome-scale high-resolution detection of long-range chromatin interactions. Nature Protocols. 2013;**8**:509-524

[61] Dostie J, Dekker J. Mapping networks of physical interactions between genomic elements using 5C technology. Nature Protocols. 2007;**2**:988-1002

[62] Lieberman-Aiden E, van Berkum NL, Williams L, Imakaev M, Ragoczy T, Telling A, et al. Comprehensive mapping of long-range interactions reveals folding principles of the human genome. Science. 2009;**326**:289-293

[63] Dixon JR, Selvaraj S, Yue F, Kim A, Li Y, Shen Y, et al. Topological domains in mammalian genomes identified by analysis of chromatin interactions. Nature. 2012;**485**:376-380

[64] Naumova N, Imakaev M, Fudenberg G, Zhan Y, Lajoie BR, Mirny LA, et al. Organization of the mitotic chromosome. Science. 2013;**342**(6161):948-953

[65] Giorgetti L, Galupa R, Nora EP, Piolot T, Lam F, Dekker J, et al. Predictive polymer modeling reveals coupled fluctuations in chromosome conformation and transcription. Cell. 2014;**157**(4):950-963

[66] Rao SSP, Huntley MH, Durand NC, Stamenova EK, Bochkov ID, Robinson JT, et al. A 3D map of the human genome at kilobase resolution reveals principles of chromatin looping. Cell. 2014;**159**:1-16

[67] Kolovos P, van de Werken HJ, Kepper N, Zuin J, Brouwer RW, Kockx CE, et al. Targeted chromatin capture (T2C): A novel high resolution high throughput method to detect genomic interactions and regulatory elements. Epigenetics & Chromatin. 2014;**7**:10

[68] Kolovos P, Brouwer RWW, Kockx CEM, Lesnussa M, Kepper N, Zuin J, et al. Investigation of the spatial structure and interactions of the genome at sub-kilobase-pair resolution using T2C. Nature Protocols. 2018;**13**(3):459-477

[69] Knoch TA. A guided protocol for array based T2C—A high-quality high-resolution high-throughput chromosome interaction capture. Current Protocols in Human Genetics. 2018;**99**(1):1-35

[70] Knoch TA, Lesnussa M, Kepper FN, Eussen HB, Grosveld FG. The GLOBE 3D genome platform—Towards a novel system-biological paper tool to integrate the huge complexity of genome organization and function. Studies in Health Technology and Informatics. 2009;**147**:105-116

[71] Gerlich D, Beaudouin J, Kalbfuss B, Daigle N, Eils R, Ellenberg J. Global chromosome positions are transmitted through mitosis in mammalian cells. Cell. 2003;**112**:751-764

[72] Dross N, Spriet C, Zwerger M, Muller G, Waldeck W, Langowski J. Mapping eGFP oligomer mobility in living cell nuclei. PLoS One. 2009;**4**:e5041

[73] Belmont AS. Visualizing chromosome dynamics with GFP. Trends in Cell Biology. 2001;**11**:250-257

[74] Bunde A, Havlin S. Fractals and Disordered Systems. 2nd ed. Berlin Heidelberg, New York: Springer Verlag; 1996. ISBN 3-540-56219-2

[75] Prigogine IR. From Being to Becoming. Freemann; 1980 ISBN 0-7167-1107-9

[76] Gladyshev GP. Hierarchical thermodynamics: Foundation of extended Darwinism. Imperial Journal of Interdisciplinary Research. 2017;**3**(2):1-14

[77] Mann F, Mann C. In: Fischer S, editor. Es werde Licht: Die Einheit von Geist und Materie in der Quantenphysik. 2017 ISBN 97803103972450

[78] Schrödinger E. What Is Life?/Mind and Matter. 1967th ed. Cambridge University Press; 1944

[79] Maturana HR, Varela F. Tree of Knowledge. Shambhala; 1992. ISBN 0877736421, ISBN 978-0877736424

[80] Luhmann N. Soziale Systeme: Grundriß einer allgemeinen Theorie. Suhrkamp; 2008. ISBN 3518282662, ISBN 978-3518282663

[81] Knoch TA, Baumgärtner V, de Zeeuw LV, Grosveld FG, Egger K. E-human grid ecology: Understanding and approaching the inverse tragedy of the commons in the e-grid society. Studies in Health Technology and Informatics. 2009;**147**:269-276

[82] Knoch TA. Sustained renewability: Approached by systems theory and human ecology. In: Nayeripour M, Keshti M, editors. Renewable Energy 2. Intech; 2011. pp. 21-48. ISBN: 978-953-307-573-0. https://www.intechopen.com/books/sustainable-growth-and-applications-in-renewable-energy-sources/sustained-renewability-approached-by-systems-theory-and-human-ecology

Chapter 5

Apicomplexa and Histone Variants: What's New?

Laura Vanagas, Susana Marisol Contreras and
Sergio Oscar Angel

Additional information is available at the end of the chapter

http://dx.doi.org/10.5772/intechopen.81409

Abstract

Plasmodium spp. and *Toxoplasma gondii* present a conserved nucleosome composition based on canonical H3 and variants, H4, canonical H2A and variants, and H2B. One-off, the phylum has also a variant H2B, named H2B.Z, which was shown to form a double variant nucleosome H2A.Z/H2B.Z. These histones also present conserved and unique post-translational modifications (PTMs). Histone variants have shown particular genomic localization and PTMs along euchromatin and heterochromatin, including telomere-associated sequences (TAS), suggesting fine-grained chromatin structure modulation. Several other nonhistone proteins present remarkable participation in controlling chromatin state, especially at TAS. Based on that, we discuss the role of epigenetics (PTMs and histone variants) in *Plasmodium* and *Toxoplasma* gene expression, replication, and DNA repair. We also discuss TAS structures and chromatin composition and its impact on antigenic variant expression in *Plasmodium*.

Keywords: *Plasmodium*, *Toxoplasma*, epigenetics, histone variants, H2B.Z, chromatin, antigenic variation, telomere-associated region

1. Introduction

Apicomplexa is a large phylum of unicellular obligate intracellular protozoan parasites responsible for a range of human and animal diseases with considerable medical and economic impact worldwide [1]. The phylum comprises several well-known genera such as *Cryptosporidium*, *Eimeria*, *Babesia*, and *Theileria*, but the most studied genera are *Plasmodium* and *Toxoplasma*.

Plasmodium genus is comprised by several species of which five infect humans: *P. falciparum*, *P. ovale*, *P. malariae*, *P. vivax*, and *P. knowlesi*. The infection due to *Plasmodium* genus is known as Malaria, a mosquito-borne infectious disease endemic in the tropical and subtropical zones of Asia, Africa, South, and Central America. Malaria also constitutes a serious problem for travelers as well as for people working in endemic regions. In 2016, an infection rate of 216 million cases was reported, causing some 445,000 deaths globally. Data show a stalling in declining burden of *Plasmodium* observed over the last decade (http://apps.who.int/iris/bitstream/10665/259492/1/9789241565523-eng.pdf?ua=1). In addition, the mass drug administration program has shown a limit due to the spreading of multidrug-resistant malaria [2]. The severe disease is mainly caused by *P. falciparum,* whereas *P. vivax* causes the majority of malaria morbidity outside Africa. The clinical manifestations of malaria include fever, shivering, arthralgia (joint pain), vomiting, jaundice, hemoglobinuria, convulsions, metabolic alterations, renal failure, liver and lung dysfunctions, anemia, and cerebral malaria (coma).

Toxoplasma gondii is the only one species of the *Toxoplasma* genus, and it is able to infect birds and mammals, including human, and cause toxoplasmosis. The infection occurs worldwide and the chronic stage reaches more than 500 million people [3]. During the first few weeks of infection, toxoplasmosis is either asymptomatic or causes a mild flu-like illness. However, those with a weakened immune system, such as AIDS patients, infected fetus during gestation or newborns with a congenital infection, may become seriously ill, and occasionally die. The parasite can cause encephalitis (inflammation of the brain) and neurologic diseases, and can affect the heart, liver, inner ears, and eyes (chorioretinitis). Recent research has also linked toxoplasmosis with neuropsychiatric symptoms such as attention-deficit hyperactivity disorder, obsessive compulsive disorder, bipolar disease, and schizophrenia [4–8]. The present chemotherapy for toxoplasmosis is efficient but, sometimes, it is not well tolerated by individuals with AIDS, and it is effective against the acute or active stage, but not against the chronic/latent stage.

2. Genome and nucleus

Both *Plasmodium* protozoan parasites and *T. gondii* present a highly complex life cycle, involving several stages along the cycle (**Figure 1**). The genome sizes are 23.3 Mb for *Plasmodium* and 80 Mb for *Toxoplasma*, being haploid (1 N) almost all their life cycle but diploid during sexual replicative stages (2 N) (**Figure 1**). *Plasmodium* genus and *T. gondii* were the first apicomplexan parasites to be included in genome projects [9, 10]. Since then, several other apicomplexan parasites genome projects were taken forward and the data uploaded at EuPathDB (https://eupathdb.org/eupathdb/). From these databases, it could be observed that *T. gondii* and *P. falciparum* present about 5300 (https://protists.ensembl.org/Plasmodium_falciparum/Info/Index) and 8172 (http://protists.ensembl.org/Toxoplasma_gondii/Info/Annotation/#assembly) gene transcripts, respectively, organized as single copy genes along the chromosomes, and in general, they are not clustered by function, pathway, or stage of expression. All of these data suggest that these genes are finely regulated throughout the cell cycle and the life cycle by transcriptional regulators and chromatin. Regarding transcription factors, they are poorly represented in apicomplexan, with only one large family of transcriptional regulators, with 24 AP2 family factors in *P. falciparum* and 68 in *T. gondii* [11, 12]. In this context, epigenetic

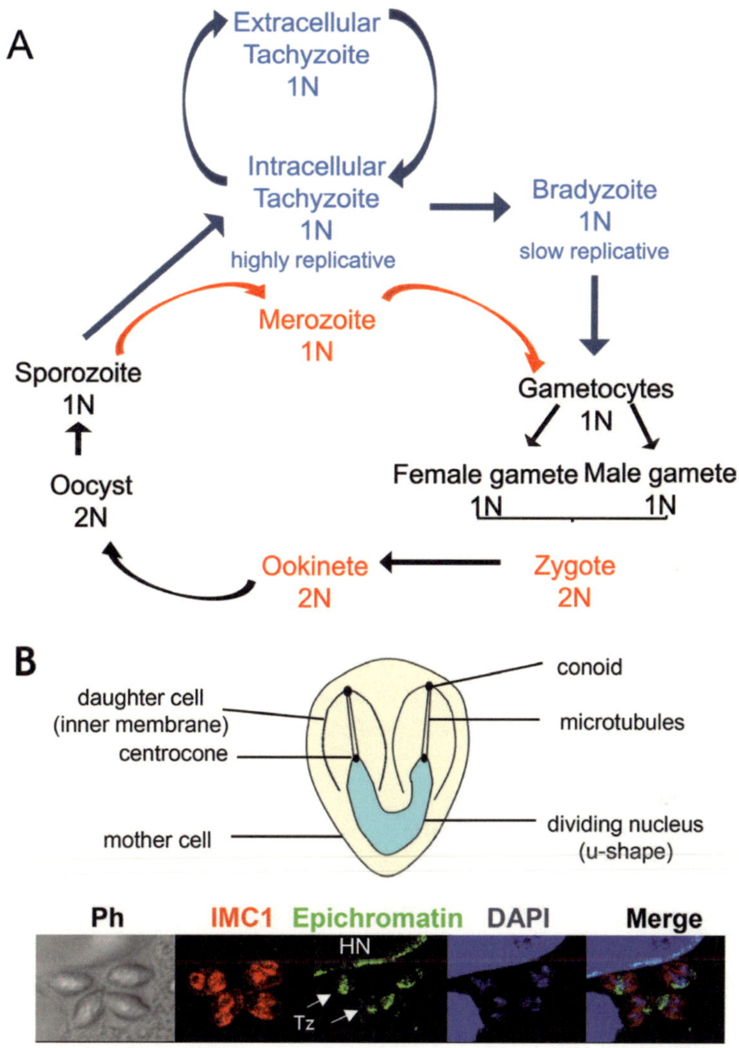

Figure 1. (A) Life cycles of T. gondii and P. falciparum. Haploid and diploid DNA content is referred to as 1 and 2 N. Black arrows and letters represent common parasite stages, blue arrows and letters represent T. gondii specific stages, and red arrows and letters represent P. falciparum-specific stages. (B) Epichromatin labeling in T. gondii. Epichromatin is a conformational epitope formed by DNA and histones H2A and H2B localized only at the exterior chromatin surface. During tachyzoite replication, the 2 N nucleus is divided, entering one genome to each budding cell, forming a typical U shape. The nuclear envelope does not disappear, and chromosomes do not present high level of condensation. Epichromatin labels predominantly one side of the nucleus, suggesting a nonhomogeneous organization of chromosomes and nuclear envelope interaction. IMC1: inner membrane complex 1.

control was proposed as a key element to facilitate parasite gene expression, DNA replication, and DNA repair [13–15].

An interesting aspect of apicomplexan parasites is that they never lose the nuclear envelope during cell division, and their chromosomes do not present the higher order level of condensation observed in metaphase chromosomes of higher eukaryotes [16]. So, the nucleus presents the same aspect along the cell cycle. However, it seems to be not homogenous: *Toxoplasma gondii* nuclear envelope and chromosomes seem to have a dynamic relocalization and/or rotation inside the nucleus during parasite budding as observed by epichromatin localization (**Figure 1B**). Epichromatin is a conformational epitope formed by DNA and histones H2A and H2B localized only at the exterior chromatin surface [17, 18]. More recently, it was observed that epichromatin forms superbead domains associated to DNA-A at the nuclear envelope [19]. A 3D analysis also shows that *P. falciparum* nucleus presents a polarization of the nuclear pore complex: in the early multinucleated schizont, it clusters in the nucleus region facing the mother plasma membrane, whereas in the late stages, when prepared for budding, it clusters toward the cytoplasm of the incipient merozoite [20].

In addition to putative polarization of the genome inside the nucleus of Apicomplexan parasites, in *T. gondii*, it was observed that the centromeres (CenH3, see below) are localized at a single spot at the apical region of the nucleus, indicating that all of them are attached to the centrocone, a structure associated to the nuclear envelope, which is traversed by microtubules coordinating the cell division [21]. Similarly, Chromo1, a *T. gondii* protein that binds to the telomere, presents a focalized localization in the nucleus, also suggesting a certain degree of chromosome organization within the parasite nucleus [22]. In *P. falciparum* prior to replication, in late ring stages and young trophozoites, CenH3 localizes to a single nuclear focus suggesting that centromeres are clustered in a single spot that most likely continues to be attached to the mitotic spindle until the end of schizogony and the intraerythrocytic developmental cycle, similar to that observed in *T. gondii* [23].

3. H3 histones: a multivariant family

H3 histone family presents canonical forms: H3, H3.1, H3.2 and variants: H3.3 and cenH3 [24]. H3.3 differs from canonical H3s in various aspects. Canonical H3s are expressed and associate to chromatin during the S-phase of cell cycle. Canonical H3s and H3.3 are highly identical differing in only four to five amino acids. The CAF-1 complex is involved in the incorporation of canonical H3s whereas CHD1/ATRX remodelers as well as HIRA chaperone complex are involved in the incorporation of H3.3 [25–30]. In addition, H3.3 is enriched in transcribed genes, enhancers, regulatory elements, and also heterochromatic repeats, including telomeres and pericentromeric regions [31–34]. In general, H3.3 is linked to gene activation or open chromatin. Moreover, it has been found to be methylated at K4, K36, and K79 and acetylated at K9 and K14, all being marks of active chromatin [32, 35]. H3.3 and H2A.Z were detected at active promoters generating nucleosomes that promote gene transcription [36–38]. Recently, it was found that H3.3 plays an essential biological role during mammal development since mice that lack H3.3 presented developmental retardation and early embryonic lethality [39]. Rather than gene expression troubles, H3.3 depletion causes genome instability due to dysfunction of heterochromatin structures at telomeres, centromeres, and pericentromeric regions of chromosomes, leading to mitotic defects.

There is little information regarding H3 histone family and the variants H3.3 in Apicomplexan parasites. The first approach is from W.J Sullivan [40] who was able to clone the entire ORFs encoding H3 and H3.3 in *Toxoplasma gondii* and also in *Plasmodium falciparum*. In this work, it was confirmed that, like in most other organisms, there is not much difference between the two variants: only four amino acids in *T. gondii* and eight between the *P. falciparum* variants. In most other species, the critical residues that differ between H3 and H3.3, resulting in different roles of these histones, are a motif, which contains SAVM in H3 canonical histone, but changes to AAIG in H3.3 [24]. However, while PfH3 has the typical SAVM motif, it changes to QAVL in PfH3.3, whereas in TgH3, the motif is SAVL and changes to QAIL in TgH3.3 [40]. Besides, there is another difference in Apicomplexa, which seems to be exclusive: KF changes for RY at position 54–55 in H3.3 [40].

In *Plasmodium*, H3.3 had a similar expression pattern to another important histone variant, H2A.Z, namely localization to active chromatin [41] (see **Figure 2**). As observed in other eukaryotic cells, it has been recently demonstrated by ChiP-seq experiments that euchromatic regions in the genome are demarcated by the presence of the H3.3 variant histone [42]. However, in *P. falciparum*, there is a particular AT versus GC content along the genome with euchromatic intergenic regions richer in AT-content compared to coding sequences with less AT content [23]. Fraschka et al. [42] have seen a particular correlation between enrichment in PfH3.3 histone variant and GC content, with this variant mainly located not only in euchromatic GC-rich sequences, but also in subtelomeric GC-rich repetitive regions. Interestingly, this correlation with the nucleotide composition is also observed with the double-variant nucleosome H2A.Z-H2B.Z (see below), but in this case, it is just the contrary: the regions with more AT content show abundance of this nucleosome [42]. However, GC-poor intergenic regions show the lowest H3.3 coverage, but the authors still argue that the incorporation of this variant to coding regions is more dependent on GC content than transcriptional activity.

It is well documented that *P. falciparum* depends on the *var* multigene family, encoding for a highly variable cytoadherence protein called *P. falciparum* erythrocyte membrane protein 1 (PfEMP1) to avoid host immunity [43–46]. This is due to the expression of only one of the ~60 *var* gene family members in any given parasite.

Regarding this important gene family, H3.3 stably occupies the promoter region and coding sequence of the active *var* gene but is evidently less incorporated into the promoter and coding sequence of silenced *var* genes [42] (see **Figure 3**). Additionally, it has been demonstrated that the PTMs affecting histone H3 are extremely important in the regulation of *var* expression. Data from fluorescence *in situ* hybridization (FISH) suggest that the *P. falciparum* SETvs (*P. falciparum variant-silencing SET* gene), which encodes an ortholog of *Drosophila melanogaster* ASH1 and controls histone H3 lysine 36 trimethylation (H3K36me3) on *var* genes, is specifically involved in *var* gene silencing, and its *knock-out* results in the transcription of virtually all *var* genes in the single parasite nuclei [47]. Besides, ChIP-qPCR analysis showed that the TSS occupancy of H3K36me3 is considerably higher in the silent *var* genes compared to the active one (see **Figure 3**) [47].

A detailed mass spectrometry study has been accomplished for *P. falciparum* histone PTMs by Trelle et al. [48], and it has been established that lysines 4, 9, 14, 18, 23, and 27 of both H3 and H3.3 are capable of being modified by acetylations and/or methylations. Also, arginine in position 17 may be mono or bimethylated. Some of these modifications had already been

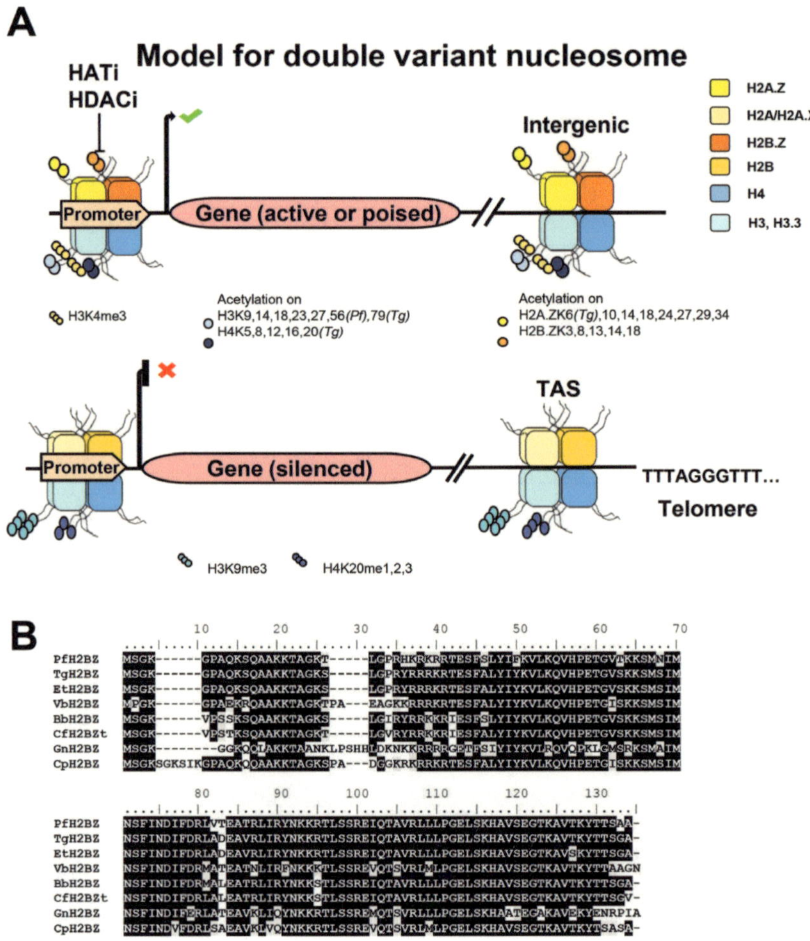

Figure 2. (A) Schematic model for double-variant nucleosome in T. gondii and P. falciparum. Hypothetical active (or poised) and silenced genes are represented with the nucleosome composition in each promoter and also downstream the genes. PTM marks are represented as well and listed below with the differences found in the literature between both the parasites. Specific HATs and HDACs are speculated to acetylate/deacetylate the unique H2B.Z histone variant. (B) Sequence alignment of histone variant H2B.Z in Apicomplexa. Pf: Plasmodium falciparum, Tg: Toxoplasma gondii, Et: Eimeria tenella, Vb: Vitrella brassicaformis, Bv: Babesia bovis, Cf: Cytauxzoon felis, Gn: Gregarina niphandrodes and Cp: Cryptosporidium parvum. All sequences can be found in EuPathDB (https://eupathdb.org/eupathdb/).

identified for H3 and H3.3 also by Miao et al. [41]. More recently, a lysine residue in the core of H3, K56, was also indicated as a site of acetylation [49, 50]. In the same way, *T. gondii* histone H3 has many lysines and also arginines capable of being modified: lysines in the positions 4, 9, 14, 23, 27 and also 36, 37, 56, 115 and 122 can be acetylated, methylated and besides some of them receive formylation, ubiquitination, or succinylation [51]. Besides, arginines 2, 17, 26, 40, and 83 can be methylated [51].

But not only acetylations and methylations are marking histones; with the development of improved acid and high-salt purification methods for *P. falciparum* histone phosphoprotein analysis, multiple phosphorylation sites have been found mostly at the N-terminal region of

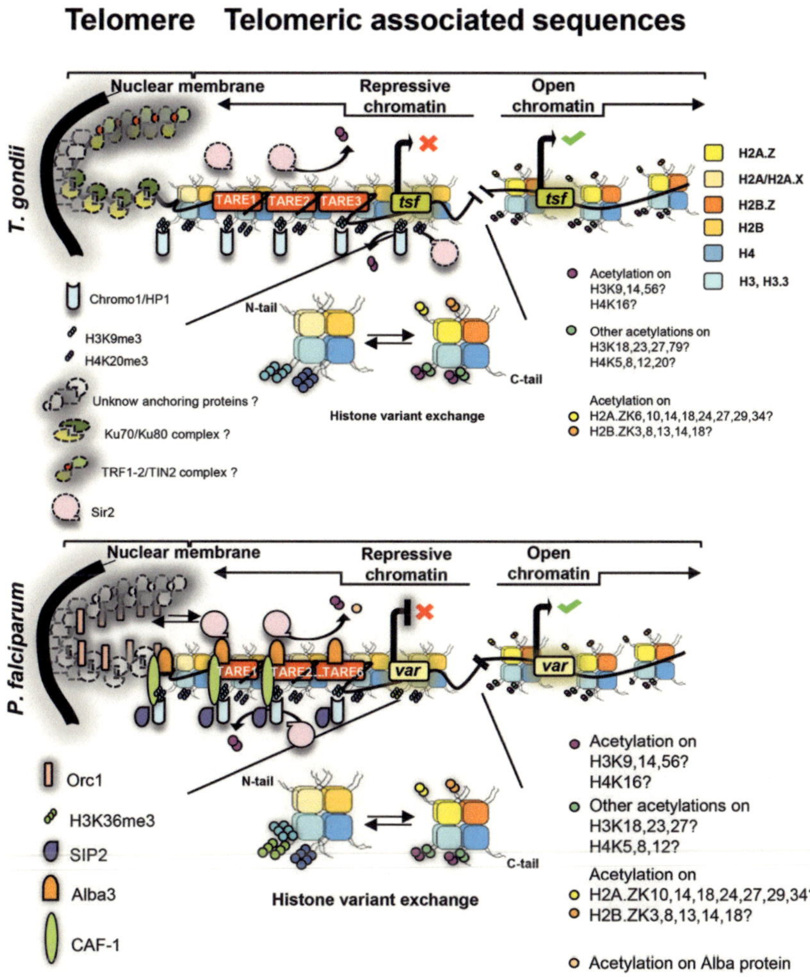

Figure 3. Hypothetical telomeric and subtelomeric structure in T. gondii (upper panel) and P. falciparum (lower panel). Here, we show the TgTAS and PfTAS composition and the possible protein interactors listed in **Table 1** and described in Section 6. In P. falciparum, we only illustrate three of the six TAREs associated. The punctuated lines define proteins described in other organisms, but some of them present an ortholog in any of those parasites. The proteins represented in continuous lines have already been described for those parasites. Proteins illustrated in grey are unknown proteins. We also show the possible histone variant exchange on silenced or active *tsf* or *var* genes. Specific PTMs on silenced or active genes are described in the figure. The questions are open because these proteins and PTMs have not been confirmed yet in these TAS regions.

most histones, including H3 and H3.3 [52]. These marks are frequently seen in combination with neighboring lysine acetylation (and methylation). In this work, they also described a Pf14-3-3 as a phosphohistone mark binding protein.

In parasites, among the most conserved modifications is histone 3 trimethylation of lysine 4 (H3K4me3), a marker of potentially active promoters. Opposed to that is H3K9 methylation, associated with silent genes and densely packed heterochromatin, although protozoan parasite histones are more highly enriched in the activation marks associated with euchromatin with lower abundance of histone modifications associated with heterochromatin [53]. However, it has been shown that the epigenome in *P. falciparum* is highly dynamic, and dependent on the stage, and, for example, H3K4me3 and H3K9ac are cycle regulated at *P. falciparum* genes [49]. This could also probably be true for *T. gondii*, where the tachyzoite to bradyzoite conversion is regulated at an epigenetic level. In this sense, it has been speculated that the H3R17me2 mark may have significance during the tachyzoite to bradyzoite differentiation process, as it was found only restricted to a subset of promoters, and taking into account the importance of arginine methylation during early development of mouse embryo [54]. In this study, using ChIP-on-chip technique, they found that H3K9ac, H4ac, and H3K4me3 modifications co-localize at focused loci in the *T. gondii* genome and correlate with significant gene expression, while the H3K4me1 and the H3K4me2 modifications were found at equal amounts in active and inactive chromatin [54].

4. Centromeric H3

CenH3, the centromere-specific H3, has been observed in animals, fungi, and plants [24] and also in Apicomplexa, including *T. gondii*, *Plasmodium spp.*, and *N. caninum* [55]. This fact was recently confirmed by Fraschka et al. [42] who found the centromeres depleted of PfH3.3 and PfH3, but occupied by PfCenH3. In *T. gondii*, this histone variant was characterized with the aim to understand the way in which chromosomes are delivered to the daughter cells after mitosis, a process that is still intriguing [21]. In this work, the authors labeled all the histone H3 variants, and used TgCenH3 as a marker of centromeres, to perform ChIP and microarray assays [21]. They found a particular combination of histone PTMs surrounding centromeres; this region had a huge concentration of H3K9 di- and trimethylation, marks usually associated to heterochromatin and found in subtelomeric regions in *P. falciparum* but not in *T. gondii.* In this parasite, these modifications concentrate in two peaks directly flanking the center of the centromere in each chromosome, while H3K4me3 or H3K9ac are not present [21, 52]. In contrast, H3K9me3 and heterochromatin protein 1 (HP1, chromodomain protein that binds to H3K9me3) were not associated with centromeres in *P. falciparum* [23], but rather found in islands of the genome that contain transcriptionally silent members of multigene families [56]. In this parasite, the enrichment of PfCenH3 on centromeres of all the chromosomes has also been demonstrated by genome-wide ChIP-seq analysis [23]. Besides, it has been characterized that a region within the carboxy-terminal histone fold domain, which is also named CENP-A targeting domain (CATD), is essential for mediating centromere targeting, while the N-terminus is not [57].

5. H2A.Z-H2B.Z: the double variant nucleosome

H2A family also has a canonical H2A and several variants: H2A.Z, H2A.X, both exchangeable by H2A.Z-H2B or H2A.X-H2B, allowing the modulation of gene transcription, DNA replication, and/or DNA damage repair [58, 59]. In vertebrates, the H2A family has two more variants: H2Abd and macro-H2A. When talking about H2A-H2B and the incorporation of variants into such nucleosomes, there are vast differences if we take a glance at Apicomplexan parasites compared to most other eukaryotes. One of the most surprising discoveries in these parasites was the presence of a novel H2B variant (formerly named H2Bv, but recently reclassified as H2B.Z [60]), a histone, which is known to be deficient in variants, similar to H4 [58, 61]. Variants of this histone family, though, are not only found in these parasites, but also in *Trypanosomatids* (even though they are not evolutionary related), and some rare testis-specific variants in human and other mammalian species (reviewed in [55]).

Different studies performed in *Toxoplasma* have shown a nucleosome composition in which H2A.Z, but not H2A.X, dimerizes with H2B.Z, while H2A.X dimerizes with canonical H2B (H2Ba in *T. gondii*), but never with H2B.Z [62, 63]. This fact is also seen in *P. falciparum*, although this parasite lacks H2A.X variant [41] and has driven the hypothesis of a new double variant nucleosome exclusive of parasites with particular characteristics that will be described in this section [64, 65] (**Figure 2A**). As it can be observed in the sequence alignment of H2B.Z in many Apicomplexan species, this histone variant is quite conserved (**Figure 2B**), suggesting that this histone, and likely the double variant nucleosome H2A.Z-H2B.Z, may have had an important role in the expansion of the phylum.

Since H2B.Z is not represented in yeast, insects, or mammals, almost all the current knowledge about the double-variant nucleosome relies on H2A.Z studies. H2A.Z is so widespread that has been catalogued as "universal" because of its origin before the divergence of eukaryotes [66]. The first observation that appears is the hyperacetylation of its N-terminal tail in most species [48–50, 67–69]. It is thought that this possibility gives H2A.Z the faculty of mediating responsiveness to the environmental changes, with so varied and seemingly contradictory effects as gene activation, heterochromatic silencing, transcriptional memory, and others, depending on the binding of activating or repressive complexes [66]. H2A.Z containing nucleosomes mark active and bivalent promoters as well as enhancers, correlating with open chromatin [70, 71]. However, acetylation of H2A.Z is necessary for gene induction and is most often associated with active gene transcription [67, 68, 70, 71], whereas ubiquitylation, which can occur at the C-terminal tail, is linked to transcriptional repression and polycomb silencing [72–75]. Acetylated H2A.Z composes nucleosomes flanking the nucleosome-depleted regions [76]. Regulation of gene expression by acetylation of H2A.Z histone tail may be a result of the participation of other proteins as "readers" in the histone code; for example, the SWR-C chromatin remodeling enzyme and related INO80 family are well characterized to catalyze chromatin incorporation of the histone variant from yeast to human, and the acetylation of histone H3 on lysine 56 (H3-K56Ac) was said to lead to promiscuous dimer exchange in which either H2A.Z or H2A can be exchanged from nucleosomes, although this is in discussion [77–82]. NuA4 acetylation activity, which is homologous to the TIP60/p400 complex, was found to be associated with SWR1-driven incorporation of H2A.Z into chromatin [83]. Besides, bromodomain-containing proteins are known to be implicated in "reading" the acetylation patterns of H2A.Z: acetylated lysines in histones, and other proteins are recognized by this motif, common in remodelers [77, 78, 84, 85]. In fact, for SWR1, bromodomains have been studied to recognize a pattern of acetylation (including H3K14ac), which may influence the deposition of H2A.Z-H2B variant dimers into the appropriate nucleosome [77, 78]. By using *Tetrahymena* as a model, it could be observed that these protozoa cannot survive with all acetylatable lysines replaced by arginines, indicating that H2A.Z acetylation modulates a charge patch with an essential function in chromatin regulation [69, 75]. Unlike the histone code, these changes need not to be site-specific. If this hypothesis is true, modulation of the charge at any one of a number of clustered sites could inhibit nucleosome condensation, facilitating transcription [86].

T. gondii H2A.Z, together with H2B.Z, was enriched in the promoters of active genes in tachyzoites, while repressed genes were enriched with H2A.X-H2Ba nucleosomes [63] (**Figure 2A**). In addition, H2A.Z-H2B.Z was also recruited within the coding region of silent bradyzoite-specific genes and within promoter regions but not coding regions of actively expressed genes [87]. It is tempting to speculate that the enrichment at active promoters or poised regions could be ruled by different PTM stages of these histone variants. In agreement with this, H2A.Z and H2B.Z have shown to be highly acetylated at the amino-terminal tail, in contrast to canonical H2A and H2B histones and the H2A.X variant [51]. Considering that H2A.Z has shown to be essential in regulating the changing gene expression program during differentiation [79–81, 88–90], and recently, it was observed that overexpression of mutated version of H2A.Z, where all five potential acetylatable lysines on H2A.Z-GFP (K4, 7, 11, 13, and 15) were mutated to arginines, blocked myoblast differentiation through disruption of myoD expression [91], it may be that the H2B variant is involved in the *T. gondii* cell differentiation process as part of H2A.Z-H2B.Z nucleosome. Whether through a patch charge modulation and/or histone code remains an open question, considering that *T. gondii* presents several bromodomain-containing proteins that can recognize some of the acetylated lysine [92].

As stated above, the sequence alignment of H2B.Z in many Apicomplexan species reveals a high degree of conservation for this histone variant (**Figure 2B**). Interestingly, every lysine that has been proved to be acetylated in *T. gondii* and *P. falciparum*, H2B.Z was detected in the other Apicomplexan species, which is also true for H2A.Z [50, 53]. Maybe, the double-variant nucleosome is present in the phylum with same PTMs and similar biological role.

5.1. Double-variant nucleosome in var genes

In *P. falciparum*, H2A.Z-containing nucleosomes were proposed to demarcate intergenic/regulatory regions of the genome, serving as a scaffold for stage specific as well as transcription-coupled recruitment of histone modifying enzymes [93]. H3K9ac and H3K4me3 were found preferentially placed/retained on or next to H2A.Z-containing nucleosomes [49]. However, it was observed that *P. falciparum* intergenic regions, including promoters, display a global nucleosome depletion, while telomeres harbored the highest nucleosomal occupancy, except for the *var* gene with the highest expression level, which again showed the lowest nucleosomal occupancy [94]. Apparently, the little amount of nucleosomes in these areas is composed largely of variant nucleosomes. Petter et al. [95] also showed an enrichment of PfH2A.Z in the promoter of a set of developmentally regulated genes in the euchromatin compartment, although not correlated with transcription levels nor with acetylation status. *P. falciparum* H2A.Z-H2B.Z promoter occupancy in *var* genes was found to be strongly associated with transcriptional activity, whereas silent or poised *var* genes would be depleted of double-variant nucleosome (see **Figure 3**) [65, 86]. The authors have speculated that it may function as a similar physical switch to control gene expression in response to temperature change (for example, during fever or as *P. falciparum* is transmitted between its two hosts), as a thermosensory response that was seen in *Arabidopsis thaliana* and yeast [96]. This could be due to reduced DNA wrapping of H2A.Z containing nucleosomes at higher temperatures, resulting in a relaxed chromatin structure, although this variant histone has also been associated with a tighter relationship with DNA, especially in heterotypic H2A.Z/H2A nucleosomes [97]. While heterochromatic intergenic regions showed to contain low levels of histone variant H2B.Z [64], it is interesting that double-variant nucleosomes are depleted from silent *var* gene promoters but not from silent promoters of heterochromatic invasion gene families, which have similar patterns of variegated expression [65]. Besides, this correlation between double-variant nucleosome presence and expression was only seen in *var* genes, while this nucleosome was also found enriched in intergenic regions across the genome, associated with euchromatic histone modifications and not necessarily associated with transcription [64, 65, 86]. Moreover, long promoter containing intergenic regions that maintain higher variant histone levels as compared with 3′UTR containing regions, which are considerably shorter, presents higher AT content, so this correlation could simply be due to the minimal length of the AT-rich content in these short 3′UTR regions [64]. As it was previously observed for H3.3 variant histone (see Section 3), a correlation between nucleosome occupancy and GC/AT content in the genome was observed, although contrary to H3.3 that was correlated with rich GC regions [42], here both H2B.Z and H2A.Z histone variant occupancy displayed a clear positive correlation toward genomic AT content [64]. In **Figure 2A**, a schematic representation of *P. falciparum* and *T. gondii* nucleosome occupancy is proposed.

6. Heterochromatin, telomeres, and subtelomeres

The telomere-associated sequences (TAS), also named subtelomeres, are heterochromatic regions adjacent to the telomeric-end looking toward the centromere. The telomeres and the TAS regions are the final structures at the chromosomes and integrate with the centromere the constitutive heterochromatin in the genome. These TAS regions have been described in *Plasmodium* and *Toxoplasma* with a size of 20–40 and near 30 Kpb, respectively (**Figure 3**) [98–100]. In *T. gondii*, the structure contains three tandem repeated elements (TARE), separated by noncoding DNA and flanked at one end by the telomere and at the other, downstream TARE 3, by a *Toxoplasma*-specific gene family, the *tsf* gene, of unknown function [100]. In general, there is only one *tsf* gene per TAS. Interestingly, based on predicted amino acidic sequence, TSF proteins present a high degree of conservation in the N-tail and middle regions while being highly variable at the C-terminal end. Up to now, only few studies were performed on chromatin modulation of *T. gondii* TAS.

The TAS element in *Plasmodium*, instead, has been deeply studied because of the presence of different families of genes associated to virulence and pathogenicity with a clonal pattern of expression [101, 102]. Telomeres are spatially restricted to nuclear periphery, where they form clusters of three to seven heterologous chromosome ends [103–105]. *Plasmodium* TAS is composed of six different TAREs, and the coding part of the genome is localized directly downstream TARE 6, and is characterized by members of multiple antigen gene families including *var*, *rif*, *stevor*, and *pfmc-2tm* genes [94, 95].

The telomeres and TAS regions are dynamic structures associated to a plethora of specific factors that not only give it structure, but also configures all the regions as constitutive heterochromatin that participates in an epigenetic way to regulate subtelomeric genes expression (**Figure 3**). This epigenetic mechanism is carried out by proteins that introduce, recognize, and implement a repressive state over the gene expression under normal environmental conditions. It has been reported that under nutritional or environmental stress, the repressed subtelomeric genes activate their expression in response to events promoting growth and survival [87, 106–109].

It is important to highlight that the *T. gondii* TAS regions show a nucleosome composition enriched in H2A.X and heterochromatin markers [100]. An *in silico* analysis using the *Plasmodium* and *Toxoplasma* databases reveals the presence of only some orthologs to the yeast and mammal's telomeric-subtelomeric proteins as TRF1-2, HP1, KU70/KU80, and Sir2 proteins (**Table 1**). But interestingly, the principal actor in this scenario would be the histone deacetylase type III -Sir2. This NAD+ deacetylase-dependent has also been implicated in different signaling pathways. *P. falciparum* has two Sir2 paralogues, Sir2A and B; with overlapping but distinct roles that regulate different subsets of *var* genes[110], binding reversibly with the promoter regions of silent but not active subtelomeric *var* genes [111]. PfSir2A is implicated in telomere length regulation [112]. In *T. gondii*, two deacetylases containing the Sir2-domain were identified: TgSir2A and TgSir2B, but their function has not been characterized yet. Another protein that had been described in *Plasmodium* is PfOrc1 (origin recognition complex 1), which together with Sir2 promotes the epigenetic silencing in *P. falciparum* TAS [113]. PfOrc1 has a role in DNA replication but also cooperates with Sir2 to coordinate the spreading of heterochromatin and regulation of *var* gene expression [114]. In general, Sir2 proteins act by removing acetyl groups in cytosolic targets and at the nuclear level at H3K9, K14 and K56, but it also was described to act on the histone mark H4K16 promoting the deposition of methyl groups on H4K20, H4K20me3 being a chromatin mark associated with heterochromatin [115]. Thus, Sir2 seems to play a very important role in linking signaling processes to gene expression and chromosome architecture.

Additionally, a member of the Alba protein family (PfAlba3) was demonstrated via ChIP assays to bind to telomeric and subtelomeric regions co-localizing with Sir2A in the periphery of the nucleus. PfAlba3 inhibits transcription *in vitro* by binding to DNA. PfSir2A was shown to interact with PfAlba3 deacetylating the lysine residue of N-terminal peptide of PfAlba3 specific for DNA binding [116] (**Figure 3**). In archaea, this interaction had been reported, in which Sir2 regulates silencing through deacetylation of the major archaeal chromatin protein Alba, highlighting an ancestrally conserved mechanism of gene regulation [117].

As stated above, heterochromatin protein 1 (HP1) is a very important protein that has been described to recognize the trimethylation on H3K9, a critical mark for the establishment, maintenance and silencing of centromeric and telomeric heterochromatic regions in various model organisms. In *P. falciparum*, it has been identified as PfHP1 [118] and the H3K9me3 mark was mainly associated with *var* genes at TAS regions, as said before [119]. Moreover, high levels of H3K9me3 correlate with genes localized to the nuclear periphery, implying chromosome loop formation. In addition, an association between PfSir2 and H3K9me3 was found, since the lack of the sirtuin deacetylases causes changes in H3K9me3 localization at the chromosome and generates disruption of the monoallelic transcription of *var* genes, suggesting the existence of perinuclear repressive centers associated with control of expression of malaria parasite genes involved in phenotypic variation and pathogenesis [119].

Flueck et al. [120] described the presence of an ApiAP2 family member in *P. falciparum*, designated as SIP2 that binds to TARE-2 and TARE-3 regions and the upstream regions of *var* upsB *in vivo*. Immunofluorescence and genome-wide high-resolution ChIP analyses demonstrated that *P. falciparum* SIP2 and HP1 proteins co-localize and associate with the same subtelomeric region, suggesting that both proteins participate in the assembly of telomeric heterochromatin. A recent report from Gupta et al. [121] has demonstrated that the protein CAF-1, a chaperone that loads the H3-H4 to the nucleosome assembly after DDR, co-localizes with PfHP1

Yeast	Mammals	*T. gondii* (ToxoDB number)	*P. falciparum* (PlasmoDB PF3D7 number)
Sir2 (P06700)	Sir2 (Q8IXJ6)	Sir2A (227020)	Sir2A (1328800)
		Sir2B (267360)	Sir2B (1451400)
Sir3 (P06701)	Sir3 (Q9NTG7)	ATPase, AAA family protein (283900)*	Orc1/Sir3 like activity (1203000)*
Sir4 (P11978)	Sir4 (Q9Y6E7)	NF	NF
RAP1 (P11938)	RAP1 (Q9NYB0)	NF	NF
RIF1 (P29539)	RIF1 (Q5UIP0)	NF	NF
RIF2 (Q06208)	NF	NF	NF
Ku70 (P32807)	XRCC6 (B1AHC9)	Ku70 (248160)	NF
Ku80 (Q04437)	XRCC5 (P13010)	Ku80 (312510)	NF
Taz1 (P79005)	TRF1 (P54274)	NF	NF
NF	TRF2 (Q15554)	NF	NF
NF	TIN2 (Q9BSI4)	NF	NF
NF	HP1 (Q13185)	Chromo1 (268280)	HP1 (1220900)
Stn1 (P38960)	NF	NF	NF
Ten1 (Q07921)	NF	NF	NF
Cdc13 (P32797)	NF	NF	NF
NF	TPP1 (Q96AP0)	NF	NF
NF	POT1 (Q9NUX5)	NF	NF
Pif1 (P07271)	Pif1 (Q9H611)	NF	NF

NF: Not found.*P. falciparum Orc1 complement yeast Sir3 activity [113]. T. gondii counterpart was detected by searching ToxoDB with PfOrc1 amino acidic sequence by BlastP.

Table 1. Telomeric proteins.

at the same subtelomeric localization, in the nuclear periphery, and also demonstrated its binding to TARE1-3 and co-localization with H3K56ac, a signal of completion on chromatin reassembly after DDR [122]. Interestingly, immunoprecipitation with PfCAF1 followed by LC-MS/MS analysis demonstrated that this protein would be interacting not only with PfHP1 but also with PfAlba3 among others [121].

In *T. gondii*, an HP1 protein was identified as TgChromo1, linked to the sequestration of chromosomes at the nuclear periphery and the process of cell division of the parasite [22]. TgChromo1 has shown to localize at *T. gondii* telomeres but not subtelomeres. However, by that time, subtelomeric regions had not yet been described and, in some cases, the sequences in these regions were not correctly assembled. Also, the presence of H4K20me3 and H2A.X at some TARE sequences and a region near *tsf* gene, previously named TgIRE, was observed [62, 63, 100, 123].

7. Double-strand break repair: H2A.X and chromatin

Cells are exposed to DNA lesions produced by exogenous (e.g., chemicals, UV-irradiation, and ionization) or endogenous factors (e.g., DNA replication stress, meiotic recombination). One of the most deleterious forms of DNA damage is the double-strand break (DSB) [124]. DSBs activate the signal transduction pathway to induce DNA damage checkpoints that delay cell cycle progression, which allows the cell to activate DNA repair mechanism [125]. The phosphorylation of SQE/DΦ motif (where Φ represents a hydrophobic residue) on histone H2A.X (referred to as γH2A.X) is one of the earliest responses to DSB [126, 127]. H2A.X seems to be incorporated randomly in the genome of resting cells [128], whereas γH2A.X is clearly observed forming foci, labeling the DSB and replication fork sites, spreading along the chromosome up to 2 Mb from the damaged site. In addition, chromatin is subjected to several changes at damage sites playing an important role in regulating DNA repair [129]. DSB can be produced by various events, either external as ionizing and UV radiations or internal such as collapse of replication forks and transcription-associated damage, among others [130, 131]. DSB can be repaired by two main mechanisms: nonhomologous end joining (NHEJ) and homologous recombination repair (HRR); the first is an error-prone mechanism available along the cell cycle, and the second is an error-free mechanism active at S/G2 phases of cell cycle because of the requirement of sister chromatid as template [131–134]. Both mechanisms were described in *T. gondii* [14, 135], but *Plasmodium* genus is thought to rely only on HRR [136–138].

Before the election of NHEJ or HRR mechanism, DSB triggers a cascade of events that starts with Mre11-RAD50-Nbs1/Xrs2 (MRN in mammals and MRX in yeast) complex binding to the damaged site, which recruits and activates ATM kinase (**Figure 4A**) [139]. ATM is able to phosphorylate H2A.X at SQE motif as well as other DSB repair enzymes allowing the spreading of γH2A.X and a correct DNA damage response (DDR) at DSB site (**Figure 4A**). ATM kinase is present in *T. gondii* and *P. falciparum* [14]. In *T. gondii*, the MYST family lysine acetyltransferase TgMYST-B has shown to mediate DDR induced by methyl methanesulfonate (MMS) and to stimulate the ATM expression at gene level [140]. In addition to this finding, histone acetyltransferases (HATs) have a predominant role in DDR on the basis of chromatin modulation. Chromatin responds to DSB first by increasing the compaction stage by replacing H2A/H2A.X with the H2A.Z variant and by methylating H3K9 by suv39h1 methyltransferase, which is recruited after spreading the DDR response at both sides of DSB sites (**Figure 4B**) [141, 142]. The arrival of H3K9me3 allows its interaction with the HAT Tip60 and the acetylation of H4 on K16 together with the acetylation of ATM kinase, an important PTM for the activation of autophosphorylation and subsequent activation of ATM (**Figure 4**) [143, 144]. The H3K9me3 and H4K16ac marks were identified in *T. gondii* and *P. falciparum* by mass spectrometry analysis [48–51]. However, in the case of *T. gondii*, an acetylated residue was also detected in H3K9 in a more frequent fashion than H3K9me1,2,3, suggesting that chromatin is preferentially in an open state and that this lysine PTM can be regulated [51]. As it was stated before, H3K9me2/3 is also enriched in centromeres in *T. gondii* [21]. In addition, *T. gondii* H4K16ac was one of the most abundant PTMs found in the mass spectrometry analysis [51]. In the case of *P. falciparum*, the treatment with MMS has increased the level of H4K8ac and H4K16ac and reduction of H3K9ac [145]. Both, *T. gondii* and *P. falciparum* present H3K9me1,2,3 and H3K9ac in normal conditions suggesting a conserved mechanism of chromatin modulation [51]. The role of these histone marks on Apicomplexan histones and the connections with DNA repair remain to be elucidated.

As mentioned above, γH2A.X spreading is a crucial step to initiate a correct DDR at DSB sites. In *T. gondii*, this PTM mark is accompanied by other DDR marks such as H3K9me2,3 and H4K16 in normal conditions of growth, opening the question whether DSBs are being produced during parasite replication [51, 135]. The *T. gondii* tachyzoite replicates at high rates, in a range of 5–9 hours [146]. So, a putative collapse of replication fork could be occurring in this stage. However, *T. gondii* ATM kinase could not be detected in normal conditions by Western blot, but it was detected by tachyzoites overexpressing MYST-B HAT [140].

The chromatin compaction that occurs early during DDR includes the remodeling of chromatin at DSB sites in which the H2A-H2B dimer is replaced by H2A.Z-H2B [142, 147]. This event is transient, allowing the recruitment of repressive kap-1(TRIM28)/HP1/suv39h1 complex that can be important to inhibit transcription. The presence of H2A-H2B dimer in the nucleosomal core particle produces a unique negatively charged region on the surface of the nucleosome, called the "acidic patch," which is extended in H2A.Z (**Figure 4B**) [148–150]. The acidic patch favors the binding of H4 N-tail, resulting in an increase in the interaction between nucleosomes and chromatin compaction [150]. Interestingly, this seems a necessary step to continue with a relaxed chromatin state, since this compaction and recruitment of kap-1(TRIM28)/HP1/suv39h1 complex lead to methylation of H3K9 and phosphorylation of KAP-1 by ATM kinase, which in turn promote H4K16 acetylation by Tip60 and release kap-1(TRIM28)/HP1/suv39h1 (**Figure 4B**) (see [142]). *T. gondii* and *P. falciparum* have the novel H2A.Z-H2B.Z double-variant nucleosome (see Section 5). However, *T. gondii* and *P. falciparum* H2A-H2B and variants conserve the acidic patch (**Figure 4B**). To note, *T. gondii* and *P. falciparum* do not appear to have KAP-1 protein at ToxoDB and PlasmoDB databases [151].

In higher eukaryotes, another important PTM mark associated to DDR is ubiquitination by E3 ubiquitin ligases RNF168 and RNF8 at DSB site after γH2A.X and MDC1 protein foci

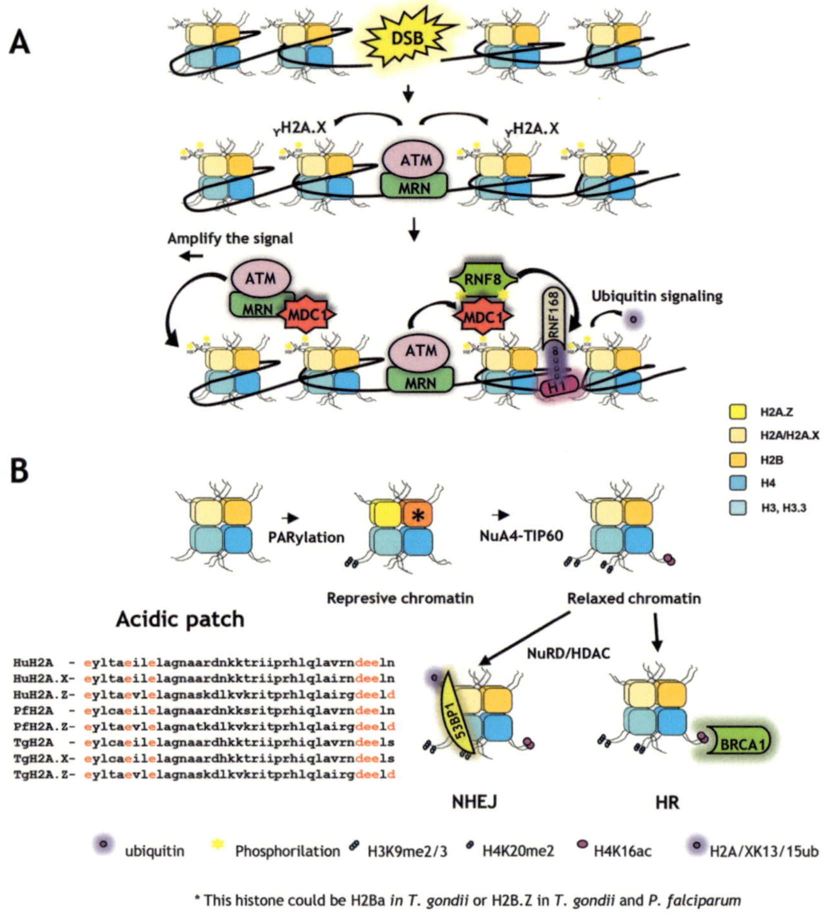

Figure 4. (A) Recognition of DSB and initial steps of DDR pathways. MRN complex and ATM kinase are recruited to a DSB. ATM phosphorylates several DDR proteins and checkpoint kinases. Phosphorylation by ATM allows the spreading of γH2A.X and DDR foci. RNF8 E3 ubiquitin ligase complexes ubiquitinate H2A and H1 histones to compact and generate histone marks, which will be read by other DDR factors such as MDC1 and E3 ubiquitin ligase RNF168. (B) Model of chromatin modulation after DSB. During PARylation, the p400 ATPase (NuA4-Tip60 complex) replaces H2A-H2B dimers by H2A.Z containing nucleosomes. Since T. gondii and P. falciparum present a double-variant nucleosome, we speculate that H2A-H2B or H2A.X-H2Ba dimers may be replaced by H2A.Z-H2B.Z dimers. This exchange might increase the interaction of the acidic patch with the N-tail of H4 increasing the chromatin compaction (see the sequence alignment: in red letters are written the acidic residues involved in generating the acidic patch). PARylation also produces an increase in H3K9me2/3 and histone deacetylation repressive marks. After a short time, the H2A.Z containing nucleosome is replaced by H2A-H2B dimer and N-tail of H4 is acetylated leading to a relaxed chromatin and recruitment of different DDR factors, among them those related to the DDR pathway choice, such as 53BP1 or BRCA1.

spreading (**Figure 4A**). MDC1 is also phosphorylated by ATM kinase allowing the recruitment of RNF168 and RNF8 [152]. Ubiquitination on H1 and H2A recruits several BRCT domain containing proteins such as BRCA1 and 53BP1 [129]. In the case of 53BP1, its binding requires the H2AK13/15ub and H4K20me2 and addresses the DDR to NHEJ pathway (**Figure 4B**). By contrast, the presence of H4K16ac impairs the 53BP1 binding to the nucleosome allowing the recruitment of BRCA, which addresses the DDR to HRR (**Figure 4B**) (see [141, 142]). As stated above, in *T. gondii* and *P. falciparum*, the mark H4K20me1,2,3 was found [48, 51, 53]. However, *T. gondii* and *P. falciparum* H2As did not contain H2AK15ub and lysine 13 ubiquitylation was not detected either [51]. In addition, *T. gondii* and *P. falciparum* did not show the presence of orthologs of BRCA1 and/or 53BP1, though *T. gondii* presents three different BRCT domain containing proteins [14].

T. gondii and *P. falciparum* conserve several histone marks present in chromatin-associated DDR to DBS, as well as histone variants—in the case of *T. gondii*, the DDR, well studied H2A.X, is present, whereas *Plasmodium* has only canonical H2A [55], involved in the recruitment of several factors that spread and choose the DDR pathway in higher eukaryotes. Although, *T. gondii* and *P. falciparum* lack some key DDR regulators such as KAP-1, 53BP1, BRCA1, MDC1, RNF168 and RNF8 [14], both parasites present the HRR mechanism of DNA repair, whereas NHEJ is present only in *T. gondii*. So, the modulation of both DDR pathways is still an intriguing issue.

8. Concluding remarks

In protozoan parasites, the modulation of chromatin seems to be a key biological process to regulate gene expression, pathogenicity and DNA repair, the latter probably associated to DNA replication, ergo, the cell cycle. In Apicomplexa, highly evident in *Plasmodium* genus, the TAS or subtelomeric regions play an important role in the control of a group of genes essential in parasite pathogenicity. This fact suggests that subtelomeres have not a trivial impact in the evolution of these organisms, and their structure can influence the features of the cell. How this genomic domain has evolved within the Apicomplexa phylum remains to be elucidated. *T. gondii*, in which to date a scenario of variant antigenicity was not detected, has shown a someway conserved structure with the presence of tandem repeated boxes and a gene family of unknown function (*tsf*). Different from *Plasmodium*, which variant antigen-associated gene is represented by hundreds of members, *T. gondii* has only one gene per TAS. However, the predicted protein sequences show conserved N-tail and middle regions, with highly variable C-terminal ends. We believe that the elucidation of the localization, role, and antigenic potential of these gene family proteins will be of high impact in our knowledge of this parasite. Also, it could be interesting to know if the members of this gene family show a regulation of gene expression similar to *Plasmodium* variable antigen gene family.

In addition to the presence of PTM marks similar to other organisms but with currently less-well characterized readers and erasers, Apicomplexa chromatin presents a double-variant nucleosome based on the new histone variant H2B.Z. If considering the partitioned knowledge in these parasites, specially *P. falciparum*, where H3.3 variant has been found in the same regions as this double-variant nucleosome, but in different studies, it would be possible that a triple-variant nucleosome exists in Apicomplexa. Since the presence of H2B.Z arose early in Apicomplexa evolution, it is expected that the double-variant nucleosome could have been important in the expansion of the phylum, maybe modulating chromatin structure during the execution of different biological processes. Interestingly, the genome-wide analyses seem to indicate that *Plasmodium* and *Toxoplasma* double-variant nucleosomes do not have the same behavior. In *T. gondii*, it is enriched in active and poised genes, whereas in *P. falciparum*, it is localized in active and silent promoters, excepting the *var* genes, in which the presence of the double-variant nucleosome is associated to active promoters. The analysis of this novel nucleosome in the different genera of the phylum can give more information to elucidate the reason of the presence of this H2B variant.

Chromatin is also important to the DDR and has an important role in determining the different pathways of DNA repair after DSB. *T. gondii* seems to have every histone variant and histone mark as well as important proteins associated to every DDR pathway to repair a DSB: NHEJ (e.g., Ku70/Ku80) and HRR (RAD51). Different from *T. gondii, Plasmodium* does not present the histone variant H2A.X, whose phosphorylation (γH2A.X) is linked to the localization of DSB on DNA. Moreover, *T. gondii* has shown a basal level of γH2A.X, even without damage. Not expected, the proteins associated to the DDR pathway choice (NHEJ or HRR), which read the chromatin, were not detected in *T. gondii* nor in *Plasmodium*. So, it is unknown if these marks are associated to other proteins (*T. gondii* has three BRCT domain containing proteins) with similar roles and/or chromatin modulates DDR in another way.

Taken all together, these differences are not only interesting at the light of evolution but also can be analyzed in the context of the identification of new parasite-specific drug targets. Gene regulation, DNA replication, pathogenicity, and DNA repair are crucial biological processes, and all of them may offer new targets to exploit as future treatments against Apicomplexan pathogens.

Acknowledgements

SO Angel (Researcher), L. Vanagas (Researcher), and S.M. Contreras (Fellow) are members of National Research Council of Argentina (CONICET). SO Angel and L. Vanagas are professors of Universidad Nacional General San Martin (UNSAM). This chapter was supported by ANPCyT PICT 2015 1288 (S.O.A.) and National Institute of Health (NIH, NIAID 1R01AI083162) grants.

Author details

Laura Vanagas, Susana Marisol Contreras and Sergio Oscar Angel*

*Address all correspondence to: sangel@intech.gov.ar

Instituto Tecnológico de Chascomús (INTECH), CONICET/UNSAM, Chascomús, Provincia de Buenos Aires, Argentina

References

[1] Black MW, Boothroyd JC. Lytic cycle of *Toxoplasma gondii*. Microbiology and Molecular Biology Reviews. 2000;**64**(3):607-623

[2] Zuber JA, Takala-Harrison S. Multidrug-resistant malaria and the impact of mass drug administration. Infection and Drug Resistance. 2018;**11**:209-336

[3] Tenter AM, Heckeroth AR, Weiss LM. *Toxoplasma gondii*: From animals to humans. International Journal for Parasitology. 2000;**30**(12-13):1217-1258

[4] Fuller Torrey E, Bartko JJ, Yolken RH. *Toxoplasma gondii* and other risk factors for schizophrenia: An update. 2012;**38**(3):642-647

[5] Fuglewicz AJ, Piotrowski P, Stodolak A. Relationship between toxoplasmosis and schizophrenia: A review. Advances in Clinical and Experimental Medicine. 2017;**26**(6):1031-1036

[6] Flegr J, Horá Ček J. Toxoplasma-infected subjects report an obsessive-compulsive disorder diagnosis more often and score higher in obsessive-compulsive inventory. European Psychiatry. 2017;**40**:82-87

[7] Sutterland AL et al. Beyond the association. *Toxoplasma gondii* in schizophrenia, bipolar disorder, and addiction: Systematic review and meta-analysis. Acta Psychiatrica Scandinavica. 2015;**132**(3):161-179

[8] Yolken R, Fuller Torrey E, Dickerson F. Evidence of increased exposure to *Toxoplasma gondii* in individuals with recent onset psychosis but not with established schizophrenia. PLoS Neglected Tropical Diseases. 2017;**6**(11):e0006040

[9] Kissinger JC, Gajria B, Li L, Paulsen IT, Roos DS. ToxoDB: Accessing the *Toxoplasma gondii* genome. Nucleic Acids Research. 2003;**31**(1):234-236

[10] Gardner MJ et al., Genome sequence of the human malaria parasite Plasmodium falciparum. Nature. 2002;**419**(6906):498-511

[11] De Silva EK et al. Specific DNA-binding by apicomplexan AP2 transcription factors. Proceedings of the National Academy of Sciences of the United States of America. 2008;**105**(16):8393-8398

[12] Behnke MS et al. Coordinated progression through two subtranscriptomes underlies the tachyzoite cycle of *Toxoplasma gondii*. PLoS One. 2010;**5**(8):e12354

[13] Dixon SE, Bhatti MM, Uversky VN, Dunker AK, Sullivan WJ. Regions of intrinsic disorder help identify a novel nuclear localization signal in *Toxoplasma gondii* histone acetyltransferase TgGCN5-B. Molecular and Biochemical Parasitology. 2011;**175**(2):192-195

[14] Fenoy IM, Bogado SS, Contreras SM, Gottifredi V, Angel SO. The knowns unknowns: Exploring the homologous recombination repair pathway in *Toxoplasma gondii*. Frontiers in Microbiology. 2016;**7**(May):1-15

[15] Batugedara G, Lu XM, Bunnik EM, Le Roch KG. The role of chromatin structure in gene regulation of the human malaria parasite. Trends in Parasitology. 2017;**33**:364-377

[16] Striepen B, Jordan CN, Reiff S, van Dooren GG. Building the perfect parasite: Cell division in apicomplexa. PLoS Pathogens. 2007;**3**(6):e78

[17] Olins AL, Ernst A, Zwerger M, Herrmann H, Olins DE. An in vitro model for Pelger-Huët anomaly. Nucleus. 2010;**1**(6):506-512

[18] Vanagas L, Dalmasso MC, Dubremetz JF, Portiansky EL, Olins DE, Angel SO. Epichromatin is conserved in *Toxoplasma gondii* and labels the exterior parasite chromatin throughout the cell cycle. Parasitology. 2013;**140**(9):1-1110

[19] Erenpreisa J, Krigerts J, Salmina K, Selga T, Sorokins H, Freivalds T. Differential staining of peripheral nuclear chromatin with Acridine orange implies an A-form epichromatin conformation of the DNA. Nucleus. 2018;**9**(1):171-181

[20] Weiner A et al. 3D nuclear architecture reveals coupled cell cycle dynamics of chromatin and nuclear pores in the malaria parasite *Plasmodium falciparum*. Cellular Microbiology. 2011;**13**(7):967-977

[21] Brooks CF, Francia ME, Gissot M, Croken MM, Kim K, Striepen B. *Toxoplasma gondii* sequesters centromeres to a specific nuclear region throughout the cell cycle. Proceedings of the National Academy of Sciences of the United States of America. 2011; **108**(9):3767-3772

[22] Gissot M, Walker R, Delhaye S, Huot L, Hot D, Tomavo S. *Toxoplasma gondii* chromodomain protein 1 binds to heterochromatin and colocalises with centromeres and telomeres at the nuclear periphery. PLoS One. 2012;**7**(3):e32671

[23] Hoeijmakers WAM et al. *Plasmodium falciparum* centromeres display a unique epigenetic makeup and cluster prior to and during schizogony. Cellular Microbiology. 2012; **14**(9):1391-1401

[24] Ahmad K, Henikoff S. Histone H3 variants specify modes of chromatin assembly. Proceedings of the National Academy of Sciences USA. 2002;**99**(Suppl 4):16477-16484

[25] Smith S, Stillman B. Purification and characterization of CAF-I, a human cell factor required for chromatin assembly during DNA replication in vitro. Cell. 1989;**58**(1):15-25

[26] Verreault A, Kaufman PD, Kobayashi R, Stillman B. Nucleosome assembly by a complex of CAF-1 and acetylated histones H3/H4. Cell. 1996;**87**(1):95-104

[27] Tagami H, Ray-Gallet D, Ve Almouzni G, Nakatani Y. Histone H3.1 and H3.3 complexes mediate nucleosome assembly pathways dependent or independent of DNA synthesis. Cell. 2004;**116**:51-61

[28] Konev AY et al. The CHD1 motor protein is required for deposition of histone variant H3.3 into chromatin in vivo Science. 2007;**317**(5841):1087-1090

[29] Drané P, Ouararhni K, Depaux A, Shuaib M, Hamiche A. The death-associated protein DAXX is a novel histone chaperone involved in the replication-independent deposition of H3.3 Genes and Development. 2010;**24**(12):1253-1265

[30] Lewis PW, Elsaesser SJ, Noh K-M, Stadler SC, Allis CD. Daxx is an H3.3-specific histone chaperone and cooperates with ATRX in replication-independent chromatin assembly at telomeres. Proceedings of the National Academy of Sciences. 2010;**107**(32):14075-14080

[31] Wirbelauer C, Bell O, Schübeler D. Variant histone H3.3 is deposited at sites of nucleosomal displacement throughout transcribed genes while active histone modifications show a promoter-proximal bias. Genes and Development. 2005;**19**(15):1761-1766

[32] Loyola A, Almouzni G. Marking histone H3 variants: How, when and why? Trends in Biochemical Sciences. 2007;**32**(9):425-433

[33] Goldberg AD et al. Distinct factors control histone variant H3.3 localization at specific genomic regions. Cell. 2010;**140**(5):678-691

[34] Szenker E, Ray-Gallet D, Almouzni G. The double face of the histone variant H3.3. Cell Research. 2011;**21**(3):421-434

[35] Mckittrick E, Gafken PR, Ahmad K, Henikoff S. Histone H3.3 is enriched in covalent modifications associated with active chromatin. Proceedings of the National Academy of Sciences USA. 2004;**101**(6):1525-1530

[36] Jin C et al. H3.3/H2A.Z double variant-containing nucleosomes mark 'nucleosome-free regions' of active promoters and other regulatory regions in the human genome. Nature Genetics. 2009;**41**(8):941-945

[37] Jin C et al. H3.3/H2A.Z double variant–containing nucleosomes mark 'nucleosome-free regions' of active promoters and other regulatory regions. Nature Genetics. 2009;**41**(8):941-945

[38] Chen PB et al. *Plasmodium falciparum* PfSET7: Enzymatic characterization and cellular localization of a novel protein methyltransferase in sporozoite, liver and erythrocytic stage parasites. Scientific Reports. 2016;**6**:21802

[39] Jang C-W, Shibata Y, Starmer J, Yee D, Magnuson T. Histone H3.3 maintains genome integrity during mammalian development. Genes & Development. 2015;**29**(13):1377-1392

[40] Sullivan WJ. Histone H3 and H3.3 variants in the protozoan pathogens *Plasmodium falciparum* and *Toxoplasma gondii*. DNA Sequence. 2003;**14**(3):227-231

[41] Miao J, Fan Q, Cui L, Li J, Cui L. The malaria parasite *Plasmodium falciparum* histones: Organization, expression, and acetylation. Gene. 2006;**369**:53-65

[42] Anne-Kristin Fraschka S, Wilhelmus Maria Henderson R, Bártfai R. H3.3 demarcates GC-rich coding and subtelomeric regions and serves as potential memory mark for virulence gene expression in *Plasmodium falciparum*. Scientific Reports. 2016;**6**:31965

[43] Scherf A, Lopez-Rubio JJ, Riviere L. Antigenic Variation in *Plasmodium falciparum*. Annual Review of Microbiology. 2008;**62**(1):445-470

[44] Baruch DI et al. Cloning the gene encoding PfEMPl, a malarial variant antigen and adherence receptor on the surface of parasitized human erythrocytes. Cell. 1995;**92**:77-97

[45] Smith JD et al. Switches in expression of *Plasmodium falciparum* var genes correlate with changes in antigenic and cytoadherent phenotypes of infected erythrocytes. Cell. 1995;**82**(1):101-110

[46] Su XZ et al. The large diverse gene family var encodes proteins involved in cytoadherence and antigenic variation of *Plasmodium falciparum*-infected erythrocytes. Cell. 1995;**82**(1):89-100

[47] Jiang L et al. PfSETvs methylation of histone H3K36 represses virulence genes in *Plasmodium falciparum*. Nature. 2013;**499**(7457):223-227

[48] Trelle MB, Salcedo-Amaya AM, Cohen AM, Stunnenberg HG, Jensen ON. Global histone analysis by mass spectrometry reveals a high content of acetylated lysine residues in the malaria parasite *Plasmodium falciparum*. Journal of Proteome Research. 2009;**8**(7):3439-3450

[49] Salcedo-Amaya AM et al. Dynamic histone H3 epigenome marking during the intraerythrocytic cycle of *Plasmodium falciparum*. Proceedings of the National Academy of Sciences. 2009;**106**(24):9655-9660

[50] Cui L, Miao J. Chromatin-mediated epigenetic regulation in the malaria parasite *Plasmodium falciparum*. Eukaryotic Cell. 2010;**9**(8):1138-1149

[51] Nardelli SC et al. The histone code of *Toxoplasma gondii* comprises conserved and unique posttranslational modifications. MBio Journal. 2013;**4**(6):e00922-13

[52] Dastidar EG et al. Comprehensive histone phosphorylation analysis and identification of Pf14-3-3 protein as a histone H3 phosphorylation reader in malaria parasites. PLoS One. 2013;**8**(1):e53179

[53] Croken MM, Nardelli SC, Kim K. Chromatin modifications, epigenetics, and how protozoan parasites regulate their lives. Trends in Parasitology. 2012;**28**(5):202-213

[54] Gissot M, Kelly KA, Ajioka JW, Greally JM, Kim K. Epigenomic modifications predict active promoters and gene structure in *Toxoplasma gondii*. PLoS Pathogens. 2007;**3**(6):e77

[55] Dalmasso MC, Sullivan WJ, Angel SO. Canonical and variant histones of protozoan parasites. Frontiers in Bioscience (Landmark Ed.). 2011;**16**:2086-2105

[56] Fischer T et al. Diverse roles of HP1 proteins in heterochromatin assembly and functions in fission yeast. Proceedings of the National Academy of Sciences of the United States of America. 2009;**106**(22):8998-9003

[57] Verma G, Surolia N. *Plasmodium falciparum* CENH3 is able to functionally complement Cse4p and its, C-terminus is essential for centromere function. Molecular and Biochemical Parasitology. 2013;**192**(1-2):21-29

[58] Talbert PB, Henikoff S. Histone variants—Ancient wrap artists of the epigenome. Nature Reviews. Molecular Cell Biology. 2010;**11**(4):264-275

[59] Henikoff S, Smith MM. Histone variants and epigenetics. Cold Spring Harbor Perspectives in Biology. 2015;**7**(1):a019364

[60] Talbert PB et al. A unified phylogeny-based nomenclature for histone variants. Epigenetics & Chromatin. 2012;**5**:7

[61] Malik HS, Henikoff S. Phylogenomics of the nucleosome. Nature Structural Biology. 2003;**10**(11):882-891

[62] Dalmasso MC, Onyango DO, Naguleswaran A, Sullivan WJ, Angel SO. Toxoplasma H2A variants reveal novel insights into nucleosome composition and functions for this histone family. Journal of Molecular Biology. 2009;**392**:33-47

[63] Bogado SS et al. Canonical histone H2Ba and H2A.X dimerize in an opposite genomic localization to H2A.Z/H2B.Z dimers in *Toxoplasma gondii*. Molecular and Biochemical Parasitology. 2014;**197**(1-2):36-42

[64] Hoeijmakers WAM et al. H2A.Z/H2B.Z double-variant nucleosomes inhabit the AT-rich promoter regions of the *Plasmodium falciparum* genome. Molecular Microbiology. 2013; **87**(5):1061-1073

[65] Petter M et al. H2A.Z and H2B.Z double-variant nucleosomes define intergenic regions and dynamically occupy *var* gene promoters in the malaria parasite *Plasmodium falciparum*. Molecular Microbiology. 2013;**87**(6):1167-1182

[66] Talbert PB, Henikoff S. Environmental responses mediated by histone variants. Trends in Cell Biology. 2014;**24**(11):642-650

[67] Millar CB, Xu F, Zhang K, Grunstein M. Acetylation of H2AZ Lys 14 is associated with genome-wide gene activity in yeast. Genes & Development. 2006;**20**(6):711-722

[68] Bruce K et al. The replacement histone H2A.Z in a hyperacetylated form is a feature of active genes in the chicken. Nucleic Acids Research. 2005;**33**(17):5633-5639

[69] Ren Q, Gorovsky MA. Histone H2A.Z acetylation modulates an essential charge patch. Molecular Cell. 2001;**7**(6):1329-1335

[70] Hartley PD, Madhani HD. Mechanisms that specify promoter nucleosome location and identity. Cell. 2009;**137**(3):445-458

[71] Saeed S et al. Chromatin accessibility, p300, and histone acetylation define PML-RAR and AML1-ETO binding sites in acute myeloid leukemia. Blood. 2012;**120**(15):3058-3068

[72] Valdés-Mora F et al. Acetylation of H2A.Z is a key epigenetic modification associated with gene deregulation and epigenetic remodeling in cancer. Genome Research. 2012;**22**(2):307-321

[73] Valdés-Mora F et al. Acetylated histone variant H2A.Z is involved in the activation of neo-enhancers in prostate cancer. Nature Communications. 2017;**8**(1):1346

[74] Sarcinella E, Zuzarte PC, Lau PNI, Draker R, Cheung P. Monoubiquitylation of H2A.Z distinguishes its association with euchromatin or facultative heterochromatin. Molecular and Cellular Biology. 2007;**27**(18):6457-6468

[75] Draker R, Sarcinella E, Cheung P. USP10 deubiquitylates the histone variant H2A.Z and both are required for androgen receptor-mediated gene activation. Nucleic Acids Research. 2011;**39**(9):3529-3542

[76] Talbert PB, Henikoff S. Histone variants on the move: Substrates for chromatin dynamics. Nature Reviews. Molecular Cell Biology. 2017;**18**(2):115-126

[77] Clapier CR, Cairns BR. The biology of chromatin remodeling complexes. Annual Review of Biochemistry. 2009;**78**(1):273-304

[78] Van C, Williams JS, Kunkel TA, Peterson CL. Deposition of histone H2A.Z by the SWR-C remodeling enzyme prevents genome instability. DNA Repair (Amst). 2015;**25**:9-14

[79] Ruhl DD et al. Purification of a human SRCAP complex that remodels chromatin by incorporating the histone variant H2A.Z into nucleosomes. Biochemistry. 2006;**45**(17): 5671-5677

[80] Watanabe S, Radman-Livaja M, Rando OJ, Peterson CL. A histone acetylation switch regulates H2A.Z deposition by the SWR-C remodeling enzyme. Science (80-). 2013; **340**(6129):195-199

[81] Wang F, Ranjan A, Wei D, Wu C. Comment on "A histone acetylation switch regulates H2A.Z deposition by the SWR-C remodeling enzyme". Science (80-). 2016;**353**(6297): 358-358

[82] Watanabe S, Peterson CL. Response to comment on "A histone acetylation switch regulates H2A.Z deposition by the SWR-C remodeling enzyme". Science (80-). 2016; **353**(6297):358-358

[83] Altaf M et al. NuA4-dependent acetylation of nucleosomal histones H4 and H2A directly stimulates incorporation of H2A.Z by the SWR1 complex. The Journal of Biological Chemistry. 2010;**285**(21):15966-15977

[84] Perell GT, Mishra NK, Sudhamalla B, Ycas PD, Islam K, Pomerantz WCK. Specific acetylation patterns of H2A.Z form transient interactions with the BPTF bromodomain. Biochemistry. 2017;**56**(35):4607-4615

[85] Draker R, Ng MK, Sarcinella E, Ignatchenko V, Kislinger T, Cheung P. A combination of H2A.Z and H4 acetylation recruits Brd2 to chromatin during transcriptional activation. PLoS Genetics. 2012;**8**(11):e1003047

[86] Ren Q, Gorovsky MA. The nonessential H2A N-terminal tail can function as an essential charge patch on the H2A.Z variant N-terminal tail. Molecular and Cellular Biology. 2003;**23**(8):2778-2789

[87] Jeffers V, Tampaki Z, Kim K, Sullivan WJ. A latent ability to persist: Differentiation in *Toxoplasma gondii*. Cellular and Molecular Life Sciences. 2018;**75**(13):2355-2373

[88] Whittle CM et al. The genomic distribution and function of histone variant HTZ-1 during *C. elegans* embryogenesis. PLoS Genetics. 2008;**4**(9):e1000187

[89] Updike DL, Mango SE. Temporal regulation of foregut development by HTZ-1/H2A.Z and PHA-4/FoxA. PLoS Genetics. 2006;**2**(9):e161

[90] Ridgway P, Brown KD, Rangasamy D, Svensson U, Tremethick DJ. Unique residues on the H2A.Z containing nucleosome surface are important for *Xenopus laevis* development. The Journal of Biological Chemistry. 2004;**279**(42):43815-43820

[91] Law C, Cheung P. Expression of non-acetylatable H2A.Z in myoblast cells blocks myoblast differentiation through disruption of MyoD expression. The Journal of Biological Chemistry. 2015;**290**(21):13234-13249

[92] Jeffers V, Yang C, Huang S, Sullivan WJ. Bromodomains in protozoan parasites: Evolution, function, and opportunities for drug development. Microbiology and Molecular Biology Reviews. 2017;**81**(1):e00047-e00016

[93] Bártfai R et al. H2A.Z demarcates intergenic regions of the *Plasmodium falciparum* epigenome that are dynamically marked by H3K9ac and H3K4me3. PLoS Pathogens. 2010;**6**(12):e1001223

[94] Westenberger SJ, Cui L, Dharia N, Winzeler E, Cui L. Genome-wide nucleosome mapping of *Plasmodium falciparum* reveals histone-rich coding and histone-poor intergenic regions and chromatin remodeling of core and subtelomeric genes. BMC Genomics. 2009;**10**(1):610

[95] Petter M et al. Expression of *P. falciparum* var genes involves exchange of the histone variant H2A.Z at the promoter. PLoS Pathogens. 2011;**7**(2):e1001292

[96] Kumar SV, Wigge PA. H2A.Z-containing nucleosomes mediate the thermosensory response in arabidopsis. Cell. 2010;**140**(1):136-147

[97] Horikoshi N, Arimura Y, Taguchi H, Kurumizaka H. Crystal structures of heterotypic nucleosomes containing histones H2A.Z and H2A. Open Biology. 2016;**6**(6):160127

[98] Scherf A, Figueiredo LM, Freitas-junior LH. Plasmodium telomeres: A pathogen's perspective. Current Opinion in Microbiology. 2001;**4**(4):409-414

[99] Figueiredo LM, Freitas-Junior LH, Bottius E, Olivo-Marin J-C, Scherf A. A central role for *Plasmodium falciparum* subtelomeric regions in spatial positioning and telomere length regulation. The EMBO Journal. 2002;**21**(4):815-824

[100] Dalmasso MC, Carmona SJ, Angel SO, Agüero F. Characterization of *Toxoplasma gondii* subtelomeric-like regions: Identification of a long-range compositional bias that is also associated with gene-poor regions. BMC Genomics. 2014;**15**(1):21

[101] Kaviratne M, Khan SM, Jarra W, Preiser PR. Small variant STEVOR antigen is uniquely located within Maurer's clefts in *Plasmodium falciparum*-infected red blood cells. Eukaryotic Cell. 2002;**1**(6):926-935

[102] Petter M, Haeggström M, Khattab A, Fernandez V, Klinkert M-Q, Wahlgren M. Variant proteins of the *Plasmodium falciparum* RIFIN family show distinct subcellular localization and developmental expression patterns. Molecular and Biochemical Parasitology. 2007;**156**(1):51-61

[103] Freitas-Junior LH et al. Frequent ectopic recombination of virulence factor genes in telomeric chromosome clusters of *P. falciparum*. Nature. 2000;**407**(6807):1018-1022

[104] Vernick KD, McCutchan TF. A novel class of supercoil-independent nuclease hypersensitive site is comprised of alternative DNA structures that flank eukaryotic genes. Journal of Molecular Biology. 1998;**279**(4):737-751

[105] Gardner MJ et al. Genome sequence of the human malaria parasite *Plasmodium falciparum*. Nature. 2002;**419**(6906):498-511

[106] Ai W, Bertram PG, Tsang CK, Chan TF, Zheng XFS. Regulation of subtelomeric silencing during stress response. Molecular Cell. 2002;**10**(6):1295-1305

[107] Buck MJ, Lieb JD. A chromatin-mediated mechanism for specification of conditional transcription factor targets. Nature Genetics. 2006;**38**(12):6-1451

[108] Tomar RS, Zheng S, Brunke-Reese D, Wolcott HN, Reese JC. Yeast Rap1 contributes to genomic integrity by activating DNA damage repair genes. The EMBO Journal. 2008; **27**(11):1575-1584

[109] Harari Y, Romano G-H, Ungar L, Kupiec M. Nature vs nurture: Interplay between the genetic control of telomere length and environmental factors. Cell Cycle. 2013;**12**(22): 3465-3470

[110] Tonkin CJ et al. Sir2 paralogues cooperate to regulate virulence genes and antigenic variation in *Plasmodium falciparum*. PLoS Biology. 2009;**7**(4):0771-0788

[111] Freitas LH et al. Telomeric heterochromatin propagation and histone acetylation control mutually exclusive expression of antigenic variation genes in malaria parasites. Cell. 2005;**121**(1):25-36

[112] Merrick CJ et al. Functional analysis of sirtuin genes in multiple *Plasmodium falciparum* strains. PLoS One. 2015;**10**(3):e0118865

[113] Mancio-Silva L, Rojas-Meza AP, Vargas M, Scherf A, Hernandez-Rivas R. Differential association of Orc1 and Sir2 proteins to telomeric domains in *Plasmodium falciparum*. Journal of Cell Science. 2008;**121**:2046-2053

[114] Deshmukh AS et al. The role of N-terminus of *Plasmodium falciparum* ORC1 in telomeric localization and var gene silencing. Nucleic Acids Research. 2012;**40**(12):5313-5331

[115] Serrano L et al. The tumor suppressor SirT2 regulates cell cycle progression and genome stability by modulating the mitotic deposition of H4K20 methylation. Genes & Development. 2013;**27**(6):639-653

[116] Goyal M et al. Identification and molecular characterization of an Alba-family protein from human malaria parasite *Plasmodium falciparum*. Nucleic Acids Research. 2012; **40**(3):1174-1190

[117] Bell SD. The interaction of Alba, a conserved archaeal chromatin protein, with Sir2 and its regulation by acetylation. Science (80-). 2002;**296**(5565):148-151

[118] Pérez-Toledo K et al. *Plasmodium falciparum* heterochromatin protein 1 binds to trimethylated histone 3 lysine 9 and is linked to mutually exclusive expression of var genes. Nucleic Acids Research. 2009;**37**(8):2596-2606

[119] Lopez-Rubio JJ, Mancio-Silva L, Scherf A. Genome-wide analysis of heterochromatin associates clonally variant gene regulation with perinuclear repressive centers in malaria parasites. Cell Host & Microbe. 2009;**5**(2):179-190

[120] Flueck C et al. A major role for the *Plasmodium falciparum* ApiAP2 protein PfSIP2 in chromosome end biology. PLoS Pathogens. 2010;**6**(2):e1000784

[121] Gupta MK, Agarawal M, Banu K, Reddy KS, Gaur D, Dhar SK. Role of chromatin assembly factor 1 in DNA replication of *Plasmodium falciparum*. Biochemical and Biophysical Research Communications. 2018;**495**(1):1285-1291

[122] Masumoto H, Hawke D, Kobayashi R, Verreault A. A role for cell-cycle-regulated histone H3 lysine 56 acetylation in the DNA damage response. Nature. 2005;**436**(7048):294-298

[123] Echeverria PC, Rojas PA, Martin V, Guarnera EA, Pszenny V, Angel SO. Characterisation of a novel interspersed *Toxoplasma gondii* DNA repeat with potential uses for PCR diagnosis and PCR-RFLP analysis. FEMS Microbiology Letters. 2000;**184**(1):23-27

[124] Fillingham J, Keogh M-C, Krogan NJ. GammaH2AX and its role in DNA double-strand break repair. Biochemistry and Cell Biology. 2006;**84**(4):568-577

[125] Nyberg KA, Michelson RJ, Putnam CW, Weinert TA. Toward maintaining the genome: DNA damage and replication checkpoints. Annual Review of Genetics. 2002;**36**(1): 617-656

[126] Redon C, Pilch D, Rogakou E, Sedelnikova O, Newrock K, Bonner W. Histone H2A variants H2AX and H2AZ. Current Opinion in Genetics and Development. 2002;**12**(2): 162-169

[127] Turinetto V, Giachino C. Histone variants as emerging regulators of embryonic stem cell identity. Epigenetics. 2015;**10**(7):563-573

[128] Seo J et al. Genome-wide profiles of H2AX and γ-H2AX differentiate endogenous and exogenous DNA damage hotspots in human cells. Nucleic Acids Research. 2012; **40**:5965-5974

[129] Lukas J, Lukas C, Bartek J. More than just a focus: The chromatin response to DNA damage and its role in genome integrity maintenance. Nature Cell Biology. 2011; **13**(10):1161-1169

[130] Ciccia A, Elledge SJ. The DNA damage response: Making it safe to play with knives. Molecular Cell. 2010;**40**(2):179-204

[131] Jackson SP, Bartek J. The DNA-damage response in human biology and disease. Nature. 2009;**461**(7267):1071-1078

[132] Bétermier M, Bertrand P, Lopez BS. Is non-homologous end-joining really an inherently error-prone process? PLoS Genetics. 2014;**10**(1):e1004086

[133] Guirouilh-Barbat J, Lambert S, Bertrand P, Lopez BS. Is homologous recombination really an error-free process? Frontiers in Genetics. 2014;**5**:175

[134] Sancar A, Lindsey-Boltz LA, Unsal-Kacmaz K, Linn S. Molecular mechanisms of mammalian DNA repair and the DNA damage checkpoints. Annual Review of Biochemistry. 2004;**73**(1):39-85 (0066-4154 (Print) LA–eng PT–Journal Article PT–Review RN–0 (Cross–Linking Reagents) RN–9007-49-2 (DNA) SB–IM)

[135] Smolarz B, Wilczyński J, Nowakowska D. DNA repair mechanisms and *Toxoplasma gondii* infection. Archives of Microbiology. 2014;**196**(1):1-8

[136] Kelso AA, Waldvogel SM, Luthman AJ, Sehorn MG. Homologous recombination in protozoan parasites and recombinase inhibitors. Frontiers in Microbiology. 2017;**8**:1716

[137] Fidock DA, Lee AH, Symington LS, Fidock A. DNA repair mechanisms and their biological roles in the malaria parasite *Plasmodium falciparum*. Microbiology and Molecular Biology Reviews. 2014;**3**:78

[138] Calhoun SF, Reed J, Alexander N, Mason CE, Deitsch KW, Kirkman LA. Chromosome end repair and genome stability in *Plasmodium falciparum*. MBio Journal. 2017;**8**(4): e00547-e00517

[139] Paull TT. Mechanisms of ATM activation. Annual Review of Biochemistry. 2015;**84**(1): 711-738

[140] Vonlaufen N, Naguleswaran A, Coppens I, Sullivan WJ. MYST family lysine acetyltransferase facilitates ataxia telangiectasia mutated (ATM) kinase-mediated DNA damage response in *Toxoplasma gondii*. The Journal of Biological Chemistry. 2010;**285**(15): 11154-11161

[141] Agarwal P, Miller KM. The nucleosome: Orchestrating DNA damage signaling and repair within chromatin. Biochemistry and Cell Biology. 2016;**94**(5):381-395

[142] Gursoy-Yuzugullu O, House N, Price BD. Patching broken DNA: Nucleosome dynamics and the repair of DNA breaks. Journal of Molecular Biology. 2016;**428**(9):1846-1860

[143] Sun Y, Jiang X, Chen S, Fernandes N, Price BD. A role for the Tip60 histone acetyltransferase in the acetylation and activation of ATM. Proceedings of the National Academy of Sciences of the United States of America. 2005;**102**(37):13182-13187

[144] Francia ME, Striepen B. Cell division in apicomplexan parasites. Nature Reviews Microbiology. 2014;**12**(2):125-136

[145] Gupta DK, Patra AT, Zhu L, Gupta AP, Bozdech Z. DNA damage regulation and its role in drug-related phenotypes in the malaria parasites. Scientific Reports. 2016; **6**(November):1-15

[146] Radke JR, Striepen B, Guerini MN, Jerome ME, Roos DS, White MW. Defining the cell cycle for the tachyzoite stage of *Toxoplasma gondii*. Molecular and Biochemical Parasitology. 2001;**115**(2):165-175

[147] Xu Y, Ayrapetov MK, Xu C, Gursoy-Yuzugullu O, Hu Y, Price BD. Histone H2A.Z controls a critical chromatin remodeling step required for DNA double-strand break repair. Molecular Cell. 2012;**48**(5):723-733

[148] Luger K, Rechsteiner TJ, Flaus AJ, Waye MM, Richmond TJ. Characterization of nucleosome core particles containing histone proteins made in bacteria. Journal of Molecular Biology. 1997;**272**(3):301-311

[149] Luger K, Suto RK, Clarkson MJ, Tremethick DJ. Crystal structure of a nucleosome core particle containing the variant histone H2A.Z. Nature Structural Biology. 2000; **7**(12):1121-1124

[150] Park Y-J, Dyer PN, Tremethick DJ, Luger K. A new fluorescence resonance energy transfer approach demonstrates that the histone variant H2AZ stabilizes the histone octamer within the nucleosome. The Journal of Biological Chemistry. 2004;**279**(23):24274-24282

[151] Aurrecoechea C et al. EuPathDB: The eukaryotic pathogen database. Nucleic Acids Research. 2013;**41**(D1):D684-D691

[152] Jackson SP, Durocher D. Regulation of DNA damage responses by ubiquitin and SUMO.Molecular Cell. 2013;**49**(5):795-807

Chapter 6

Epigenetic Modulation of Circadian Rhythms: *Bmal1* Gene Regulation

Tatsunosuke Tomita and Yoshiaki Onishi

Additional information is available at the end of the chapter

http://dx.doi.org/10.5772/intechopen.79975

Abstract

Circadian rhythms that function in behaviour and physiology have adaptive significance for living organisms from bacteria to humans and reflect the presence of a biological clock. The engine of circadian rhythms is a transcription-translation feedback loop that is fine-tuned by epigenetic regulation in higher eukaryotes. We elucidated the chromatin structure of the *Bmal1* gene, a critical component of the mammalian clock system, and have continued to investigate transcriptional regulation including DNA methylation. Various ailments including metabolic diseases can disrupt circadian rhythms, and many human diseases are associated with altered DNA methylation. Therefore, regulated circadian rhythms are important for human health. Here, we summarise the importance of epigenetic clock gene regulation, including DNA methylation of the *Bmal1* gene, from the viewpoint of relationships to diseases.

Keywords: molecular clock, transcriptional mechanism, cytosine methylation, chromatin, cancer, metabolic syndrome

1. Introduction

Circadian rhythms function in most living organisms and govern many behavioural and biochemical processes with 24-h periodicity regardless of changes in the cellular environment. This is closely associated with the natural rhythm of the sun, which provides light and heat with 24-h periodicity. The master clock that generates circadian rhythms in mammals is located in the suprachiasmatic nucleus (SCN) of the hypothalamus and is governed by blue-light sensing in eyes. Peripheral organs also contain molecular clocks. These biological clocks control all aspects of physiology such as sleep-wake cycles, body temperature, hormone secretion, blood pressure and metabolism [1]. Biological clocks oscillate via a mechanism based on interlocking transcriptional-translational feedback loops that have both positive and negative elements. The circadian oscillator orchestrates the rhythmic mRNA expression and output of hundreds or thousands of clock-controlled genes (CCG) that temporally coordinate many cellular functions [2]. Circadian transcriptional regulators are apparently involved in the initial stages of RNA polymerase II recruitment and initiation, as well as the histone modifications associated with these events to set the stage for gene expression [3]. The methylation of cytosine on CpG dinucleotides, which is also epigenetic regulation of gene expression, either directly interferes with the binding of transcriptional regulators or indirectly inactivates a gene by modulating chromatin to a repressive structure. About 43% of all protein-encoding genes in mice exhibit circadian rhythms of mRNA abundance somewhere in the body, largely in an organ-specific manner [4]. The temporal coordination of cellular functions is lost when circadian rhythms are disrupted by age, the environment or genetic mutation, with deleterious effects on health. For instance, the adrenal steroid hormone glucocorticoid that controls various physiological processes, such as metabolism, the immune response, cardiovascular activity and brain function, is under the control of the circadian clock [5], implying that several diseases are closely associated with disrupted circadian rhythms.

2. Transcriptional mechanism of the circadian clock

2.1. Basic regulation of circadian transcription

The engine of the mammalian molecular clock consists of a transcription-translation feedback loop initiated by the transcription factor BMAL1-CLOCK heterodimer. BMAL1 and CLOCK have paralogs, known as BMAL2 and NPAS2, respectively. Heterodimers such as BMAL1-CLOCK bind to E-box enhancer sequences and activate the transcription of three *Per* (*Per1*, *Per2* and *Per3*) and two *Cry* (*Cry1* and *Cry2*) genes. The PER and CRY proteins subsequently repress the transcription at their own promoters through negative feedback by acting on the BMAL1-CLOCK heterodimer. The cellular circadian clock mediates the rhythmic output of the hundreds or thousands of CCG transcripts that are regulated by transcription factors or coregulators with rhythmic abundance that is a part of the cellular circadian clock [3]. The prominent transcription factors activated by BMAL1-CLOCK are REV-ERB α and β, which bind to ROREs, as well as DBP and E4BP4, which bind to D-boxes. E-box motifs contain a core CANNTG sequence, which is recognised by a basic helix-loop-helix (bHLH) domain that contains transcription factors. BMAL1-CLOCK binds tandem E boxes spaced 6 or 7 nucleotides (nt) apart with high affinity [6]. The bHLH containing the oncoprotein Myc also binds to E-boxes and directly activates the expression of multiple repressors of the clock, including *Rev-erbα* and *Rev-erbβ* [7]. In addition, USF1 binds to the E-box motifs of *Dbp*, *Per1* and *Per2* [8]. The RORE motif comprises an AT-rich sequence preceding a core (G/A) GGTCA motif. ROR and REV-ERB, respectively, activate and repress the transcription of genes by binding to ROREs [9]. They co-ordinately maintain robust circadian expression of core clock proteins, such as BMAL1. D-boxes are variants of basic leucine-zipper (bZIP) motifs and are 9- or 10-bp palindromes of two GTAA (C/T) half-site sequences [10]. The D-box motif is bound by the proline- and acidic amino acid-rich bZIP (PAR-bZIP) transcription factor family, including DBP, E4BP4, HLF and TEF [11, 12]. A combination of three binding elements, E-boxes, ROREs and D-boxes, coordinates CCG transcription. **Figure 1** shows that most core clock proteins including BMAL1, CLOCK, PER, CRY, REB-ERB, ROR and E4BP4 bind to many thousands of sites in the genome in a circadian manner [13].

2.2. Epigenetic mechanism: effect of chromatin structure

Transcriptional regulation initially requires the coordinated control of chromatin and the genome structure [3]. In general, genetic information is packed into the chromatin structure, of which the nucleosome is the most basic unit; it determines the large-scale chromatin structure as a building block and influences transcription. Eukaryotic promoter regions are thought to have inactive states, assured by the tendency of nucleosomes to inhibit transcription by protecting protein-DNA interaction. Therefore, chromatin remodelling and loosening of the nucleosomal barrier including histone tail modifications are key steps in circadian modifications followed by sequence-specific, transcription factor binding that regulates gene expressions [14]. Distinct chromatin states are determined by unique histone post-translational modifications. First, histone acetylation levels fluctuate rhythmically at clock gene promoters and enhancers. Specifically, acetylated histone H3 at Lys27 (H3K27ac), a marker of active enhancers, and H3 at Lys9 (H3K9ac) are rhythmic and positively correlate with clock gene expression. For example, rhythmic BMAL1-CLOCK binding and H3K9ac are required as well as rhythmic histone H3 abundance at the start site for *Dbp* transcription [15]. Complexes of clock proteins such as PER contain various interactive partners with known catalytic activity towards chromatin [16, 17]. The acetylation of histone H3 (at Lys9 and Lys14) at *Per1*, *Per2* and *Cry1* and of H4 at *Per1* during the transcriptional activation phase has been identified [18, 19]. Rhythmic histone acetylation at clock loci is largely mediated by p300 and CBP histone

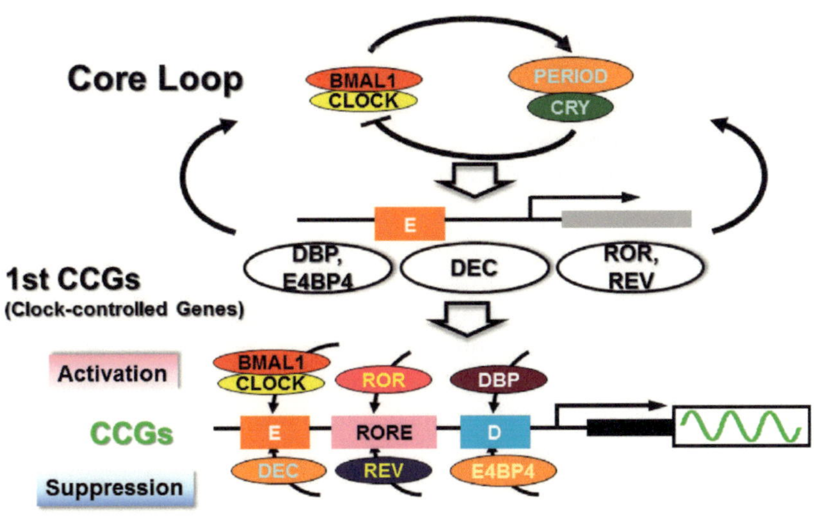

Figure 1. Hierarchical regulation mechanism of circadian transcription. E, RORE and D indicate transcription factor recognition sites: E-box, RORE and D-box, respectively.

acetyltransferases (HAT) [15, 19], and CLOCK itself might also have intrinsic HAT activity [20]. Levels of histone acetylation are also regulated by histone deacetylases (HDAC) as well as by HAT. Several HDAC are important in the control of circadian histone acetylation. For example, REV-ERBα represses transcription in part by recruiting the co-repressor complexes NCoR and/or SMRT to ROREs [21]. One major mechanism of transcriptional repression mediated by CRY and PER is the direct recruitment of the Sin3 complex, which contains HDAC1 and HDAC2 [17]. Another co-repressor complex containing HDAC1 and HDAC2 subunits, NuRD, binds PER-CRY and deacetylates nearby histones, thereby represses clock genes [22]. Sirtuins are another class of HDAC involved in the core clock mechanism that associate with the BMAL-CLOCK heterodimer, and levels of their common cofactor, nicotinamide adenine dinucleotide, are under tight circadian control in many physiological systems [23, 24]. In addition to being acetylated, lysine side chains can be methylated by methyltransferases, and their deacetylation often precedes and facilitates an acetylation-methylation switch. Histone H3 at Lys9 methylation (H3K9me) promotes heterochromatin formation and transcriptional repression. Rhythmic H3K9me near circadian E boxes is mediated by SUV39 methyltransferase and is antiphase to H3K9ac rhythms in the mouse liver [15]. The di- and trimethylation of H3 at Lys27 also proceed at *Per1* and *Per2* during the repressive phase [25]. The circadian clock regulates global transcriptional integrity and chromatin status by regulating RNA polymerase II, because circadian transcription is clustered in phase and accompanied by circadian control of RNA polymerase II recruitment and initiation [26]. The above individual mechanism is governed by the three-dimensional (3D) architecture of chromatin and its critical contributions to long-distance cis-acting mechanisms of gene regulation [27]. Regulatory elements such as enhancers, silencers and insulators built up functional 3D architectures in the nucleus and manage the transcription factory with specific properties [28]. Several looping factors, such as components of the Mediator complex, interact with clock transcription factors [29]. Deletion of one of the factors important for looping, Smc3, causes major disruptions to the clock [30]. Recently, the detailed 3D multi-loop aggregate/rosette chromatin architecture and functional dynamics have been revealed [31, 32], and this may explain how physiological functions are regulated with a tissue-specific rhythm in spite of the same core clock system. These results suggest that epigenetic regulation caused by the chromatin structure is important for circadian transcription, and further researches from the viewpoint of 3D chromatin structure are required to elucidate the physiological function with circadian rhythm in the tissue.

2.3. DNA methylation

The most common epigenetic modification is DNA methylation, which is a covalent chemical alteration that plays a crucial role in numerous biological processes. It occurs in mammals predominantly on cytosine residues in cytosine-guanine (CpG) dinucleotides, and tissue-specific genomic DNA methylation patterns play a fundamental role in establishing cell identity during differentiation. Generally, although about 70% of all CpG sequences in mouse and human genomes are methylated, CpG islands in promoter sequences are methylated at a relatively lower level [33]. Overall, DNA methylation exhibits no major rhythmic changes and the cellular function of DNA methylation depends on which gene is methylated. One of the most important issues regarding DNA methylation is how the machinery is directed towards and maintains specific genomic sequences. One mechanism might be the PML-RAT fusion protein in leukaemia, which induces DNA hypermethylation and gene silencing at specific target promoters [34]. Another is siRNA-mediated, RNA-directed DNA methylation, which is a stepwise process initiated by dsRNA that recruits DNMT to catalyse the *de novo* DNA methylation of specific regions [35]. Therefore, the susceptibility of individual CpG islands to *de novo* methylation might intrinsically differ, but the mechanism remains obscure. In any event, CpG methylation is strictly regulated and stable, and changes in methylation profiles are associated with diseases, indicating close relationships among DNA methylation sites, the mechanism of methylation and biological functions.

3. Transcriptional regulation of the *Bmal1* gene

Bmal1 was originally characterised due to its high expression levels in brain and muscle cells [36]. The activity of *Bmal1*$^{-/-}$ mice immediately becomes arrhythmic in constant darkness; therefore, BMAL1 is apparently an essential and non-redundant component of the mammalian clock [37]. Among the core clock genes, *BmalL1* expression oscillates in the SCN and in peripheral clock cells, in close association with circadian rhythms [38]. We evaluated the chromatin structure of the *Bmal1* gene and discovered a unique structure within the *Bmal1* promoter. The *Bmal1* promoter region comprises mainly a general nucleosome structure upstream of a 5′ SacI site, an open chromatin structure around RORE and a nuclear matrix-like structure at a 3′-flanking region (**Figure 2**). Oscillatory transcription of the *Bmal1* gene requires the chromatin structure to undergo rhythmic alterations *in vivo* at the region around the ROREs and at the 3′-flanking region in response to SAF-A binding, indicating cooperative alteration of the chromatin structure between the 3′-flanking region and the ROREs [39]. The methylation of DNA on CpG islands results in transcriptional repression either by interfering with transcription factor binding or by including a repressive chromatin structure [40]. The methylation of CpG adjacent to the core Sp1 motif decreases Sp1/Sp3 binding [41], which might be associated with the repression of *Bmal1* transcription by DNA methylation, because many putative Sp1-binding motifs are located around the *Bmal1* promoter. The level of DNA methylation within a ±1 kb region surrounding the transcription start site closely correlates with gene repression, and the promoter of clock genes including *Bmal1* is usually unmethylated [39]. However, the hypermethylation of CpG islands in the promoter of *Bmal1* transcriptionally silences its expression in haematological

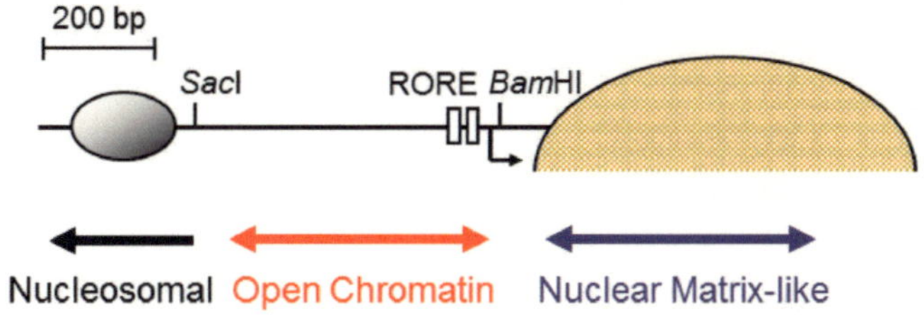

Figure 2. Chromatin structure of *Bmal1* promoter. Oval, unfilled boxes and arrow near *BamHI* indicate nucleosome, RORE and transcription start site, respectively. RORE: recognition motifs for retinoic acid receptor-related orphan receptor (ROR) and reverse Erb (REV-ERB) orphan nuclear receptors.

malignancies [42, 43]. Relationships between the DNA methylation of clock genes and diseases have been identified. The ROREs, which are critical elements for *Bmal1* oscillatory transcription [44], are embedded in a unique GC-rich open chromatin structure. We also found that DNA demethylation of the *Bmal1* promoter enhances *Bmal1*, and then *Per2* and *Cry1* transcription that function in the circadian oscillation of *Bmal1* transcription recover, suggesting that the circadian rhythm is restored [42, 43]. Furthermore, DNA methylation might contribute to the developmental expression of clock genes [45]. These lines of evidence suggest that the DNA methylation of clock genes, in particular, *Bmal1*, plays a key role in the disruption of circadian rhythms that are closely associated with various diseases.

We recently found that recovery from DNA methylation by 5-aza-2′-deoxycytidine (aza-dC) differs between the *Bmal1* and *Rpib9* genes, suggesting that the release of methylation depends on the locus/gene or sequence and that methylation status is specific to the DNA site [46]. Taken together, these findings imply that methylation is specific to gene function and that an early response to the aza-dC demethylation of sites in *Bmal1* might be functionally important for adaptation to environmental change.

4. Disease

Appropriate circadian gene expression is necessary for the normal cell development. That is, distorted clock gene expression leads to various diseases. This chapter focuses on cancers and some other diseases.

4.1. Cancer

Close relationships between clock gene expression and the initiation and progression of cancer are obvious from the findings of many studies. Clock gene expression is altered in many types of malignancies including breast, lung, haematopoietic, pancreatic and skin cancers. Clock genes are apt to be downregulated in many cancer types, as shown in **Table 1**. These phenomena imply that clock genes have some anti-tumour effects. The physiological disruption of circadian

Gene	Expression	Mechanism	Cancer type	DNA methylation	References
Per1	Downregulated	Apoptosis	Colon, lung, breast	—	[38]
Per2	Downregulated	Apoptosis	Lung, lymphocyte	—	[36, 37]
	Downregulated	MYC-downregulation	Lung, breast	—	[34, 36, 37]
	Downregulated	p53-upregulation	Lung, breast	—	[36, 37]
Cry2	Downregulated	Unknown	Breast	Hypermethylation	[39, 40]
Bmal1	Downregulated	p53 pathway	Pancreas	—	[35]
	Downregulated	p300, CAT activation	Leukaemia	Hypermethylation	[29, 33, 34]

Table 1. Clock genes and their possible functions for cancer suppression.

rhythms and the genetic loss of *Per2* or *Bmal1* promote tumorigenesis in lung cancer [47], and such disruptions are associated with upregulated *c-Myc* levels. The expression of *Bmal1* is suppressed in pancreatic cancer, and this gene activates the p53 tumour suppressor pathway, playing an important role in cancer suppression [48]. Fu et al. found that PER2 is an important factor for tumour suppression and the DNA damage response [49]. The overexpression of *Per1* or *Per2* can lead to the apoptosis of cancer cells [50, 51]. Mao et al. reported that *Cry2* expression is decreased in breast cancer, resulting in an altered methylation pattern in CpG islands [52]. The findings of another study support this observation, and CRY2 suppression is closely associated with risks for breast cancer [53]. From a mechanistic viewpoint, one of the main factors in such disrupted circadian gene expression might be MYC. According to a report by Altman et al., this gene directly activates REV-ERB, which suppresses *Bmal1*, and their constitutive expression suspends clock mechanisms [7]. These findings suggest that the appropriate expression of clock genes is necessary to maintain normal tissues. On the other hand, leukaemia stem cells in acute myeloid leukaemia (AML) have intact circadian expression. Furthermore, knockdown studies have shown that *Bmal1* and *Clock* are required for AML cell growth and that disrupted circadian rhythm machinery is an anti-leukaemic factor that leads to leukaemia stem cell differentiation [54]. In addition, upregulated *Clock* plays critical roles in the proliferation of colorectal carcinoma cells and the inhibition of apoptosis [55].

The roles of clock genes seem to differ among stages or tissues in patients with cancer. In addition to classical genetic mutations, the epigenetic landscapes of cancer cells are rather contorted. From an epigenetic perspective, clock genes functionally associate with histone modifying genes that are responsible for cancer progression and maintenance. Mixed lineage leukaemia (MLL) genes were originally discovered through detailed analyses of leukaemogenic rearrangement but they are now thought to be responsible for histone H3K4 methyltransferase activity and promoters of target gene transcription. Mutations of MLL genes literally trigger mixed lineage leukaemia and are necessary to maintain malignancy through aberrant epigenetic gene regulation [56]. The relationship between MLL genes and circadian rhythm maintenance through histone modification has been studied in detail. According to Katada et al., MLL1 has CLOCK-associated histone modifying activity, and it is necessary to generate circadian rhythms in fibroblasts [57]. Kim et al. found that MLL3 and 4 are factors that regulate circadian rhythmic homeostasis in the liver [58]. In addition, MLL3 contributes to circadian rhythm generation in mouse embryonic fibroblasts (MEFs) [59]. The histone modifying enzyme EZH2 is another histone-lysine N methyl transferase that is responsible for histone H3K27 methyl transfer. This modification results in transcription repression. *Ezh2* also promotes tumorigenesis by altering the expression of numerous tumour suppressor genes [60]. EZH2 interacts with CLOCK-BMAL1 complexes and is necessary for circadian rhythm maintenance [25]. Although CLOCK *per se* is not considered to be an oncogene, it might affect cancer cell proliferation if it is atypically expressed [55].

Considering the altered methylation patterns of the promoter regions of clock genes, the features of epigenetic abnormalities of cancer cells comprise highly methylated CpG islands of specific genes accompanied by low methylation status of other genes [61]. Some studies have indicated that this phenomenon is true for clock genes. The *Cry2* promoter tends to be highly methylated in patients with breast cancer, resulting in lower *Cry2* expression compared with controls [43]. Taniguchi et al. reported that CpG islands of the *Bmal1* promoter are hypermethylated in diffuse large B-cell lymphoma and in acute lymphocytic and myeloid leukaemia [42]. We also reported this phenomenon and that the methylation pattern of the *Per2* promoter region does not change in RPMI8402 cells [46]. The aberrant methylation pattern of the *Bmal1* promoter was restored, and the intrinsic rhythm was revived after 1 day of aza-dC treatment. These findings indicate that active mechanisms in leukaemia cells maintain the promoters of hypermethylated *Bmal1* gene status.

As noted above, many studies have emphasised close relationships between epigenetic modification and circadian clock genes in cancer proliferation and progression. However, the precise mechanisms seem highly complex and remain obscure. Further investigation is required to elucidate these mechanisms.

4.2. Other diseases

Circadian rhythms are also associated with diseases other than cancer through effects on the cardiovascular, renal, immune, endocrine, neuropsychiatric and metabolic systems [5, 62–67]. Many physiological processes cannot be harmonised when the intrinsic rhythm is aberrant and such dyssynchrony leads to many diseases.

Here, we consider neuropsychiatric disorders. Disrupted sleep-wake cycles, depression, Alzheimer's disease and mood disorders among neuropsychiatric disorders are notably linked to altered circadian rhythms. However, circadian epigenomics have received less consideration in studies of neuropsychiatric disorders compared with cancers.

Alzheimer's disease is an age-dependent neurodegenerative disorder that is associated with severe cognitive impairment, and its incidence is increasing, particularly in developed countries due to extended life spans. The typical clinical symptoms are disordered circadian rhythms and abnormal sleep patterns. Amyloid beta is a key molecule in this neurodegeneration [68], and it reportedly degrades BMAL1 protein [69]. The lack of this powerful rhythm generator disrupts circadian rhythms in many patients. Furthermore, the methylation rhythm of the *Bmal1* promoter changes in the neocortex of patients with this disease. These phenomena imply that the aberrant methylation of the *Bmal1* promoter and rapid BMAL1 degradation together affect behavioural changes or cognitive impairments. Furthermore, a methylome study of the neocortex of brains at autopsy revealed attenuated methylation rhythms in samples from patients with Alzheimer disease compared with controls [70]. The neocortex is very rare in terms of tissues with circadian methylation rhythms.

According to many studies, contorted clock gene expression patterns and mood disorders are closely associated in experimental animal models. Genetic experiments have found that CLOCK is a key factor in maniac states because *Clock* mutant mice (*Clock*Δ19) develop clear features [71–73] of mania, circadian rhythm disruption, hyperactivity and decreased sleep. The physiological features of these mutant mice include altered gene expression patterns and excited neurons due to upregulated dopamine content in the ventral tegmental area (VTA) [74]. Notably, knockdown of CLOCK in the VTA using RNA interference results in concomitant mania-like (hyperactivity and decreased anxiety) and depression-like behaviours in mice. Since patients with mania often experience depressive episodes, this knockdown mouse is a more appropriate model of mania in humans. However, precisely how these CLOCK disruptions affect the upregulated dopamine content in the VTA remains obscure. The expression of monoamine oxidase A (MAOA), which inactivates monoamine neurotransmitters including dopamine, serotonin and norepinephrine, is regulated by circadian clock genes including *Bmal1*, *Npas2* and *Per2* [75]. However, in this mechanism, CLOCK, unlike NPAS2, does not work as a transcriptional activator. Therefore, the absence of CLOCK directly results in downregulated MAOA activity, and consequent dopamine upregulation cannot be concluded. Some other CLOCK functions including histone modification activity or an indirect action of CLOCK might be involved in dopamine upregulation, and investigations into this are underway.

Patients with depression frequently have insomnia and abnormal circadian rhythms that could reasonably relate to altered clock gene expression. Circadian clock gene expression has been compared between post-mortem brain samples from patients with major depressive disorder (MDD) and age-matched controls [76]. The findings showed abnormal clock gene phasing and decreased *Bmal1* and *Per2* oscillation in most brain regions of the patients. These findings provided direct evidence that clock gene expression is altered in the central nervous system of patients with MDD. On the other hand, depression states and anti-depressant effects are often tested in experimental animal models such as laboratory mice that are suspended by the tail or forced to swim to mimic short duration stress or exposed to social defeat to mimic chronic stress [77]. The volume of the hippocampus is reduced in both patients and in a model that develops depressive pathophysiology after exposure to chronic stress, and this volume is restored by administering anti-depressant medicine. Brain-derived neurotropic factor (BDNF) plays a very important role in the hippocampus as an anti-depressant and for adaptation to stress. Anti-depressants enhance BDNF expression in the mouse brain [78]; BDNF infused into the hippocampus has anti-depressant effects in behavioural mouse models of depression [78, 79] and the action of the anti-depressant desipramine is attenuated mice with a BDNF deletion in the forebrain [78, 80]. Expression of *Bdnf* gene is rhythmic in rat brain regions including hippocampus. However, its downregulation in an animal model of depression was due to the methylation status of the promoter region of the *Bdnf* gene [81]. Tsankova et al. found that the expression of two BDNF variants, Bdnf III and IV, is downregulated and that the promoter regions of corresponding variants are hypermethylated in laboratory mice exposed to defeat stress [82]. Furthermore, chronic imipramine administration increased histone acetylation on the corresponding promoters, and this downregulation was reversed. The findings of the above studies indicate that *Bdnf* gene expression is rhythmically maintained under normal conditions but is epigenetically regulated under stress. However, precisely how *Bdnf* expression is rhythmically maintained remains unclear and awaits further investigation.

5. Assays of *Bmal1* transcription modulators

The circadian clock controls the daily oscillations of gene expression and physiological function at the cellular level, indicating that the control of circadian rhythms at the cellular level is important for human health. After we elucidated the transcriptional mechanism of the non-redundant essential unique clock gene, *Bmal1*, we developed a circadian functional assay system that consists of luminescent reporter cells and the application of *Bmal1* findings. We found that the minimal essential region of the *Bmal1* promoter for circadian transcription is embedded in an open chromatin structure, suggesting that this region can remain functional even when inserted into a reporter plasmid [39]. We then established stable reporter cell lines with which to analyse circadian clock function [83] and the effects of DNA methylation on circadian clock function [43]. **Figure 3** shows the application of these systems to further dissection of the molecular mechanisms underlying the mammalian circadian clock [46, 84, 85].

One of the most important findings was that altering the DNA configuration of the *Bmal1* promoter causes the epigenetic regulation of *Bmal1* circadian transcription [86]. Topoisomerase I (TOP1) is located at an intermediate region between two ROREs that are critical *cis*-elements of circadian transcription, which is required for transcriptional suppression in cooperation with the distal RORE. The DNA fragment between the ROREs, where the TOP1-binding site is located, behaved like a right-handed superhelical twist, and the modulation of TOP1 activity by the TOP1 inhibitor, camptothecin and *Top1* siRNA altered the footprint, indicating the modulation of the chromatin structure. These findings indicated that TOP1 modulates the chromatin structure of the *Bmal1* promoter, regulates the *Bmal1* transcription and influences the circadian period.

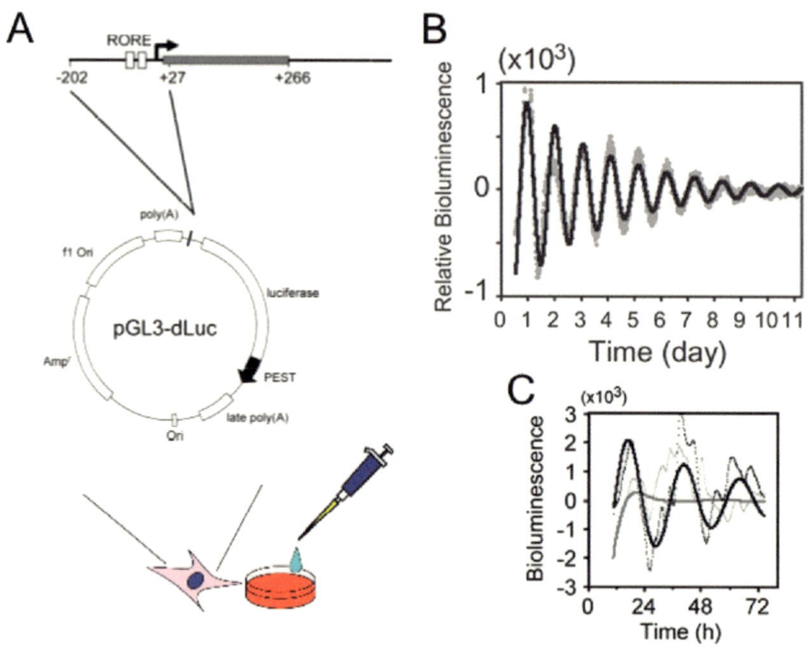

Figure 3. Monitoring cellular circadian clock system using stable reporter cell line. (A) Monitoring method. Promoter region (−202 to +27) of Bmal1 was inserted into pGL3-dluc and used to create cell lines with stable, real-time reporter gene to evaluate cellular circadian clock system. (B) Circadian oscillation monitored using host NIH3T3 cells. (C) Circadian oscillation monitored using host CPT-K cells with hypermethylated Bmal1 promoter region. Cells with stable gene expression derived from CPT-K cells were incubated with 2.5 μM aza-dC for 2 days, stimulated with 50% FBS for 2 h and then bioluminescence was measured. Detrended fit curves are representative of at least three independent experiments (control, grey; aza-dC, black). Dots, raw values; lines, fit curve data.

Another important finding was the epigenetic inactivation or DNA methylation of the *Bmal1* promoter. The methylation status of the *Bmal1* promoter is critical for the circadian system. Because the *Bmal1* gene is inactivated by the DNA hypermethylation of its promoter, the circadian oscillation of *Bmal1* transcription was absent in the haematological malignant cells. The demethylating agent aza-dC restored circadian oscillation, whereas continuous *Bmal1* expression did not. Because BMAL1 protein has distinct tissue-specific regulation and functions [87], tissue-specific regulation of BMAL1 expression might be required, and this can be introduced endogenously by aza-dC to establish the negative feedback loop system and restore circadian oscillation. Because the *Bmal1* promoter is basically hypomethylated, the methyltransferases DNMT3a and DNMT3b might be mainly responsible for introducing cytosine methylation *de novo* at unmethylated CpG sites in the promoter [40]. The methylation of DNA contributes to the expression of clock genes [45] in addition to *Bmal1*, a key player in the disruption of circadian rhythms.

6. Conclusion

Circadian rhythms control all aspects of physiology. When they are disrupted by changes in clock gene expression, various critical intracellular physiological processes become dysregulated, and this can lead to diseases that are induced partly by epigenetic effects including DNA methyation. The pathologies that are closely associated with disrupted circadian rhythms include cancer [88], dementia [89], Parkinson's disease [90] and obesity [91]. Among the clock genes, *Bmal1* is unique because the loss of BMAL1 protein in mice results in immediate and complete loss of circadian rhythmicity [33], indicating the importance of a specific amount of BMAL1 expression for circadian rhythms. In addition, DNA methylation of the *Bmal1* promoter disrupts the circadian system even when the *Per* and *Cry* gene promoters are unmethylated, indicating that the *Bmal1* gene is functionally important [43]. Epigenetic regulation, especially DNA methylation status, is specific to DNA sites and gene functions. Therefore, the finding that the epigenetic transcriptional regulation of *Bmal1* is functionally important for adaptation to environmental changes provides novel insights into clock gene functions that should affect the clinical diagnosis and treatment of diseases. Therefore, modulators of *Bmal1* transcription are needed for the human health.

Acknowledgements

This work was supported by a KAKENHI grant [number 23592756].

Conflict of interest

The authors have no conflicts of interest to declare.

Author details

Tatsunosuke Tomita and Yoshiaki Onishi*

*Address all correspondence to: y-onishi@aist.go.jp

Biomedical Research Institute, National Institute of Advanced Industrial Science and Technology (AIST), DBT-AIST International Laboratory for Advanced Biomedicine (DAILAB), Tsukuba, Japan

References

[1] Ruby NF, Brennan TJ, Xie X, Cao V, Franken P, Heller HC, O'Hara BF. Role of melanopsin in circadian responses to light. Science. 2002;**298**(5601):2211-2213. DOI: 10.1126/science. 1076701

[2] Doherty CJ, Kay SA. Circadian control of global gene expression patterns. Annual Review of Genetics. 2010;**44**:419-444. DOI: 10.1146/annurev-genet-102209-163432

[3] Koike N, Yoo S-H, Huang H-C, Kumar V, Lee C, Kim T-K, Takahashi JS. Transcriptional architecture and chromatin landscape of the core circadian clock in mammals. Science. 2012;**338**(6105):349-354. DOI: 10.1126/science.1226339

[4] Zhang R, Lahens NF, Ballance HI, Hughes ME, Hogenesch JB. A circadian gene expression atlas in mammals: Implications for biology and medicine. Proceedings of the National Academy of Sciences of the United States of America. 2014;**111**(45):16219-16224. DOI: 10.1073/pnas.1408886111

[5] Dickmeis T. Glucocorticoids and the circadian clock. The Journal of Endocrinology. 2009; **200**(1):3-22. DOI: 10.1677/JOE-08-0415

[6] Rey G, Cesbron F, Rougemont J, Reinke H, Brunner M, Naef F. Genome-wide and phase-specific DNA-binding rhythms of BMAL1 control circadian output functions in mouse liver. PLoS Biology. 2011;**9**(2):e1000595. DOI: 10.1371/journal.pbio.1000595

[7] Altman BJ, Hsieh AL, Sengupta A, Krishnanaiah SY, Stine ZE, Walton ZE, Gouw AM, Venkataraman A, Li B, Goraksha-Hicks P, Diskin SJ, Bellovin DI, Simon MC, Rathmell JC, Lazar MA, Maris JM, Felsher DW, Hogenesch JB, Weljie AM, Dang CV. MYC disrupts the circadian clock and metabolism in cancer cells. Cell Metabolism. 2015;**22**(6):1009-1019. DOI: 10.1016/j.cmet.2015.09.003

[8] Shimomura K, Kumar V, Koike N, Kim TK, Chong J, Buhr ED, Whiteley AR, Low SS, Omura C, Fenner D, Owens JR, Richards M, Yoo SH, Hong HK, Vitaterna MH, Bass J, Pletcher MT, Wiltshire T, Hogenesch J, Lowrey PL, Takahashi JS. Usf1, a suppressor of the circadian clock mutant, reveals the nature of the DNA-binding of the CLOCK:BMAL1 complex in mice. eLife. 2013;**2**:e00426. DOI: 10.7554/eLife.00426

[9] Duez H, Staels B. The nuclear receptors Rev-erbs and RORs integrate circadian rhythms and metabolism. Diabetes & Vascular Disease Research. 2008;**5**(2):82-88. DOI: 10.3132/dvdr.2008.0014

[10] Papazyan R, Zhang Y, Lazar MA. Genetic and epigenomic mechanisms of mammalian circadian transcription. Nature Structural & Molecular Biology. 2016;**23**(12):1045-1052. DOI: 10.1038/nsmb.3324

[11] Fonjallaz P, Ossipow V, Wanner G, Schibler U. The two PAR leucine zipper proteins, TEF and DBP, display similar circadian and tissue-specific expression, but have different target promoter preferences. The EMBO Journal. 1996;**15**(2):351-362. DOI: 10.1002/j.1460-2075.1996.tb00365.x

[12] Doerks T, Copley RR, Schultz J, Ponting CP, Bork P. Systematic identification of novel protein domain families associated with nuclear functions. Genome Research. 2002;**12**(1):47-56. DOI: 10.1101/gr.203201

[13] Ueda HR. Systems biology of mammalian circadian clocks. Cold Spring Harbor Symposia on Quantitative Biology. 2007;**72**:365-380. DOI: 10.1101/sqb.2007.72.047

[14] Onishi Y, Kiyama R. Enhancer activity of HS2 of the human beta-LCR is modulated by distance from the key nucleosome. Nucleic Acids Research. 2001;**29**(16):3448-3457. DOI: 10.1093/nar/29.16.3448

[15] Ripperger JA, Schibler U. Rhythmic CLOCK-BMAL1 binding to multiple E-box motifs drives circadian Dbp transcription and chromatin transitions. Nature Genetics. 2006;**38**(3):369-374. DOI: 10.1038/ng1738

[16] Brown SA, Ripperger J, Kadener S, Fleury-Olela F, Vilbois F, Rosbash M, Schibler U. PERIOD1-associated proteins modulate the negative limb of the mammalian circadian oscillator. Science. 2005;**308**(5722):693-696. DOI: 10.1126/science.1107373

[17] Duong HA, Weitz CJ. Temporal orchestration of repressive chromatin modifiers by mammalian circadian clock PERIOD complexes. Nature Structural & Molecular Biology. 2014;**21**(2):126-132. DOI: 10.1038/nsmb.2746

[18] Curtis AM, Seo SB, Westgate EJ, Rudic RD, Smyth EM, Chakravarti D, FitzGerald GA, McNamara P. Histone acetyltransferase-dependent chromatin remodeling and the vascular clock. The Journal of Biological Chemistry. 2004;**279**(8):7091-7097. DOI: 10.1074/jbc.M311973200

[19] Etchegaray JP, Lee C, Wade PA, Reppert SM. Rhythmic histone acetylation underlies transcription in the mammalian circadian clock. Nature. 2003;**421**(6919):177-182. DOI: 10.1038/nature01314

[20] Doi M, Hirayama J, Sassone-Corsi P. Circadian regulator CLOCK is a histone acetyltransferase. Cell. 2006;**125**(3):497-508. DOI: 10.1016/j.cell.2006.03.033

[21] Feng D, Liu T, Sun Z, Bugge A, Mullican SE, Alenghat T, Liu XS, Lazar MA. A circadian rhythm orchestrated by histone deacetylase 3 controls hepatic lipid metabolism. Science. 2011;**331**(6022):1315-1319. DOI: 10.1126/science.1198125

[22] Kim JY, Kwak PB, Weitz CJ. Specificity in circadian clock feedback from targeted reconstitution of the NuRD corepressor. Molecular Cell. 2014;**56**(6):738-748. DOI: 10.1016/j.molcel.2014.10.017

[23] Asher G, Gatfield D, Stratmann M, Reinke H, Dibner C, Kreppel F, Mostoslavsky R, Alt FW, Schibler U. SIRT1 regulates circadian clock gene expression through PER2 deacetylation. Cell. 2008;**134**(2):317-328. DOI: 10.1016/j.cell.2008.06.050

[24] Nakahata Y, Kaluzova M, Grimaldi B, Sahar S, Hirayama J, Chen D, Guarente LP, Sassone-Corsi P. The NAD^+-dependent deacetylase SIRT1 modulates CLOCK-mediated chromatin remodeling and circadian control. Cell. 2008;**134**(2):329-340. DOI: 10.1016/j.cell.2008.07.002

[25] Etchegaray JP, Yang X, DeBruyne JP, Peters AH, Weaver DR, Jenuwein T, Reppert SM. The polycomb group protein EZH2 is required for mammalian circadian clock function. The Journal of Biological Chemistry. 2006;**281**(30):21209-21215. DOI: 10.1074/jbc.M603722200

[26] Takahashi JS. Molecular components of the circadian clock in mammals. Diabetes, Obesity & Metabolism. 2015;**17**(Suppl 1):6-11. DOI: 10.1111/dom.12514

[27] Jhunjhunwala S, van Zelm MC, Peak MM, Cutchin S, Riblet R, van Dongen JJ, Grosveld FG, Knoch TA, Murre C. The 3D structure of the immunoglobulin heavy-chain locus: Implications for long-range genomic interactions. Cell. 2008;**133**(2):265-279. DOI: 10.1016/j.cell.2008.03.024

[28] Kolovos P, Knoch TA, Grosveld FG, Cook PR, Papantonis A. Enhancers and silencers: An integrated and simple model for their function. Epigenetics & Chromatin. 2012;**5**(1):1. DOI: 10.1186/1756-8935-5-1

[29] Lehmann R, Childs L, Thomas P, Abreu M, Fuhr L, Herzel H, Leser U, Relogio A. Assembly of a comprehensive regulatory network for the mammalian circadian clock: A bioinformatics approach. PLoS One. 2015;**10**(5):e0126283. DOI: 10.1371/journal.pone.0126283

[30] Xu Y, Guo W, Li P, Zhang Y, Zhao M, Fan Z, Zhao Z, Yan J. Long-range chromosome interactions mediated by cohesin shape circadian gene expression. PLoS Genetics. 2016;**12**(5):e1005992. DOI: 10.1371/journal.pgen.1005992

[31] Knoch TA, Wachsmuth M, Kepper N, Lesnussa M, Abuseiris A, Ali Imam AM, Kolovos P, Zuin J, Kockx CEM, Brouwer RWW, van de Werken HJG, van IJcken WFJ, Wendt KS, Grosveld FG. The detailed 3D multi-loop aggregate/rosette chromatin architecture and functional dynamic organization of the human and mouse genomes. Epigenetics & Chromatin. 2016;**9**:58. DOI: 10.1186/s13072-016-0089-x

[32] Wachsmuth M, Knoch TA, Rippe K. Dynamic properties of independent chromatin domains measured by correlation spectroscopy in living cells. Epigenetics & Chromatin. 2016;**9**:57. DOI: 10.1186/s13072-016-0093-1

[33] Law JA, Jacobsen SE. Establishing, maintaining and modifying DNA methylation patterns in plants and animals. Nature Reviews Genetics. 2010;**11**:204. DOI: 10.1038/nrg2719

[34] Di Croce L, Raker VA, Corsaro M, Fazi F, Fanelli M, Faretta M, Fuks F, Lo Coco F, Kouzarides T, Nervi C, Minucci S, Pelicci PG. Methyltransferase recruitment and DNA hypermethylation of target promoters by an oncogenic transcription factor. Science. 2002; **295**(5557):1079-1082. DOI: 10.1126/science.1065173

[35] Zaratiegui M, Irvine DV, Martienssen RA. Noncoding RNAs and gene silencing. Cell. 2007;**128**(4):763-776. DOI: 10.1016/j.cell.2007.02.016

[36] Ikeda M, Nomura M. cDNA cloning and tissue-specific expression of a novel basic helix-loop-helix/PAS protein (BMAL1) and identification of alternatively spliced variants with alternative translation initiation site usage. Biochemical and Biophysical Research Communications. 1997;**233**(1):258-264. DOI: 10.1006/bbrc.1997.6371

[37] Bunger MK, Wilsbacher LD, Moran SM, Clendenin C, Radcliffe LA, Hogenesch JB, Simon MC, Takahashi JS, Bradfield CA. Mop3 is an essential component of the master circadian pacemaker in mammals. Cell. 2000;**103**(7):1009-1017. DOI: 10.1016/S0092-8674(00)00205-1

[38] Shearman LP, Sriram S, Weaver DR, Maywood ES, Chaves I, Zheng B, Kume K, Lee CC, van der Horst GT, Hastings MH, Reppert SM. Interacting molecular loops in the mammalian circadian clock. Science. 2000;**288**(5468):1013-1019. DOI: 10.1126/science.288.5468.1013

[39] Onishi Y, Hanai S, Ohno T, Hara Y, Ishida N. Rhythmic SAF-A binding underlies circadian transcription of the Bmal1 gene. Molecular and Cellular Biology. 2008;**28**(10):3477-3488. DOI: 10.1128/mcb.02227-07

[40] Klose RJ, Bird AP. Genomic DNA methylation: The mark and its mediators. Trends in Biochemical Sciences. 2006;**31**(2):89-97. DOI: 10.1016/j.tibs.2005.12.008

[41] Zhu WG, Srinivasan K, Dai Z, Duan W, Druhan LJ, Ding H, Yee L, Villalona-Calero MA, Plass C, Otterson GA. Methylation of adjacent CpG sites affects Sp1/Sp3 binding and activity in the p21(Cip1) promoter. Molecular and Cellular Biology. 2003;**23**(12):4056-4065. DOI: 10.1128/MCB.23.12.4056-4065.2003

[42] Taniguchi H, Fernandez AF, Setien F, Ropero S, Ballestar E, Villanueva A, Yamamoto H, Imai K, Shinomura Y, Esteller M. Epigenetic inactivation of the circadian clock gene BMAL1 in hematologic malignancies. Cancer Research. 2009;**69**(21):8447-8454. DOI: 10.1158/0008-5472.CAN-09-0551

[43] Satou R, Sugihara N, Ishizuka Y, Matsukubo T, Onishi Y. DNA methylation of the BMAL1 promoter. Biochemical and Biophysical Research Communications. 2013;**440**(3):449-453. DOI: 10.1016/j.bbrc.2013.09.124

[44] Ueda HR, Chen W, Adachi A, Wakamatsu H, Hayashi S, Takasugi T, Nagano M, Nakahama K, Suzuki Y, Sugano S, Iino M, Shigeyoshi Y, Hashimoto S. A transcription factor response element for gene expression during circadian night. Nature. 2002;**418**(6897):534-539. DOI: 10.1038/nature00906

[45] Ji Y, Qin Y, Shu H, Li X. Methylation analyses on promoters of mPer1, mPer2, and mCry1 during perinatal development. Biochemical and Biophysical Research Communications. 2010;**391**(4):1742-1747. DOI: 10.1016/j.bbrc.2009.12.146

[46] Tomita T, Kurita R, Onishi Y. Epigenetic regulation of the circadian clock: Role of 5-aza-2′-deoxycytidine. Bioscience Reports. 2017;**37**(3). DOI: 10.1042/BSR20170053

[47] Papagiannakopoulos T, Bauer MR, Davidson SM, Heimann M, Subbaraj L, Bhutkar A, Bartlebaugh J, Vander Heiden MG, Jacks T. Circadian rhythm disruption promotes lung tumorigenesis. Cell Metabolism. 2016;**24**(2):324-331. DOI: 10.1016/j.cmet.2016.07.001

[48] Jiang W, Zhao S, Jiang X, Zhang E, Hu G, Hu B, Zheng P, Xiao J, Lu Z, Lu Y, Ni J, Chen C, Wang X, Yang L, Wan R. The circadian clock gene Bmal1 acts as a potential anti-oncogene in pancreatic cancer by activating the p53 tumor suppressor pathway. Cancer Letters. 2016;**371**(2):314-325. DOI: 10.1016/j.canlet.2015.12.002

[49] Fu L, Pelicano H, Liu J, Huang P, Lee CC. The circadian gene Period2 plays an important role in tumor suppression and DNA damage response in vivo. Cell. 2002;**111**(1):41-50. DOI: 10.1016/S0092-8674(02)00961-3

[50] Hua H, Wang Y, Wan C, Liu Y, Zhu B, Yang C, Wang X, Wang Z, Cornelissen-Guillaume G, Halberg F. Circadian gene mPer2 overexpression induces cancer cell apoptosis. Cancer Science. 2006;**97**(7):589-596. DOI: 10.1111/j.1349-7006.2006.00225.x

[51] Gery S, Komatsu N, Baldjyan L, Yu A, Koo D, Koeffler HP. The circadian gene per1 plays an important role in cell growth and DNA damage control in human cancer cells. Molecular Cell. 2006;**22**(3):375-382. DOI: 10.1016/j.molcel.2006.03.038

[52] Mao Y, Fu A, Hoffman AE, Jacobs DI, Jin M, Chen K, Zhu Y. The circadian gene CRY2 is associated with breast cancer aggressiveness possibly via epigenomic modifications. Tumour Biology. 2015;**36**(5):3533-3539. DOI: 10.1007/s13277-014-2989-3

[53] Hoffman AE, Zheng T, Yi CH, Stevens RG, Ba Y, Zhang Y, Leaderer D, Holford T, Hansen J, Zhu Y. The core circadian gene Cryptochrome 2 influences breast cancer risk, possibly by mediating hormone signaling. Cancer Prevention Research (Philadelphia, Pa.). 2010; **3**(4):539-548. DOI: 10.1158/1940-6207.capr-09-0127

[54] Puram RV, Kowalczyk MS, de Boer CG, Schneider RK, Miller PG, McConkey M, Tothova Z, Tejero H, Heckl D, Jaras M, Chen MC, Li H, Tamayo A, Cowley GS, Rozenblatt-Rosen O, Al-Shahrour F, Regev A, Ebert BL. Core circadian clock genes regulate leukemia stem cells in AML. Cell. 2016;**165**(2):303-316. DOI: 10.1016/j.cell.2016.03.015

[55] Wang Y, Qian R, Sun N, Lu C, Chen Z, Hua L. Circadian gene hClock enhances proliferation and inhibits apoptosis of human colorectal carcinoma cells in vitro and in vivo. Molecular Medicine Reports. 2015;**11**(6):4204-4210. DOI: 10.3892/mmr.2015.3247

[56] Li BE, Ernst P. Two decades of leukemia oncoprotein epistasis: The MLL1 paradigm for epigenetic deregulation in leukemia. Experimental Hematology. 2014;**42**(12):995-1012. DOI: 10.1016/j.exphem.2014.09.006

[57] Katada S, Sassone-Corsi P. The histone methyltransferase MLL1 permits the oscillation of circadian gene expression. Nature Structural & Molecular Biology. 2010;**17**:1414. DOI: 10.1038/nsmb.1961

[58] Kim DH, Rhee JC, Yeo S, Shen R, Lee SK, Lee JW, Lee S. Crucial roles of mixed-lineage leukemia 3 and 4 as epigenetic switches of the hepatic circadian clock controlling bile acid homeostasis in mice. Hepatology. 2015;**61**(3):1012-1023. DOI: 10.1002/hep.27578

[59] Valekunja UK, Edgar RS, Oklejewicz M, van der Horst GT, O'Neill JS, Tamanini F, Turner DJ, Reddy AB. Histone methyltransferase MLL3 contributes to genome-scale circadian transcription. Proceedings of the National Academy of Sciences of the United States of America. 2013;**110**(4):1554-1559. DOI: 10.1073/pnas.1214168110

[60] Yamagishi M, Uchimaru K. Targeting EZH2 in cancer therapy. Current Opinion in Oncology. 2017;**29**(5):375-381. DOI: 10.1097/CCO.0000000000000390

[61] Kelly TK, De Carvalho DD, Jones PA. Epigenetic modifications as therapeutic targets. Nature Biotechnology. 2010;**28**(10):1069-1078. DOI: 10.1038/nbt.1678

[62] Cao Q, Zhao X, Bai J, Gery S, Sun H, Lin D-C, Chen Q, Chen Z, Mack L, Yang H, Deng R, Shi X, Chong L-W, Cho H, Xie J, Li Q-Z, Müschen M, Atkins AR, Liddle C, Yu RT, Alkan S, Said JW, Zheng Y, Downes M, Evans RM, Koeffler HP. Circadian clock cryptochrome proteins regulate autoimmunity. Proceedings of the National Academy of Sciences. 2017; **114**(47):12548-12553. DOI: 10.1073/pnas.1619119114

[63] Firsov D, Tokonami N, Bonny O. Role of the renal circadian timing system in maintaining water and electrolytes homeostasis. Molecular and Cellular Endocrinology. 2012;**349**(1): 51-55. DOI: 10.1016/j.mce.2011.06.037

[64] Morris CJ, Yang JN, Scheer FAJL. Chapter 19—The impact of the circadian timing system on cardiovascular and metabolic function. In: Kalsbeek A et al., editors. Progress in Brain Research. Netherlands: Elsevier; 2012. pp. 337-358. DOI: 10.1016/B978-0-444-59427-3.00019-8

[65] Bollinger T, Schibler U. Circadian rhythms–from genes to physiology and disease. Swiss Medical Weekly. 2014;**144**:w13984. DOI: 10.4414/smw.2014.13984

[66] Bailey SM, Udoh US, Young ME. Circadian regulation of metabolism. The Journal of Endocrinology. 2014;**222**(2):R75-R96. DOI: 10.1530/JOE-14-0200

[67] Liu C, Chung M. Genetics and epigenetics of circadian rhythms and their potential roles in neuropsychiatric disorders. Neuroscience Bulletin. 2015;**31**(1):141-159. DOI: 10.1007/s12264-014-1495-3

[68] Van Someren EJW. Circadian and sleep disturbances in the elderly. Experimental Gerontology. 2000;**35**(9):1229-1237. DOI: 10.1016/S0531-5565(00)00191-1

[69] Song H, Moon M, Choe HK, Han D-H, Jang C, Kim A, Cho S, Kim K, Mook-Jung I. Aβ-induced degradation of BMAL1 and CBP leads to circadian rhythm disruption in Alzheimer's disease. Molecular Neurodegeneration. 2015;**10**(1):13. DOI: 10.1186/s13024-015-0007-x

[70] Lim AS, Srivastava GP, Yu L, Chibnik LB, Xu J, Buchman AS, Schneider JA, Myers AJ, Bennett DA, De Jager PL. 24-hour rhythms of DNA methylation and their relation with

rhythms of RNA expression in the human dorsolateral prefrontal cortex. PLoS Genetics. 2014;**10**(11):e1004792. DOI: 10.1371/journal.pgen.1004792

[71] Bechtel W. Circadian rhythms and mood disorders: Are the phenomena and mechanisms causally related? Frontiers in Psychiatry. 2015;**6**:118. DOI: 10.3389/fpsyt.2015.00118

[72] Roybal K, Theobold D, Graham A, DiNieri JA, Russo SJ, Krishnan V, Chakravarty S, Peevey J, Oehrlein N, Birnbaum S, Vitaterna MH, Orsulak P, Takahashi JS, Nestler EJ, Carlezon WA Jr, McClung CA. Mania-like behavior induced by disruption of CLOCK. Proceedings of the National Academy of Sciences of the United States of America. 2007; **104**(15):6406-6411. DOI: 10.1073/pnas.0609625104

[73] Coyle JT. What can a clock mutation in mice tell us about bipolar disorder? Proceedings of the National Academy of Sciences of the United States of America. 2007;**104**(15):6097-6098. DOI: 10.1073/pnas.0701491104

[74] Mukherjee S, Coque L, Cao JL, Kumar J, Chakravarty S, Asaithamby A, Graham A, Gordon E, Enwright JF 3rd, DiLeone RJ, Birnbaum SG, Cooper DC, McClung CA. Knock-down of clock in the ventral tegmental area through RNA interference results in a mixed state of mania and depression-like behavior. Biological Psychiatry. 2010;**68**(6):503-511. DOI: 10.1016/j.biopsych.2010.04.031

[75] Hampp G, Ripperger JA, Houben T, Schmutz I, Blex C, Perreau-Lenz S, Brunk I, Spanagel R, Ahnert-Hilger G, Meijer JH, Albrecht U. Regulation of monoamine oxidase A by circadian-clock components implies clock influence on mood. Current Biology. 2008; **18**(9):678-683. DOI: 10.1016/j.cub.2008.04.012

[76] Li JZ, Bunney BG, Meng F, Hagenauer MH, Walsh DM, Vawter MP, Evans SJ, Choudary PV, Cartagena P, Barchas JD, Schatzberg AF, Jones EG, Myers RM, Watson SJ, Akil H, Bunney WE. Circadian patterns of gene expression in the human brain and disruption in major depressive disorder. Proceedings of the National Academy of Sciences. 2013; **110**(24):9950-9955. DOI: 10.1073/pnas.1305814110

[77] Hollis F, Kabbaj M. Social defeat as an animal model for depression. ILAR Journal. 2014; **55**(2):221-232. DOI: 10.1093/ilar/ilu002

[78] Martinowich K, Lu B. Interaction between BDNF and serotonin: Role in mood disorders. Neuropsychopharmacology. 2008;**33**(1):73-83. DOI: 10.1038/sj.npp.1301571

[79] Shirayama Y, Chen AC-H, Nakagawa S, Russell DS, Duman RS. Brain-derived neurotrophic factor produces antidepressant effects in behavioral models of depression. The Journal of Neuroscience. 2002;**22**(8):3251-3261

[80] Monteggia LM, Barrot M, Powell CM, Berton O, Galanis V, Gemelli T, Meuth S, Nagy A, Greene RW, Nestler EJ. Essential role of brain-derived neurotrophic factor in adult hippocampal function. Proceedings of the National Academy of Sciences of the United States of America. 2004;**101**(29):10827-10832. DOI: 10.1073/pnas.0402141101

[81] Chen D, Meng L, Pei F, Zheng Y, Leng J. A review of DNA methylation in depression. Journal of Clinical Neuroscience. 2017;**43**:39-46. DOI: 10.1016/j.jocn.2017.05.022

[82] Tsankova NM, Berton O, Renthal W, Kumar A, Neve RL, Nestler EJ. Sustained hippocampal chromatin regulation in a mouse model of depression and antidepressant action. Nature Neuroscience. 2006;**9**(4):519-525. DOI: 10.1038/nn1659

[83] Onishi Y, Oishi K, Kawano Y, Yamazaki Y. The harmala alkaloid harmine is a modulator of circadian Bmal1 transcription. Bioscience Reports. 2012;**32**(1):45-52. DOI: 10.1042/bsr20110002

[84] Onishi Y, Kawano Y, Yamazaki Y. Lycorine, a candidate for the control of period length in mammalian cells. Cellular Physiology and Biochemistry. 2012;**29**(3–4):407-416. DOI: 10.1159/000338495

[85] Ogawa Y, Kawano Y, Yamazaki Y, Onishi Y. Shikonin shortens the circadian period: Possible involvement of Top2 inhibition. Biochemical and Biophysical Research Communications. 2014;**443**(1):339-343. DOI: 10.1016/j.bbrc.2013.11.116

[86] Onishi Y, Kawano Y. Rhythmic binding of Topoisomerase I impacts on the transcription of Bmal1 and circadian period. Nucleic Acids Research. 2012;**40**(19):9482-9492. DOI: 10.1093/nar/gks779

[87] McDearmon EL, Patel KN, Ko CH, Walisser JA, Schook AC, Chong JL, Wilsbacher LD, Song EJ, Hong HK, Bradfield CA, Takahashi JS. Dissecting the functions of the mammalian clock protein BMAL1 by tissue-specific rescue in mice. Science. 2006;**314**(5803):1304-1308. DOI: 10.1126/science.1132430

[88] Masri S, Kinouchi K, Sassone-Corsi P. Circadian clocks, epigenetics, and cancer. Current Opinion in Oncology. 2015;**27**(1):50-56. DOI: 10.1097/CCO.0000000000000153

[89] Liu HC, Hu CJ, Tang YC, Chang JG. A pilot study for circadian gene disturbance in dementia patients. Neuroscience Letters. 2008;**435**(3):229-233. DOI: 10.1016/j.neulet.2008.02.041

[90] Lin Q, Ding H, Zheng Z, Gu Z, Ma J, Chen L, Chan P, Cai Y. Promoter methylation analysis of seven clock genes in Parkinson's disease. Neuroscience Letters. 2012;**507**(2):147-150. DOI: 10.1016/j.neulet.2011.12.007

[91] Milagro FI, Gomez-Abellan P, Campion J, Martinez JA, Ordovas JM, Garaulet M. CLOCK, PER2 and BMAL1 DNA methylation: Association with obesity and metabolic syndrome characteristics and monounsaturated fat intake. Chronobiology International. 2012;**29**(9):1180-1194. DOI: 10.3109/07420528.2012.719967

Chapter 7

Epigenome Editing

Paul Enríquez

Additional information is available at the end of the chapter

http://dx.doi.org/10.5772/intechopen.79564

Abstract

The regulation of chromatin structure and gene expression depends substantially on a dynamic and intricate layer of biological and chemical information that constitutes the epigenome. This epigenetic layer of information holds fundamental clues to the molecular mechanisms, not yet fully understood, by which a genotype can influence and configure a specific phenotype. A profound understanding of the molecular underpinnings of epigenetic processes is, thus, essential to wield deliberate spatiotemporal control of gene activation and repression. However, only recently has the technology required to adequately probe the functional significance of specific epigenetic mechanisms become available. This chapter provides an overview of modern epigenome editing systems, including zinc finger proteins, TAL effectors, and CRISPR–Cas systems. It highlights the use of biotechnological tools to investigate the role of DNA and histone post-translational modifications as well as regulatory RNAs to manipulate specific patterns of gene expression. This chapter further discusses the technological limitations that have limited our ability to elucidate epigenetic mechanisms in local and genome-wide contexts.

Keywords: epigenome editing, epigenetic manipulation, CRISPR, CRISPR–dCas9, ZFNs, TALENs, epigenome engineering, targeted gene activation and repression, epigenetics

1. Introduction

Elucidating the underlying basis for the molecular links that bridge the gap between genotype and phenotype has propelled research and scientific discovery for decades. More than three-quarters of a century have passed since Conrad Waddington introduced the concept of "epigenetics." The term signaled an attempt to describe the causal links by which genes give rise to specific phenotypes in the context of developmental changes that drive cellular differentiation [1]. Waddington's "epigenetic landscapes" [2] laid a foundation for exploring phenotypic plasticity as a phenomenon that is not entirely dependent on genetic mechanisms, but one which can also be shaped by environmental cues that concomitantly orchestrate the process of cell differentiation from an initial totipotent state [3].

The scope of epigenetics as a scientific discipline, much like its definition, has expanded with the passage of time. Today, the field broadly refers to the study of fundamental processes related to mitotic and meiotic stable and heritable changes that emerge without alteration of DNA sequences [3–5]. Stable changes encompass durable alterations in gene expression patterns, which may be neither permanent nor heritable. In contrast, heritable changes in gene expression may constitute persistent alterations that are carried onto the progeny of cells or individual organisms [6].

In the last two decades, the emerging field of epigenetics has revealed crucial information about the regulation of chromatin states in the eukaryotic nucleus. Heterochromatic DNA is organized into compact, higher order, chromatin fibers. By contrast, euchromatic DNA comprises lightly packed chromatin that represents an active and accessible part of the genome. Both states feature an array of nucleosomes, the basic subunits of chromatin, consisting of approximately 145–147 base pairs of DNA wrapped around a core histone octamer [7]. Each histone octamer carries two copies of all of the core histone proteins—H2A, H2B, H3, H4—assembled into nucleosomes, which are subsequently stabilized into high-order chromatin structures by the presence of the linker histone H1 and linker DNA [7].

The organization and regulation of chromatin at the epigenetic level depend on dynamic and diverse combinations of covalent chemical modifications that occur on histone proteins—known as histone post-translational modifications (PTMs)—and DNA bases, as well as the expression of regulatory noncoding RNA (ncRNA) molecules [3]. Together, these epigenetic signals and ncRNAs constitute a layer of information that controls the spatiotemporal regulation of gene expression patterns by remodeling the structure of chromatin and modulating its bio-physicochemical properties.

The scientific community has made great strides in elucidating functional roles for epigenetic processes in recent years. Progress has come at the hand of biotechnologies aimed at manipulating endogenous, site-specific epigenetic targets. The technologies rely on the use of catalytically active or scaffolding epigenetic effectors fused to programmable DNA-binding proteins that target specific genetic loci. This chapter focuses on the three most important platforms for modern targeted epigenome editing: zinc finger proteins, transcription activator-like effectors (TALEs), and clustered, regularly interspaced, short palindromic repeats (CRISPR), and CRISPR-associated sequences (Cas) (CRISPR–Cas) systems.

Notwithstanding the progress made in recent years, much remains to be learned about epigenetic mechanisms in development, cellular programming, disease, and personalized medicine. Epigenome editing technologies are poised to make significant contributions to the field of epigenetics. This chapter provides an overview of each epigenome editing system, highlights their use for manipulating specific patterns of gene expression, and discusses the technological limitations that have thus far limited our ability to interrogate the mechanisms of epigenetic regulation in local and genome-wide contexts.

http://dx.doi.org/10.5772/intechopen.79564

2. Programmable epigenome editing systems

Regulating endogenous levels of gene expression by targeting specific epigenetic modifications is a relatively nascent field. Basic scientific research in the last few decades has provided insights that facilitated the development of technologies aimed at interrogating epigenetic processes. At their core, epigenome editing tools are based on the concept of fusing programmable DNA-binding proteins that target specific genetic loci, with catalytically active or scaffolding effector domains that exert some influence on epigenetic processes. The three most important molecular tools that have been developed for targeted epigenome editing are zinc finger proteins, TALEs, and CRISPR–Cas systems. All have been repurposed into epigenome editing platforms designed to manipulate gene expression patterns in particular contexts. An overview of each system is provided below. However, it should be noted that other, less common, tools have also been developed to probe epigenetic mechanisms and modulate gene expression in a sequence-dependent manner including synthetic polyamides [8] and triple helix-forming oligonucleotides (TFOs) [9].

2.1. Zinc finger-based systems

Zinc finger proteins are among the most characterized systems used for the manipulation of targeted, sequence-specific nucleic acids. Their discovery arose from research of transcriptional mechanisms in *Xenopus laevis* oocytes dating back to over three decades ago. At the time, scientists realized that a unique set of proteins, containing repetitive zinc-binding domains of roughly 30 amino acid residues, was required for transcription factor-mediated gene regulation [10]. These finger-like peptide arrangements rely on interactions between conserved pairs of cysteine and histidine residues that are anchored by a centered and tetrahedrally coordinated zinc ion (**Figure 1**). From a structural standpoint, each zinc finger

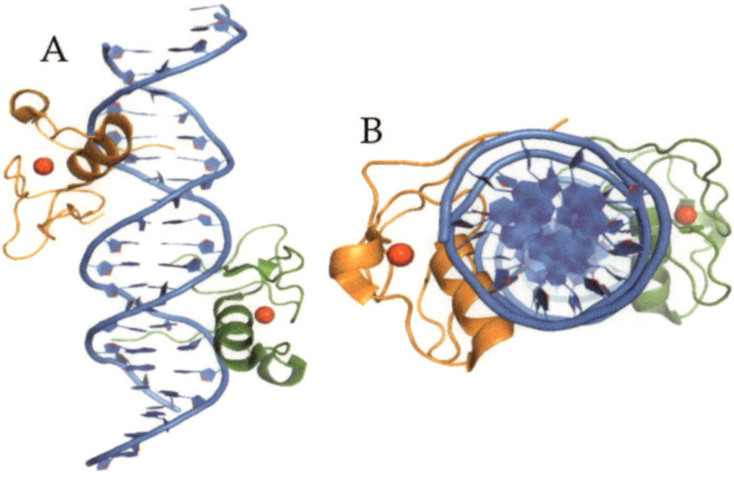

Figure 1. Crystal structure of two zinc finger proteins in complex with DNA. (A) Recognition of target DNA (blue) by two distinct zinc finger proteins (orange and green). A tetrahedrally coordinated zinc ion (red) stabilizes each zinc finger motif. (B) Orthogonal view of zinc finger-mediated DNA recognition. [PDB 3DFX].

comprises an antiparallel β sheet and α helix, which are stabilized by the zinc ion and a set of hydrophobic residues [11]. The α helix of each zinc finger binds directly to the major groove of B-DNA. Residues from the NH_2-terminal region of the helix mediate recognition of a three base-pair target site [11]. Thus, side chain residues in the α helix that can be engineered to recognize diverse nucleic acid base-pair triplets drive the double-stranded DNA sequence specificity of each zinc finger.

The intrinsic modularity of DNA recognition by zinc finger proteins led to the swift development of a range of biotechnological applications. For example, manipulation of gene expression via site-specific, DNA-binding zinc finger proteins was first established by targeting a 9 base-pair region of a *BCR-ABL* fusion oncogene [12]. Using a three-zinc finger peptide, scientists demonstrated that transcriptional repression of a leukemic oncogene in a chromosomal DNA context was possible [12].

Similarly, the development of zinc finger nucleases (ZFNs) marked an important milestone for modern genome editing biotechnologies. ZFNs are engineered chimeric fusions composed of a set of tandem zinc finger DNA-binding proteins and a nuclease domain—such as the catalytic domain of the restriction endonuclease *Fok*I—that cleaves DNA [13]. Synthetic zinc finger-*Fok*I fusions coupled the DNA sequence specificity of zinc finger proteins with the non-specific cleavage activity of *Fok*I to trigger double-stranded breaks at desired genomic loci. Importantly, repurposing zinc finger proteins into ZFNs facilitated the path for epigenome editing biotechnologies aimed at controlling transcriptional activation and repression.

Zinc finger-related epigenome editing tools are based on the concept of fusing programmable, DNA binding, zinc finger proteins designed to target diverse sequences [14, 15] with catalytically active or scaffolding effector domains. The chimeric proteins are designed to alter gene expression patterns and act as artificial transcription factors (ATFs) [15]. Pioneer studies on transcriptional repression by directing DNA cytosine methylation [16] and local histone H3K9 methylation [17] deposition at specific promoter sequences established the feasibility of the approach in the late 1990s and early 2000s. The use of zinc finger-based epigenome editing tools has contributed valuable insights into epigenetic mechanisms, as will be discussed in the next section. However, the high cost and technical expertise required to engineer and validate context-dependent specificity in zinc finger proteins [18, 19] has greatly limited their widespread adoption.

2.2. TALE-based systems

Transcription activator-like effectors (TALEs) are proteins of bacterial origin. They were first reported in the literature in 2007, after two independent research groups discovered that certain bacterial proteins can bind specific promoter sequences in eukaryotic cells [20, 21]. Upon binding to DNA, the pathogenic effector proteins induce the expression of genes that promote the spread of bacterial infection in host cells. TALEs are secreted by gram-negative bacteria of the *Xanthomonas* genus and injected into eukaryotic cells via the type III secretion system [22]. Once inside the eukaryotic cell, they translocate directly to the nucleus where they act as transcription factors to regulate the expression of genes that support bacterial infection [20, 21].

http://dx.doi.org/10.5772/intechopen.79564

The molecular basis for DNA recognition of each TALE comes from a central tandem amino acid repeat domain of approximately 33–35 residues in length [23, 24]. This tandem repeat is flanked by an N-terminal region required for type III secretion and a C-terminal region involved in nuclear localization and transcriptional activation. Each repeat folds into two left-handed α helices linked by a short loop that contains two hypervariable residues, known as the repeat variable diresidues (RVDs). The RVDs occupy positions 12 and 13 of each repeat, where the 12th residue stabilizes the RVD loop by mediating contacts with the protein backbone, while the 13th residue interacts directly with a specific nucleotide nitrogenous base (**Figure 2**). Thus, the DNA specificity of TALEs is modular and encoded in the tandem repeat sequence [25, 26].

Unlike zinc fingers, which require triplet sequence recognition sites, TALEs are able to target a single nucleotide at a time through its RVDs. This characteristic makes TALEs simpler to engineer and has enabled the rational design of artificial TALEs for biotechnological applications [26]. Indeed, borrowing from its ZFN predecessors, TALE proteins fused to the *Fok*I nuclease domain gave rise to TALE nucleases (TALENs), which have been utilized for genome editing applications [27].

Similarly, fusions of TALEs to epigenetic effectors have been repurposed for epigenome editing to manipulate gene expression. For example, TALE fusions to activation domains have shown the ability to target specific genomic loci and induce robust transcriptional activation [28]. Inactivation of enhancer regulatory elements has also been demonstrated using TALE fusions to chromatin effectors targeting specific histone modifications [29], thereby establishing a platform to interrogate the function of specific regulatory elements in diverse chromatin landscapes.

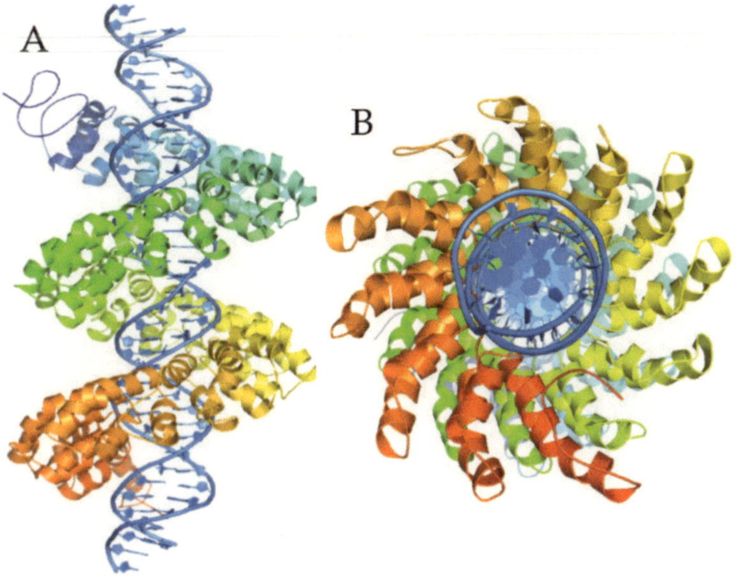

Figure 2. Structure of a TAL effector bound to its target DNA. (A) The TALE tandem amino acid repeats (multi-colored helices) associate into a right-handed superhelix that wraps around the major groove of the target DNA (blue). (B) Orthogonal view of target DNA recognition by a TALE [PDB 3UGM].

Despite the advantages of possessing modular properties compared to zinc finger proteins and the improvements made in design and assembly of custom TALE-based systems [30, 31], significant technological shortcomings remain to be addressed. The highly repetitive sequences associated with TALE tandem repeats make them susceptible to rearrangements when expressed in target cells due to recombination events that trigger deletions [32]. Moreover, the sheer size of TALE repeats creates obstacles to cellular delivery using some standard viral vectors—e.g., lentiviruses [32]. Such limitations have precluded wider adaptability of TALE-based tools for epigenome editing.

2.3. CRISPR–Cas–based systems

Clustered, regularly interspaced, short palindromic repeats (CRISPR) and CRISPR-associated sequences (Cas) constitute bacterial and archaeal endogenous adaptive immunity systems. They were first reported over three decades ago upon publication of findings of a cryptic cluster of unknown function in the bacterium *Escherichia coli* [33]. Advances in bioinformatics—alongside greater access to bacterial genomes in public databases around the turn of the century—led to the realization that such clusters exist in a wide range of bacteria and archaea [34]. Interest about the functional significance of the clusters in the scientific community grew steadily, which eventually led researchers to empirical evidence that CRISPR–Cas systems provide adaptive immunity against attack by viruses and plasmids [35] **Figure 3**.

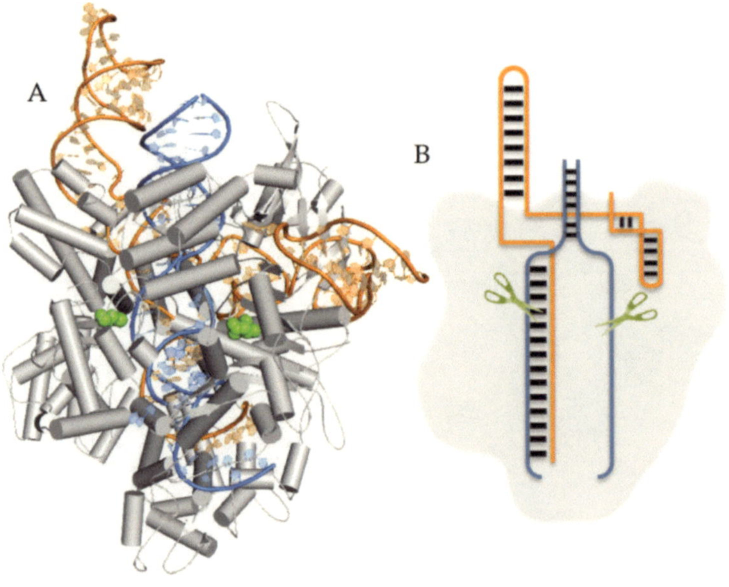

Figure 3. Structural representation of CRISPR–Cas9 in complex with target double-stranded DNA and sgRNA. (A) Crystal structure of the CRISPR–Cas9 nuclease (gray) bound to a sgRNA (orange) and double-stranded target DNA (blue) primed for cleavage. Mutation of two Cas9 residues within the RuvC and HNH nuclease domains—D10A and H840A (green spheres)—abolishes the catalytic activity of the enzyme and transforms it into dCas9, which is able to bind its target sequence without triggering DNA cleavage. (B) Schematic cartoon representation of the CRISPR–Cas9–sgRNA–DNA complex (colors as shown in A). [PDB 5F9R].

CRISPR–Cas systems are currently classified into class 1 and class 2 groups, which are further divided into six different types and multiple subtypes [36]. Adaptive immunity to foreign nucleic acids is encoded in CRISPR arrays, which contain DNA sequences derived from the integration of invasive DNA as a new CRISPR spacer. CRISPR arrays are transcribed into precursor CRISPR RNAs (pre-crRNAs) and subsequently processed into mature CRISPR RNAs (crRNAs). After processing takes place, crRNAs mediate interference by acting as guides that recruit a variety of Cas effector proteins to complementary nucleic acids from invasive genetic elements [36]. Unlike zinc finger- and TALE-based systems, which rely on protein-DNA interactions to target specific genetic loci, CRISPR–Cas systems are DNA-encoded and RNA-guided. Thus, CRISPR–Cas systems harness nature's principles of Watson-Crick base pairing of nucleic acids to identify the target genetic loci. This property confers significant advantages to CRISPR–Cas systems because they bypass the need for complex protein engineering to mediate DNA recognition.

The type II CRISPR–Cas9 system is the most characterized CRISPR–Cas system to date. This complex requires dual RNA molecules—a crRNA and a trans-activating crRNA (tracrRNA)—that hybridize together and direct Cas9, a DNA endonuclease, to a target DNA sequence. Upon target recognition, Cas9 triggers double-stranded breaks a few nucleotides away from a protospacer adjacent motif (PAM) site [37]. The RNA-guided Cas9 endonuclease cleaves double-stranded DNA via conserved RuvC and HNH nuclease domains, which form a compact catalytic core [38].

In recent years, the CRISPR–Cas9 system has been repurposed into an efficient genome editing tool by engineering a chimeric single guide RNA (sgRNA) that fuses the crRNA and tracrRNA into a programmable RNA molecule capable of sequence-specific DNA targeting [37]. Genome editing biotechnological applications take advantage of Cas9-induced double-stranded DNA breaks to mediate DNA repair via one of two pathways: nonhomologous end joining (NHEJ), which introduces insertion or deletion mutations, or homology-directed repair (HDR), which introduces specific mutations or DNA sequences by recombination with a donor template. Importantly, adapting the CRISPR–Cas9 system for gene editing has also paved the path for repurposing the system for epigenome editing applications.

CRISPR-Cas9–mediated epigenome editing is based on the use of an engineered Cas9 protein, which has been purposefully stripped of its catalytic activity into a nuclease-null or "dead" Cas9 (dCas9). Mutating two residues—D10A and H840A—located in the RuvC and HNH nuclease domains is necessary and sufficient to deactivate the nuclease activity of the enzyme [38]. The mutations allow CRISPR–Cas9 to target specific DNA sequences and bind to them without cleaving DNA. Epigenetic effector proteins can then be fused to dCas9 to trigger desired local epigenetic changes, in much the same way mentioned earlier using zinc finger and TALE fusions to epigenetic effectors.

CRISPR–dCas9 has recently been used to demonstrate the feasibility of targeting specific epigenetic modifications to trigger transcriptional activation or repression. For instance, tethering dCas9 to an epigenetic effector that acetylates histone tail residues led to transcriptional activation in one study [39], and a dCas9 fusion to a repression domain showed the CRISPR–dCas9 system's ability to recruit a heterochromatin-forming complex to silence gene expression [40]. The next section discusses these and other epigenome editing studies in more detail.

3. Epigenome editing to regulate gene expression

Manipulation of endogenous gene expression at target loci has thus far been achieved by coupling DNA-binding proteins with (1) transcriptional activators and repressors, which modulate gene expression by acting as site-specific ATFs, or (2) epigenetic effectors, which catalyze the deposition or removal of specific epigenetic modifications at target loci. In addition, the CRISPR–dCas9 system has recently enabled engineering of modular sgRNA molecules that carry RNA aptamers as scaffolding platforms to recruit molecules for site-specific regulation.

3.1. Transcriptional activation

Transcriptional activator effector domains were first used to upregulate levels of gene expression by fusing the herpes simplex virus protein VP16—involved in recruitment of chromatin remodeling factors that increase chromatin accessibility—and its VP64 tetrameric form to zinc finger proteins [41, 42]. For instance, an early study demonstrated that zinc finger proteins targeted to 18 base-pair sequences of the 5′ untranslated region of the proto-oncogene *erbB-2* could upregulate transcription in an endogenous context [41]. Upregulation was also shown to occur even under control of an exogenous chemical inducer molecule [41]. Similarly, zinc finger fusions to the VP64 activation domain have been used to reactivate dormant genes. In one study, a dormant mammary serine protease inhibitor (*maspin*) tumor suppressor gene was reactivated by targeting the *maspin* promoter region, which is often epigenetically silenced in aggressive epithelial tumors [43]. Other transcriptional activator domains, such as the p65 subunit of the NF-κB complex, have also been coupled to zinc finger proteins for targeted transcriptional activation [44].

The proof-of-concept studies using zinc fingers fused to transcriptional activation domains to modulate gene expression patterns served as blueprints for the subsequent use of activator domain fusions to TALEs and CRISPR–dCas9 systems. TALE activators targeted to promoter sites of the endogenous human *VEGF-A, NTF3,* and the microRNA *miR-302/367* cluster have been reported to induce increased target gene expression by fivefold or more [28]. Endogenous levels of gene expression of select human pluripotency factors have also been shown to increase by twofold to fivefold using TALE fusions to VP64 [45].

Furthermore, although single TALE fusions to transcriptional activators have demonstrated the ability to upregulate gene expression, combinations of TALE-VP64 fusions targeting the promoter regions of genes implicated in inflammation, immunomodulation, and cancer pathways have established the occurrence of synergistic activation effects at target sites, which presents opportunities to develop tunable transcriptional networks [46]. Interestingly, it appears that different types of transcriptional activators induce varying levels of gene expression. In one study, the mean-fold activation induced with TALE-p65 activators was lower than that with TALE-VP64 activators [28]. The combination of VP64 and p65 activators exhibits synergistic effects on endogenous gene expression, which can provide a platform not only to design tools for targeted gene expression, but also to fine-tune the fold activation induced by different combinations of activator domains fused to DNA-binding proteins [28].

Recently, researchers have built upon work on zinc finger and TALE fusions to develop technologies for targeted gene expression utilizing the CRISPR–dCas9 system. RNA-guided transcriptional upregulation using CRISPR–dCas9-VP64 and CRISPR–dCas9-p65 fusions has been reported by several research teams in the last 5 years [47–50]. Introduction of individual sgRNAs for dCas9 targeting can induce transcriptional activity at desired loci, but the use of multiple sgRNAs to target a single locus can act synergistically to induce multifold gene expression [48–50]. The CRISPR–dCas9 system offers several advantages over zinc fingers and TALEs for targeted gene activation. Because CRISPR–dCas9 is an RNA-guided, DNA-targeting system, it does not require complex protein engineering and could open additional research paths for spatiotemporal control of gene expression alongside complex chemical and optogenetic inducible systems. However, some studies have reported that dCas9-based activation tools induce weaker levels of transcription compared to TALE-based activators [50, 51].

In addition to the use of transcriptional activators for targeted gene expression, research teams have demonstrated the efficacy of using epigenetic effectors fused to DNA targeting proteins to manipulate epigenetic landscapes. In 2015, a study reported the use of CRISPR–dCas9, TALE, and zinc finger fusions to the p300 histone acetyltransferase (HAT) catalytic domain, which deposits H3K27 and other acetylation marks on histone tails, to trigger transactivation of genes [39]. Notably, the study showed that CRISPR–dCas9-p300 Core fusions led to more efficient and robust activation from proximal and distal enhancer regions compared to synthetic transcription factors engineered to carry activation domains that target promoters. Other epigenetic effectors that catalyze covalent modifications on DNA have also been fused to DNA-binding proteins to induce gene expression. For example, ten-eleven translocation (TET) demethylase enzymes that catalyze the sequential oxidation of 5-methyldeoxycytosine (m^5dC) to form 5-hydroxymethylcytosine (hm^5C), 5-formylcytosine (f^5C), and 5-carboxylcytosine (ca^5C) have been used with zinc fingers [52–54], TALEs [55], and CRISPR–dCas9 [56, 57] to direct promoter-specific DNA demethylation and, thus, upregulate gene expression.

As mentioned earlier, CRISPR–dCas9 systems represent promising tools for the next generation of complex inducible systems. More recently, a novel system comprising CRISPR–dCas9 fused to a tripartite activator featuring the constituent activation domains VP64, p65, and Rta (VPR) was shown to activate endogenous coding and noncoding regions while targeting several genes simultaneously [58]. Another platform built with CRISPR–dCas9 and a protein scaffold—namely, a repeating peptide array called the SUperNova tagging system (SunTag)—capable of recruiting multiple copies of an antibody fusion protein showed robust levels of endogenous gene activation at target sites [59]. Lastly, it has been demonstrated that CRISPR–Cas sgRNAs can be engineered to carry protein-binding cassettes, artificial aptamers, and other ncRNAs directly into the sgRNA stem-loop structures, thereby creating modular CRISPR–dCas9 complexes that enable locus targeting, multiplexing, and highly inducible regulatory action to upregulate patterns of gene expression [60–62].

3.2. Transcriptional repression

Silencing endogenous gene expression at target loci has been achieved in similar ways to gene activation by fusing repressor domains or epigenetic effectors to DNA-binding proteins.

The domain most commonly used to trigger gene silencing is the Krüppel-associated box (KRAB). KRAB repression is mediated by the recruitment of complexes that elicit formation of heterochromatin via interactions between the KRAB-associated protein 1 (KAP1) corepressor and other factors that catalyze histone methylation and deacetylation [17, 63]. KRAB fusions to zinc finger proteins [41, 64], TALEs [65], and CRISPR–dCas9 [40, 47] have demonstrated robust inactivation from promoter as well as proximal and distal regulatory elements.

Although the KRAB repressor is the most common domain used for recruitment of heterochromatin-forming factors, researchers have also reported efficient gene silencing with other repressor domains such as the mSin interaction domain (SID) [65] and Mxi1 [47]. One study comparing dCas9-KRAB and dCas9-Mxi1 showed that the Mxi1 fusion could induce more repression by nearly threefold levels [47]. A separate study found that TALE-SID repressors could lead to 26% more transcriptional repression than its TALE-KRAB counterparts [65]. Moreover, concatenating four SID (SID4X) domains—much like combining VP16 units into VP64 activator proteins—and fusing them to DNA-binding proteins can induce greater repression than single SID domain fusions [66].

In contrast to transcriptional repressors, which aptly recruit heterochromatin-forming complexes to desired loci, epigenetic effectors can directly catalyze specific modifications on histones and DNA that trigger targeted epigenetic repression. Several studies have thus far tested and confirmed robust repression efficiency rates of histone methyltransferases [17], histone demethylases [29, 67], and DNA methyltransferases [68–70] with zinc finger, TALE, and CRISPR–dCas9 proteins at target promoter and enhancer regions.

Notably, a study featuring a CRISPR–dCas9 fusion to the catalytic domain of DNMT3A, a de novo DNA methyltransferase, found that multiplexing with several sgRNAs exhibits synergistic effects on site-specific CpG methylation [70]. Thus, CRISPR–dCas9-DNMT3A multiplexing can lead to a greater increase of methylation levels at target regions. Despite the synergistic effects seen with DNMT3A repressor fusions, it appears that dCas9-DNMT3A epigenetic editing of target CpG methylation sites is not wholly stable and does not persist beyond the loss of expression of the editing construct [70]. To overcome this type of transient epigenetic editing, a study recently reported a system for inheritable silencing of endogenous genes by transiently expressing combinations of DNA-binding proteins fused to transcriptional repressors, including KRAB and DNMT3 domains, to prompt long-term memory of repressive epigenetic states [71]. Lastly, the modular nature of sgRNAs has enabled the creation of scaffolding systems to achieve multiplexing and multimerization of epigenetic effectors, which will be useful to study endogenous and synthetic gene repression [62, 72].

3.3. Technological limitations

Epigenome editing directed by catalytically active or scaffolding epigenetic effectors fused to programmable DNA binding proteins has contributed important insights about the causal relationships between epigenetic states and gene regulation. However, despite the progress made in recent years to uncover the molecular basis of epigenetic processes, epigenome editing

http://dx.doi.org/10.5772/intechopen.79564

remains a nascent field that must overcome many technological challenges related to editing efficiency, delivery, cytotoxicity, and specificity.

A major current limitation concerns the spatio-temporal specificity of epigenome editing, which may involve multiple factors including, but not limited to, non-specific deposition or removal of select epigenetic marks on target DNA loci or histones, mitigation of off-target DNA binding effects, and temporal expression of epigenetic effectors in different cell types and chromatin contexts [3]. For example, enzymes that catalyze the deposition of histone PTMs—e.g., p300's HAT domain, which catalyzes acetylation of multiple residues on all four core histones—have various degrees of histone substrate specificity. Thus, the inherent promiscuity of an enzyme may pose challenges to the study of direct functional roles for unique PTMs in specific contexts [3].

Similarly, the expression levels and times at which an epigenetic effector is present in a cell type can exert effects on the efficiency of DNA binding specificity. Following target site saturation, off-target binding activity is likely to increase if the effector remains, or continues to be expressed, in the cell. This could partly explain the extensive global off-target effects reported with the use of dCas9–methyltransferase fusions in one study [73]. Given the lack of target site specificity previously reported for dCas9–sgRNA complexes [74], further research is needed to establish whether other dCas9–effector fusions exhibit similar off-target activities. The use of zinc finger and TALE fusions to epigenetic effectors offers an alternative to mitigate possible dCas9-related off-target effects. However, zinc finger and TALE epigenome editing biotechnologies are limited by the complex protein engineering required to mediate DNA recognition.

Progress in other areas associated with challenges in delivery, efficiency, and stability of epigenetic states has come at a rapid pace [75–77]. Chemical and optogenetic molecules have expanded the toolbox for reversible and inducible epigenome editing in endogenous contexts [3, 66, 78]. And emerging technologies for RNA editing [79] could soon provide new insights about the epigenetic roles of ncRNAs. As the field matures, epigenome editing will undoubtedly make significant contributions to the advancement of basic and applied research.

4. Conclusion

Recent technological breakthroughs in epigenome editing have expanded our understanding of the underlying mechanisms responsible for the regulation of chromatin structure and the spatiotemporal control of gene activation and repression. As detailed in this chapter, modern epigenome editing systems are revolutionizing the field of epigenetics. Programmable DNA-binding proteins fused to catalytically active or scaffolding epigenetic effectors represent invaluable tools to uncover the functional significance of site-specific epigenetic mechanisms in a myriad of contexts including development, cell differentiation, and disease. Undoubtedly, progress in this emerging field will offer great contributions to translational medicine and biotechnology in the near future.

Conflict of interest

The author declares no conflicts of interest.

Acronyms and abbreviations

PTM	post-translational modification
ncRNA	noncoding RNA
ZFN	zinc finger nuclease
ATF	artificial transcription factor
TALEN	transcription activator-like effector nuclease
RVD	repeat variable diresidue
CRISPR	clustered, regularly interspaced, short palindromic repeats
Cas	CRISPR-associated sequences
pre-RNA	precursor CRISPR RNA
crRNA	CRISPR RNA
tracrRNA	trans-activating crRNA
sgRNA	single-guide RNA
NHEJ	nonhomologous end joining
HDR	homology-directed repair
TFO	triple helix-forming oligonucleotide
MASPIN	mammary serine protease inhibitor
HAT	histone acetyltransferase
TET	ten-eleven translocation
m^5dC	5-methyldeoxycytosine
hm^5C	5-hydroxymethylcytosine
f^5C	5-formylcytosine
ca^5C	5-carboxylcytosine
VPR	VP64, p65, and Rta

SunTag	SUperNova tagging
KRAB	Krüppel-associated box
KAP1	KRAB-associated protein 1
SID	mSin interaction domain

Author details

Paul Enríquez

Address all correspondence to: penriqu@ncsu.edu

North Carolina State University, Raleigh, NC, United States of America

References

[1] Waddington CH. The Epigenotype. Endeavor. 1942;1:18-20. International Journal of Epidemiology. 2012;**41**:10-13

[2] Waddington CH. The Strategy of the Genes. London: George Allen & Unwin; 1957

[3] Enríquez P. CRISPR-mediated epigenome editing. Yale Journal of Biology and Medicine. 2016;**89**:471-486

[4] Goldberg AD, Allis CD, Bernstein E. Epigenetics: A landscape takes shape. Cell. 2007; **128**:635-638

[5] Berger SL, Kouzarides T, Shiekhattar R, Shilatifard A. An operational definition of epigenetics. Genes & Development. 2009;**23**:781-783

[6] Roadmap Epigenomics Project. Overview of the Roadmap Epigenomics Project [Internet]. 2010. Available from: http://www.roadmapepigenomics.org/overview [Accessed: 15-04-2018]

[7] Luger K, Mäder AW, Richmond RK, Sargent DF, Richmond TJ. Crystal structure of the nucleosome core particle at 2.8 Å resolution. Nature. 1997;**389**:251-260

[8] Hochhauser D, Kotecha M, O'Hare C, Morris PJ, Hartley JM, Taherbhai Z, Harris D, Forni C, Mantovani R, Lee M, Hartley JA. Modulation of topoisomerase IIalpha expression by a DNA sequence-specific polyamide. Molecular Cancer Therapeutics. 2007;**6**:346-354

[9] van der Gun BT, Maluszynska-Hoffman M, Kiss A, Arendzen AJ, Ruiters MH, McLaughlin PM, Weinhold E, Rots MG. Targeted DNA methylation by a DNA methyltransferase coupled to a triple helix forming oligonucleotide to down-regulate the epithelial cell adhesion molecule. Bioconjugate Chemistry. 2010;**21**:1239-1245

[10] Miller J, McLachlan AD, Klug A. Repetitive zinc-binding domains in the protein transcription factor IIIA from Xenopus oocytes. EMBO Journal. 1985;**4**:1609-1614

[11] Pavletich NP, Pabo CO. Zinc finger-DNA recognition: Crystal structure of a Zif268-DNA complex at 2.1 Å. Science. 1991;**252**:809

[12] Choo Y, Sanchez-Garcia I, Klug A. In vivo repression by a site-specific DNA-binding protein designed against an oncogenic sequence. Nature. 1994;**372**:642-645

[13] Kim YG, Cha J, Chandrasegaran S. Hybrid restriction enzymes: Zinc finger fusions to Fok I cleavage domain. Proceedings of the National Academy of Sciences of the United States of America. 1996;**93**:1156

[14] Sera T, Uranga C. Rational Design of Artificial Zinc-Finger Proteins Using a nondegenerate recognition code table. Biochemistry. 2002;**41**:7074-7081

[15] Pabo CO, Peisach E, Grant RA. Design and selection of novel Cys2His2 zinc finger proteins. Annual Review of Biochemistry. 2001;**70**:313-340

[16] Xu GL, Bestor TH. Cytosine methylation targetted to pre-determined sequences. Nature Genetics. 1997;**17**:376-378

[17] Snowden AW, Gregory PD, Case CC, Pabo CO. Gene-specific targeting of H3K9 methylation is sufficient for initiating repression in vivo. Current Biology. 2002;**12**:2159-2166

[18] Rebar EJ, Pabo CO. Zinc finger phage: Affinity selection of fingers with new DNA-binding specificities. Science. 1994;**263**:671-673

[19] Choo Y, Isalan M. Advances in zinc finger engineering. Current Opinion in Structural Biology. 2000;**10**:411-416

[20] Römer P, Hahn S, Jordan T, Strauß T, Bonas U, Lahaye T. Plant pathogen recognition mediated by promoter activation of the pepper *Bs3* resistance gene. Science. 2007;**318**:645-648

[21] Kay S, Hahn S, Marois E, Hause G, Bonas U. A bacterial effector acts as a plant transcription factor and induces a cell size regulator. Science. 2007;**318**:648-651

[22] Galán JE, Wolf-Watz H. Protein delivery into eukaryotic cells by type III secretion machines. Nature. 2006;**444**:567-573

[23] Mak AN-S, Bradley P, Cernadas RA, Bogdanove AJ, Stoddard BL. The crystal structure of TAL effector PthXo1 bound to its DNA target. Science. 2012;**335**:716-719

[24] Deng D, Yan C, Pan X, Mahfouz M, Wang J, Zhu J-K, Shi Y, Yan N. Structural basis for sequence-specific recognition of DNA by TAL effectors. Science. 2012;**335**:720-723

[25] Boch J, Scholze H, Schornack S, Landgraf A, Hahn S, Kay S, Lahaye T, Nickstadt A, Bonas U. Breaking the code of DNA binding specificity of TAL-type III effectors. Science. 2009;**326**:1509-1512

[26] Moscou MJ, Bogdanove AJ. A simple cipher governs DNA recognition by TAL effectors. Science. 2009;**326**:1501

[27] Christian M, Cermak T, Doyle EL, Schmidt C, Zhang F, Hummel A, Bogdanove AJ, Voytas DF. TAL effector nucleases create targeted DNA double-strand breaks. Genetics. 2010;**186**:757-761

[28] Maeder ML, Linder SJ, Reyon D, Angstman JF, Fu Y, Sander JD, Joung JK. Robust, synergistic regulation of human gene expression using TALE activators. Nature Methods. 2013;**10**:243-245

[29] Mendenhall EM, Williamson KE, Reyon D, Zou JY, Ram O, Joung JK, Bernstein BE. Locus-specific editing of histone modifications at endogenous enhancers. Nature Biotechnology. 2013;**31**:1133-1136

[30] Cermak T, Doyle EL, Christian M, Wang L, Zhang Y, Schmidt C, Baller JA, Somia NV, Bogdanove AJ, Voytas DF. Efficient design and assembly of custom TALEN and other TAL effector-based constructs for DNA targeting. Nucleic Acids Research. 2011;**39**:e82

[31] Zhang Z, Li D, Xu H, Xin Y, Zhang T, Ma L, Wang X, Chen Z, Zhang Z. A simple and efficient method for assembling TALE protein based on plasmid library. PLoS One. 2013;**8**:e66459

[32] Holkers M, Maggio I, Liu J, Janssen JM, Miselli F, Mussolino C, Recchia A, Cathomen T, Gonçalves MA. Differential integrity of TALE nuclease genes following adenoviral and lentiviral vector gene transfer into human cells. Nucleic Acids Research. 2013;**41**:e63

[33] Ishino Y, Shinagawa H, Makino K, Amemura M, Nakata A. Nucleotide sequence of the iap gene, responsible for alkaline phosphatase isozyme conversion in *Escherichia coli*, and identification of the gene product. Journal of Bacteriology. 1987;**169**:5429-5433

[34] Mojica FJ, Diez-Villasenor C, Soria E, Juez G. Biological significance of a family of regularly spaced repeats in the genomes of archaea, bacteria and mitochondria. Molecular Microbiology. 2000;**36**:244-246

[35] Barrangou R, Fremaux C, Deveau H, Richards M, Boyaval P, Moineau S, Romero DA, Horvath P. CRISPR provides acquired resistance against viruses in prokaryotes. Science. 2007;**315**:1709-1712

[36] Shmakov S, Abudayyeh OO, Makarova KS, Wolf YI, Gootenberg JS, Semenova E, Minakhin L, Joung J, Konermann S, Severinov K, Zhang F, Koonin EV. Discovery and functional characterization of diverse class 2 CRISPR-Cas systems. Molecular Cell. 2015;**60**:385-397

[37] Jinek M, Chylinski K, Fonfara I, Hauer M, Doudna JA, Charpentier E. A programmable dual-RNA-guided DNA endonuclease in adaptive bacterial immunity. Science. 2012;**337**:816-821

[38] Jinek M, Jiang F, Taylor DW, Sternberg SH, Kaya E, Ma E, Anders C, Hauer M, Zhou K, Lin S, Kaplan M, Iavarone AT, Charpentier E, Nogales E, Doudna JA. Structures of Cas9 endonucleases reveal RNA-mediated conformational activation. Science. 2014;**343**:1247997

[39] Hilton IB, D'Ippolito AM, Vockley CM, Thakore PI, Crawford GE, Reddy TE, Gersbach CA. Epigenome editing by a CRISPR/Cas9-based acetyltransferase activates genes from promoters and enhancers. Nature Biotechnology. 2015;**33**:510-517

[40] Thakore PI, D'Ippolito AM, Song L, Safi A, Shivakumar NK, Kabadi AM, Reddy TE, Crawford GE, Gersbach CA. Highly specific epigenome editing by CRISPR-Cas9 repressors for silencing of distal regulatory elements. Nature Methods. 2015;**12**:1143-1149

[41] Beerli RR, Dreier B, Barbas CF. Positive and negative regulation of endogenous genes by designed transcription factors. Proceedings National Academy of Sciences USA. 2000;**97**:1495-1500

[42] Zhang L, Spratt SK, Liu Q, Johnstone B, Qi H, Raschke EE, Jamieson AC, Rebar EJ, Wolffe AP, Case CC. Synthetic zinc finger transcription factor action at an endogenous chromosomal site. Activation of the human erythropoietin gene. Journal of Biological Chemistry. 2000;**275**:33850-33860

[43] Beltran A, Parikh S, Liu Y, Cuevas B, Johnson GL, Futscher BW, Blancafort P. Re-activation of a dormant tumor suppressor gene maspin by designed transcription factors. Oncogene. 2007;**26**:2791-2798

[44] Liu P-Q, Rebar EJ, Zhang L, Liu Q, Jamieson AC, Liang Y, Qi H, Li P-X, Chen B, Mendel MC, Zhong X, Lee YL, Eisenberg SP, Spratt SK, Case CC, Wolffe AP. Regulation of an endogenous locus using a panel of designed zinc finger proteins targeted to accessible chromatin regions. Activation of vascular endothelial growth factor A. Journal of Biological Chemistry. 2001;**276**:11323-11334

[45] Zhang F, Cong L, Lodato S, Kosuri S, Church G, Arlotta P. Programmable sequence-specific transcriptional regulation of mammalian genome using designer TAL effectors. Nature Biotechnology. 2011;**29**:149-153

[46] Perez-Pinera P, Ousterout DG, Brunger JM, Farin AM, Glass KA, Guilak F, Crawford G, Hartemink AJ, Gersbach CA. Synergistic and tunable human gene activation by combinations of synthetic transcription factors. Nature Methods. 2013;**10**:239-242

[47] Gilbert LA, Larson MH, Morsut L, Liu Z, Brar GA, Torres SE, Stern-Ginossar N, Brandman O, Whitehead EH, Doudna JA, Lim WA, Weissman JS, Qi LS. CRISPR-mediated modular RNA-guided regulation of transcription in eukaryotes. Cell. 2013;**154**:442-451

[48] Maeder ML, Linder SJ, Cascio VM, Fu Y, Ho QH, Joung JK. CRISPR RNA-guided activation of endogenous human genes. Nature Methods. 2013;**10**(10):977-979

[49] Mali P, Aach J, Stranges PB, Esvelt KM, Moosburner M, Kosuri S, Yang L, Church GM. CAS9 transcriptional activators for target specificity screening and paired nickases for cooperative genome engineering. Nature Biotechnology. 2013;**31**:833-838

[50] Perez-Pinera P, Kocak DD, Vockley CM, Adler AF, Kabadi AM, Polstein LR, Thakore PI, Glass KA, Ousterout DG, Leong KW, Guilak F, Crawford GE, Reddy T, Gersbach CA. RNA-guided gene activation by CRISPR-Cas9-based transcription factors. Nature Methods. 2013;**10**:973-976

[51] Gao X, Tsang JC, Gaba F, Wu D, Lu L, Liu P. Comparison of TALE designer transcription factors and the CRISPR/dCas9 in regulation of gene expression by targeting enhancers. Nucleic Acids Research. 2014;**42**:e155

[52] Huisman C, van der Wijst MG, Falahi F, Overkamp J, Karsten G, Terpstra MM, Kok K, van der Zee AG, Schuuring E, Wisman GBA, Rots MG. Prolonged re-expression of the hypermethylated gene EPB41L3 using artificial transcription factors and epigenetic drugs. Epigenetics. 2015;**10**:384-396

[53] Chen H, Kazemier HG, de Groote ML, Ruiters MH, Xu G-L, Rots MG. Induced DNA demethylation by targeting ten-eleven translocation 2 to the human ICAM-1 promoter. Nucleic Acids Research. 2014;**42**:1563-1574

[54] Huisman C, Van Der Wijst MG, Schokker M, Blancafort P, Terpstra MM, Kok K, Van Der Zee AG, Schuuring E, Wisman GBA, Rots MG. Re-expression of selected epigenetically silenced candidate tumor suppressor genes in cervical cancer by TET2-directed demethylation. Molecular Therapy. 2016;**24**:536-547

[55] Maeder ML, Angstman JF, Richardson ME, Linder SJ, Cascio VM, Tsai SQ, Ho QH, Sander JD, Reyon D, Bernstein BE, Costello JF, Wilkinson MF, Joung JK. Targeted DNA demethylation and activation of endogenous genes using programmable TALE-TET1 fusion proteins. Nature Biotechnology. 2013;**31**:1137-1142

[56] Choudhury SR, Cui Y, Lubecka K, Stefanska B, Irudayaraj J. CRISPR-dCas9 mediated TET1 targeting for selective DNA demethylation at BRCA1 promoter. Oncotarget. 2016;**7**:46545-46556

[57] Xu X, Tao Y, Gao X, Zhang L, Li X, Zou W, Ruan K, Wang F, Xu GL, Hu R. A CRISPR-based approach for targeted DNA demethylation. Cell Discovery. 2016;**2**:16009

[58] Chavez A, Scheiman J, Vora S, Pruitt BW, Tuttle M, Iyer EP, Lin S, Kiani S, Guzman CD, Wiegand DJ, Ter-Ovanesyan D, Braff JL, Davidsohn N, Housden BE, Perrimon N, Weiss R, Aach J, Collins JJ, Church GM. Highly efficient Cas9-mediated transcriptional programming. Nature Methods. 2015;**12**:326-328

[59] Tanenbaum ME, Gilbert LA, Qi LS, Weissman JS, Vale RD. A protein-tagging system for signal amplification in gene expression and fluorescence imaging. Cell. 2014;**159**:635-646

[60] Konermann S, Brigham MD, Trevino AE, Joung J, Abudayyeh OO, Barcena C, Hsu PD, Habib N, Gootenberg JS, Nishimasu H, Nureki O, Zhang F. Genome-scale transcriptional activation by an engineered CRISPR-Cas9 complex. Nature. 2015;**517**:583-588

[61] Shechner DM, Hacisuleyman E, Younger ST, Rinn JL. Multiplexable, locus-specific targeting of long RNAs with CRISPR-display. Nature Methods. 2015;**12**:664-670

[62] Zalatan JG, Lee ME, Almeida R, Gilbert LA, Whitehead EH, La Russa M, Tsai JC, Weissman JS, Dueber JE, Qi LS, Lim WA. Engineering complex synthetic transcriptional programs with CRISPR RNA scaffolds. Cell. 2015;**160**:339-350

[63] Friedman JR, Fredericks WJ, Jensen DE, Speicher DW, Huang XP, Neilson EG, Rauscher FJ III. KAP-1, a novel corepressor for the highly conserved KRAB repression domain. Genes & Development. 1996;**10**:2067-2078

[64] Stolzenburg S, Rots MG, Beltran AS, Rivenbark AG, Yuan X, Qian H, Strahl BD, Blancafort P. Targeted silencing of the oncogenic transcription factor SOX2 in breast cancer. Nucleic Acids Research. 2012;**40**:6725-6740

[65] Cong L, Zhou R, Y-c K, Cunniff M, Zhang F. Comprehensive interrogation of natural TALE DNA binding modules and transcriptional repressor domains. Nature Communications. 2012;**3**:968

[66] Konermann S, Brigham MD, Trevino AE, Hsu PD, Heidenreich M, Cong L, Platt RJ, Scott DA, Church GM, Zhang F. Optical control of mammalian endogenous transcription and epigenetic states. Nature. 2013;**500**:472-476

[67] Kearns NA, Pham H, Tabak B, Genga RM, Silverstein NJ, Garber M, Maehr R. Functional annotation of native enhancers with a Cas9-histone demethylase fusion. Nature Methods. 2015;**12**:401-403

[68] McDonald JI, Celik H, Rois LE, Fishberger G, Fowler T, Rees R, Kramer A, Martens A, Edwards JR, Challen GA. Reprogrammable CRISPR/Cas9-based system for inducing site-specific DNA methylation. Biology Open. 2016;**5**:866-874

[69] Siddique AN, Nunna S, Rajavelu A, Zhang Y, Jurkowska RZ, Reinhardt R, Rots MG, Ragozin S, Jurkowski TP, Jeltsch A. Targeted methylation and gene silencing of VEGF-A in human cells by using a designed Dnmt3a–Dnmt3L single-chain fusion protein with increased DNA methylation activity. Journal of Molecular Biology. 2013;**425**:479-491

[70] Vojta A, Dobrinić P, Tadić V, Bočkor L, Korać P, Julg B, Klasić M, Zoldoš V. Repurposing the CRISPR-Cas9 system for targeted DNA methylation. Nucleic Acids Research. 2016;**44**:5615-5628

[71] Amabile A, Migliara A, Capasso P, Biffi M, Cittaro D, Naldini L, Lombardo A. Inheritable silencing of endogenous genes by hit-and-run targeted epigenetic editing. Cell. 2016; **167**:219-232

[72] Cheng AW, Jillette N, Lee P, Plaskon D, Fujiwara Y, Wang W, Taghbalout A, Wang H. Casilio: A versatile CRISPR-Cas9-Pumilio hybrid for gene regulation and genomic labeling. Cell Research. 2016;**26**:254-257

[73] Galonska C, Charlton J, Mattei AL, Donaghey J, Clement K, Gu H, Mohammad AW, Stamenova EK, Cacchiarelli D, Klages S, Timmermann B, Cantz T, Schöler HR, Gnirke A, Ziller MJ, Meissner A. Genome-wide tracking of dCas9-methyltransferase footprints. Nature Communications. 2018;**9**:597

[74] Wu X, Scott DA, Kriz AJ, Chiu AC, Hsu PD, Dadon DB, Cheng AW, Trevino AE, Konermann S, Chen S, Jaenisch R, Zhang F, Sharp PA. Genome-wide binding of the CRISPR endonuclease Cas9 in mammalian cells. Nature Biotechnology. 2014;**32**:670-676

[75] Ehrke-Schulz E, Schiwon M, Leitner T, Dávid S, Bergmann T, Liu J, Ehrhardt A. CRISPR/Cas9 delivery with one single adenoviral vector devoid of all viral genes. Scientific Reports. 2017;**7**:17113

[76] Zuris JA, Thompson DB, Shu Y, Guilinger JP, Bessen JL, Hu JH, Maeder ML, Joung JK, Chen ZY, Liu DR. Cationic lipid-mediated delivery of proteins enables efficient protein-based genome editing in vitro and in vivo. Nature Biotechnology. 2015;**33**:73-80

[77] Kroll C, Rathert P. Stable Expression of Epigenome Editors via Viral Delivery and Genomic Integration. Methods in Molecular Biology. 2018;**1767**:215-225

[78] Braun SMG, Kirkland JG, Chory EJ, Husmann D, Calarco JP, Crabtree GR. Nature Communications. 2017;**8**:560

[79] Abudayyeh OO, Gootenberg JS, Essletzbichler P, Han S, Joung J, Belanto JJ, Verdine V, Cox DBT, Kellner MJ, Regev A, Lander ES, Voytas DF, Ting AY, Zhang F. Nature. 2017;**550**:280-284

Chapter 8

Resetting Cell Fate by Epigenetic Reprogramming

Molly Pui Man Wong and Ray Kit Ng

Additional information is available at the end of the chapter

http://dx.doi.org/10.5772/intechopen.81015

Abstract

Epigenetic modifications and their regulations govern the identity of every cell type in an organism. Cell differentiation involves a switch in gene expression profile that is accompanied by heritable changes of epigenetic signatures in the differentiated cell type. Differentiation is generally not reversible, thereby conferring cell fate decisions once an altered epigenetic pattern is set. Nevertheless, attempts have been made to reverse a differentiation cell fate to a pluripotent state by various experimental approaches, such as somatic cell nuclear transfer, cell fusion and ectopic expression of defined transcription factors. The fundamental basis of all these strategies is to mediate epigenetic reprogramming, which allows a permanent and completed conversion of cell fate. A comprehensive understanding of the dynamic of epigenetic changes during cell differentiation would provide a more precise and efficient way of reprogramming cell fate. Here we summarize the epigenetic aspects of different reprogramming strategies and discuss the possible mechanisms underlying these epigenetic reprogramming events.

Keywords: epigenetic, reprogramming, somatic cell nuclear transfer, cell fusion, transcription factors, pluripotency, differentiation, cell fate

1. Introduction

Development is a complex process that involves a series of cell differentiation pathways starting from totipotent embryonic cells. According to Waddington's concept of epigenetic landscape, a cell has to interact with surrounding stimuli and respond by giving a phenotype which defines its identity during development [1]. Each cell experiences different inter/intra-cellular signals and hence has its epigenetic signature of cell identity, which in turn directs its own specific gene expression pattern without alteration of DNA sequences (with the exception of the immunoglobulin genes in B and T cells). It is now clear that the diversity of cell type specific gene expression pattern is mediated by means of epigenetic mechanisms, such as DNA methylation, covalent histone modifications and chromatin remodeling. Once a cell's identity is set, it is difficult to convert it to other unrelated lineages, thus leading to stable and irreversible differentiated cell states. Nevertheless, there are a few exceptions of cell fate conversion during embryo development and adult tissue/organ regeneration, e.g. vascular endothelium to smooth muscle cells [2], which involve changing a cell's epigenetic signature into another unrelated kind.

Apart from a few exceptions of natural cell fate conversion events, different strategies have been developed aiming to reprogram differentiated somatic cell fate to a pluripotent state (**Figure 1**). A historical strategy of reprogramming is by somatic cell nuclear transfer (SCNT) experiments. SCNT involves transplantation of a cell nucleus into an enucleated egg/oocyte in order to generate a "cloned" animal with an equivalent genetic composition as the donor individual. It is possible to derive embryonic stem (ES) cells from nuclear transplanted (NT) embryos (ntES cells), which shows indistinguishable pluripotent gene expression profiles

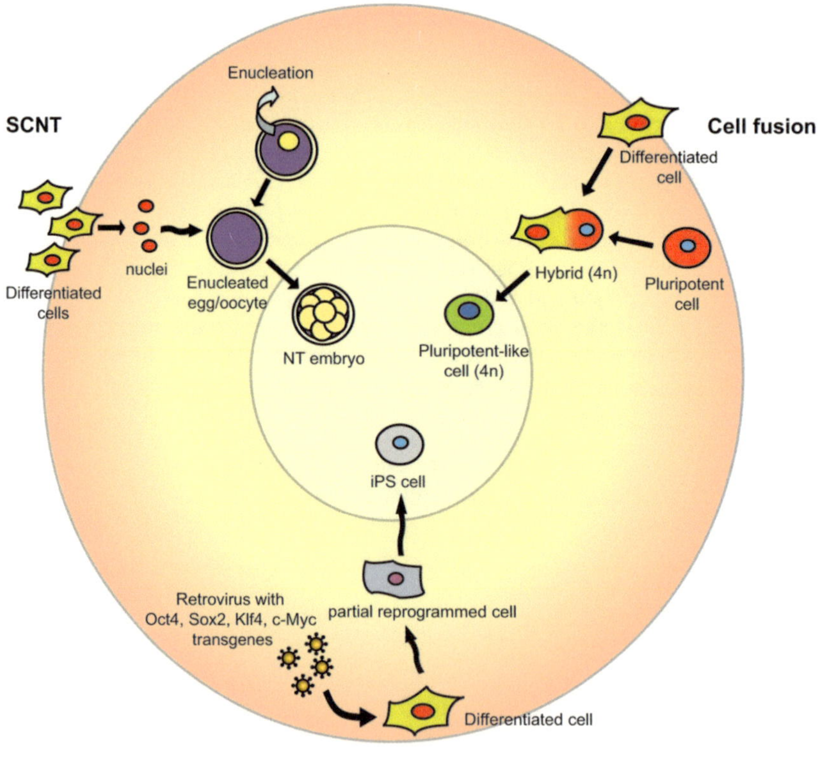

Figure 1. Strategies of reprogramming cell fate. Differentiated cells can be reprogrammed to pluripotent state by somatic cell nuclear transfer (SCNT), cell fusion, and ectopic expression of defined transcription factors. SCNT involves transplantation of a single differentiated cell nucleus into an enucleated egg/oocyte, which develops as a nuclear transplanted (NT) embryo. Cell fusion involves artificial fusion of a differentiated and a pluripotent cell to form a tetraploid pluripotent-like cell. Defined transcription factors (Oct4, Sox2, Klf4, c-Myc) can be ectopically expressed in differentiated cells and convert them to induced pluripotent stem (iPS) cells.

when compared to the normal ES cells derived from fertilized embryos. Another strategy of reprogramming is achieved by fusion of a differentiated cell with a pluripotent cell in order to generate a pluripotent-like tetraploid hybrid cell [3–5]. It has been proposed that cellular components, such as transcription factors, in the pluripotent cell are able to reprogram the differentiated cell nucleus. This idea aligns with the use of cell extracts from pluripotent cell types to revert differentiated cells into a pluripotent-like state [6, 7]. Presumably the cytoplasmic "reprogramming factors" from the pluripotent cells can be isolated and concentrated to achieve a higher reprogramming efficacy. A third strategy involves ectopic expression of defined transcription factors in somatic differentiated cells to generate induced pluripotent stem (iPS) cells. Delivery of the ectopic transcription factors can be achieved by viral approaches, such as the use of retrovirus, lentivirus, adeno-associated virus or Sendai virus, or by using episomal vesicles, or by direct mRNA or protein transfection. This technique has been successfully applied to reprogram a vast number of differentiated somatic cell types. Importantly, iPS cells can also be generated by using combinations of microRNAs (miRNAs) or small chemical molecules without the needs of ectopic expression of reprogramming factors [8]. The three reprogramming strategies show different reprogramming kinetics and efficiencies, which can be associated with the distinct epigenetic mechanisms in the erasure of somatic cell epigenetic signature and re-establishment of the pluripotent one. In this review, we focus on the dynamic changes of epigenetics mediated by different reprogramming strategies and how the modulation of epigenetic status improves the reprogramming efficiency.

2. Epigenetic reprogramming by SCNT

SCNT was first done by Briggs and Kings in 1952, who transplanted a blastula nucleus into an enucleated egg of the amphibian *Rana pipiens* [9]. Few years later, Gurdon *et al.* succeeded by using differentiated *Xenopus* intestinal epithelial donor nuclei for SCNT [10]. In 1997, the first cloned mammal, Dolly the sheep, was generated [11], and since then, more than 23 other mammalian species have now been successfully cloned [12]. Normal development of nuclear transplanted (NT) embryos requires recapitulation of the gene expression profile that supports the embryogenesis process by the differentiated donor nucleus. This involves re-activation of pluripotency genes, in particular *Oct4*, *Nanog* and *Sox2*, and repression of somatic lineage genes. In fact, the efficiency of reprogramming by SCNT is generally very low and less than 1% of NT embryos can develop into normal adults [13–15]. The cloned newborns often suffer from developmental abnormalities owing to incomplete reprogramming. It has been observed that *Oct4* was aberrantly expressed in cloned mouse blastocysts derived from cumulus donor nuclei [16, 17]. Besides, continuous expression of other somatic donor marker genes was demonstrated in some *Xenopus* NT embryos [18]. Some imprinted genes in donor cells were found to be aberrantly expressed in cloned embryos, presumably owing to the incomplete epigenetic reprogramming of the regulatory regions of imprinting loci [19–21]. Dysregulation of imprinted genes, such as *Igf2*, *Igf2r*, *H19*, and *Xist*, in cloned embryos can lead to both fetal and placental overgrowth and result in embryonic lethality or an abnormal growth condition called "large offspring syndrome", which is commonly found in cloned mammals [22–24]. Since the SCNT process does not increase the frequency of genetic alterations, it is suggested that the variable phenotypes observed in cloned embryos are associated with the reprogrammed epigenetic status of the donor nuclei [25]. This is supported by the findings that the developmental defects in cloned animals were not transmittable to the next offspring generation, indicating the presence of aberrant epigenetic reprogramming [26]. Aberrant DNA methylation patterns were indeed observed in NT embryos [27]. It was also demonstrated that the bovine NT blastocysts lack asymmetric patterns of both H3K9 methylation and acetylation between the inner cell mass and trophectoderm [28], which may account for abnormal cloned embryo development.

Although the rate of successful SCNT is very low, the reprogramming ability of factors in the egg/oocyte is highly efficient as the transplanted nuclei take less than 1 day to initiate cell division and trigger the "normal" developmental program. The donor cell epigenetic status has to be reprogrammed in order to support the embryonic program of development. In fact, genome-wide demethylation was observed in the cloned blastocysts [29]. It has been shown that the *Oct4* promoter of somatic cells undergoes DNA demethylation after nuclear transplantation into the germinal vesicle of *Xenopus* oocytes [30]. This demethylation of the *Oct4* promoter was found to be mediated by Tet3 in a mouse SCNT study [31], which is essential for the reactivation of *Oct4* expression for successful SCNT. In addition, chromatin remodeling factors, such as ISWI and BRG1, are documented in facilitating the reprogramming of cell fate. It has been shown that ISWI, which is a chromatin remodeling ATPase, is able to dissociate TATA binding protein in somatic nuclei after incubation in the *Xenopus* egg extract, suggesting that ISWI-containing complexes are facilitating epigenetic reprogramming in an egg environment [32]. Besides, *Xenopus* egg extract depleted of BRG1 protein showed an abolishment of the reprogramming ability and hence failed to induce *Oct4* expression in the somatic nuclei [33]. Therefore, chromatin structure remodeling is believed to be one of the reprogramming mechanisms. In addition, the maternal-derived histone H3 variant H3.3 in the enucleated egg was found to replace the canonical histone H3 in the donor nuclei after SCNT, leading to the reactivation of key pluripotent genes that are originally associated with repressive histone marks [34]. Histone H2A variant, macroH2A, also plays an important role in the reactivation of female donor cell's inactive X chromosome during reprogramming. It was shown that knockdown of macroH2A facilitates X-reactivation and the expression of pluripotent genes in cloned *Xenopus* embryos [35]. More recently, it has been demonstrated that the H3K9 tri-methylation (H3K9me3) of the donor cell genome is a major epigenetic barrier to SCNT. Ectopic expression of H3K9 demethylases *Kdm4b* or *Kdm4d* in the mouse donor cells de-repressed the genomic regions that are resistant to reprogramming and thus significantly improving SCNT efficiency [36, 37]. Similarly, removal of H3K9me3 by ectopic expression of other H3K9 demethylases also demonstrated improved reprogramming efficiency in human and bovine SCNT experiments [38, 39].

Interestingly, it has been shown that the epigenetic state of a differentiated cell is directly correlated with its reprogrammability [40]. SCNT with ES cell nuclei demonstrated a much higher efficiency of generation of NT blastocysts than using other somatic cell types [41]. This could be associated with a more relaxed chromatin configuration in ES cells that may render their epigenome more susceptible for reprogramming [42]. Alternatively, it has been demonstrated that the cloning efficiency can be significantly improved by pre-treating the more differentiated and condensed chromatin state of somatic nuclei with epigenetic modifying agents, e.g. 5-aza-deoxycytidine and trichostatin A (TSA), that facilitate chromatin relaxation [43–45]. Interestingly, the effect of TSA treatment in improving SCNT is associated with the reactivation of a subset of genes that are repressed by H3K9me3 in the somatic cells [46], presumably through introducing histone hyper-acetylation at their promoters. Altogether, SCNT provides a quick route of epigenetic reprogramming for a differentiated cell to a pluripotent state. Identification of the responsible reprogramming factors in the egg and oocyte cytoplasm will be one of the key future directions to improve the efficiency of SCNT and therapeutic cloning.

3. Epigenetic reprogramming by cell fusion

Cell fusion is a natural event that is crucial for fertilization and in various organs such as placenta, skeletal muscles and bones [47]. It has been proposed that the fusion of bone marrow-derived stem cells and tissue cells, e.g. hepatocytes, is one of the mechanisms of tissue repair [48, 49]. Cell fusion experiments using pluripotent cells, e.g. an ES or embryonic germ (EG) cell, were shown to be able to reprogram a differentiated cell type [3–5]. Both ES and EG cells possess reprogramming ability and are able to reactivate pluripotent genes and silence differentiation genes in the somatic cell nucleus within a tetraploid hybrid cell after cell fusion. It is indeed the case that the new transcription profile of a hybrid cell is partly contributed by the reprogrammed somatic nucleus to a pluripotent-like state. Moreover, injection of hybrid cells into normal diploid blastocysts demonstrated their contribution to all three germ layers in the chimeras [4, 50], indicating the pluripotent nature of hybrid cells. Similar to the SCNT, different somatic cell types show different kinetics of reprogramming by the cell fusion approach, which could be associated with the somatic chromatin accessibility status [51, 52].

Cell fusion with a pluripotent cell can trigger extensive epigenetic reprogramming in the differentiated cell nucleus. It has been shown that reactivation of *Oct4* from the somatic nucleus occurs before DNA replication after cell fusion [53], suggesting the involvement of active demethylation process [54]. This was further supported by the functional roles of Tet1 and Tet2 in the demethylation of *Oct4* and imprinted control regions by fusing somatic cells with EG cells [55]. Besides, in a cell fusion experiment using thymocytes and ES cells from two different mouse strains, it was observed that the epigenetic profile of the somatic cell nucleus was reprogrammed to a similar pattern to that of the ES cell. Global histone H3 and H4 acetylation and H3K4 di- and tri-methylation were increased in the hybrid cell to a level comparable to ES cells, whereas these modifications are weak in the parental somatic thymocytes. Examination of gene specific loci showed that the *Oct4* promoter was enriched with H3 acetylation and the promoter of the thymocyte marker *Thy-1* was enriched with H3K27 tri-methylation in both ES cell and hybrid cell chromatin, whereas these epigenetic modifications are missing in the thymocyte [56]. Hence, the somatic genome has undergone epigenetic reprogramming triggered by fusion with the ES cell, suggesting that the process of cell fusion mediates a transcription activation-permissive chromatin state in the hybrid genome. In addition, silencing the somatic differentiation genes was shown to be associated with polycomb repressive complexes in the cell fusion experiment using ES cells [57].

Reprogramming to pluripotency by cell fusion approach requires lengthy selection of the successfully reprogrammed hybrid cells. The reprogramming efficiency of cell fusion is usually less than 0.001%, depending on the somatic cell types [50, 58]. The low reprogramming efficiency in hybrid cells can be largely enhanced by manipulation of key pluripotency-associated genes like *Nanog* [58, 59] and *Sall4* [60], or by activation of the Wnt signaling pathway [61], emphasizing the importance of these factors in cell fusion reprogramming. Overexpression of *Nanog* or *Sall4* in ES cells demonstrated a several hundred-fold increase in reprogramming efficiency after cell fusion. Similarly, treatment of ES cells with Wnt3a for 24–48 hours enhanced the reprogramming of somatic cells by 20-fold. However, owing to the low reprogramming efficiency and the tetraploid genome of the resulting hybrid cells, reprogramming by the cell fusion approach becomes less promising in regenerative medicine.

4. Epigenetic reprogramming by defined transcription factors

In a groundbreaking discovery, Yamanaka *et al.* demonstrated that somatic cell state can be reprogrammed to a pluripotent state by the introduction of only four transcription factors; Oct4, Sox2, Klf4 and c-Myc, which are now also known as the Yamanaka factors. The first generation of iPS cells was obtained using a Fbx15-driven selection construct (Fbx15-iPS) and displayed a gene expression pattern very similar to that of normal pluripotent ES cells. However, the somatic epigenetic signature was only partially reprogrammed; the *Oct4* promoter, for example, retained some DNA methylation and no germline transmission was observed for these cells in chimeric mice. Hence these first-generation iPS cells were not fully pluripotent in nature [62]. Given the potential reprogramming capacity of these four factors, the second generation of iPS cells was generated by selection with a Nanog reporter construct (Nanog-iPS) [63, 64]. During the reprogramming process, the virally delivered transgenes were silenced, but on the contrary, the endogenous *Oct4* and *Sox2* loci were re-activated for the maintenance of pluripotency in iPS cells. In contrast to the Fbx15-iPS cells, these Nanog-iPS cells were able to undergo germline transmission in chimeric mice, and thus share this crucial feature of pluripotency with normal ES cells. These landmark studies that pioneered the derivation of mouse iPS cells led to the possibility of using the same strategy to generate human iPS cells. An initial study was performed by Thomas *et al.* in which a different combination of factors, OCT4, SOX2, NANOG and LIN28, was used to reprogram human fibroblasts into iPS cells [65]. Thereafter, Yamanaka and other groups succeeded in generating human iPS cells by using the same 4 Yamanaka factors as in the mouse iPS systems [66, 67]. To date, a number of different somatic cell types have been successfully reprogrammed into iPS cells, e.g. neural stem cells, keratinocytes, hepatocytes, gastric epithelia cells, pancreatic β cells, terminally differentiated B and T cells [8].

By using the defined transcription factor approach to reprogram cell fate, about 0.1–3% of the somatic starting cell population can be converted into iPS cells in around 2–3 weeks. The reprogramming efficiency is believed to be correlated with the differentiation state of the starting somatic cells. It has been shown that hematopoietic stem and progenitor cells can be reprogrammed to iPS cells 300 times more efficient than the terminally differentiated B and T cells [68]. Interestingly, partially de-differentiating mature B cells by either knockdown of

Pax5 or forced expression of *C/EBPα* resulted in efficient reprogramming by the Yamanaka factors [69], suggesting that the epigenetic status of differentiated cells is crucial in successful reprogramming. The dynamics of transcription profile and epigenetic patterns during the reprogramming process from somatic to iPS cells were studied in details. The reprogramming of somatic cell fate is a sequential stochastic event which involves a gradual silencing of the somatic lineage genes and the viral transgenes, and a sequential expression of alkaline phosphatase and SSEA1 in partially reprogrammed cells, whereas endogenous *Oct4* and *Nanog* are only activated in fully reprogrammed iPS cells. Induction of the four Yamanaka factors results in an immediate cellular response of inactivation of thousands of somatic lineage distal enhancers and, to a smaller extent, the H3K4me3-enriched somatic gene promoters [70, 71], leading to down-regulation of somatic identity genes. This initial phase of reprogramming is also accompanied by a global reduction of H3K27me3 resulting in loss of heterochromatin [72]. Meanwhile, mesenchymal transcription factors, such as *Snail1/2, Zeb1/2,* are repressed [70, 73, 74], whereas epithelial transcription factors, such as *Cdh1, Epcam,* are activated [75, 76], resulting in mesenchymal-epithelial transition (MET). This is associated with an increase in H3K4me2 at epithelial genes, but a decrease in H3K4me2 and H3K79me2 at mesenchymal genes [71, 77]. Despite an increase in global H3K36me2/3 level, loss of H3K36me2/3 was observed at the *Ink4-Arf* locus, leading to enhanced cell proliferation during reprogramming [78]. The following phase of reprogramming is marked by upregulation of endogenous pluripotency genes to establish transcriptional program, which is independent of transgene expression [73, 79]. The final phase of reprogramming involves elongation of telomeres, X-reactivation in female iPS cells, and upregulation of DNA methylation genes [79]. This coincides with loss of DNA methylation and downregulation of *Xist* expression in the somatic inactive X chromosome [80, 81].

A number of epigenetic remodeling factors are involved in the reprogramming events. Both polycomb (PcG) and trithorax (TrxG) group proteins were found to be crucial in the derivation of iPS cell colonies. Upon knockdown of *Wdr5,* which is a core component of TrxG protein complex, cells failed to establish H3K4me3 at the pluripotent genes, like *Oct4* and *Nanog,* for their reactivation [82]; whereas inhibition of the core components of the polycomb repressive complex 1 and 2 reduced reprogramming efficiency [83], partly because of the dysregulation of genes involved in the MET process [84]. This is similar to the findings that inhibition of H3K79 methyltransferase DOT1L facilitates the loss of H3K79me2 mark at the mesenchymal genes to promote MET during reprogramming [83]. The H3K27 demethylase Kdm6a (also known as Utx) can directly interact with Oct4/Sox2/Klf4 to remove the repressive H3K27me3 mark from the early pluripotent genes in somatic cells for their reactivation [85]. This is in agreement with the findings that depletion of histone H2A variant, macroH2A, enhances reprogramming, owing to its co-occupancy with H3K27me3 to repress pluripotent genes [86, 87]. Besides, the H3K36 demethylase Kdm2b (also known as Jhdm1b) enhances the activation of early responsive genes (*Cdh1, Epcam, Dsg2, Dsp, Irf6*) during reprogramming through the removal of H3K36me2 at their promoters [88]. Interestingly, H3K9me3 was also found to be one of the major epigenetic barriers in the generation of iPS cells [89], similar to the findings in SCNT experiments. Depletion of the H3K9 methyltransferases SUV39H1/H2, Ehmt1/2 and Setdb1 or inhibition of Cbx3, a protein that recognizes H3K9 methylation, enhances reprogramming by de-repressing *Nanog* and abolishing the cellular responses to BMP signaling [89–91]. Although

Dnmt3a/b were found to be dispensable [92], DNA demethylation of key pluripotency loci mediated by Tet proteins is required for efficient reprogramming [93–95].

Previous studies demonstrated that the partially reprogrammed iPS cells contained significantly fewer genes marked by the bivalent chromatin signature (co-existence of both H3K4 and H3K27 methylation) and an enrichment of DNA hyper-methylated loci when compared to the wild-type ES cells and the fully reprogrammed iPS cells [70]. Therefore, it is proposed that completion of the epigenetic reprogramming process is pre-requisite for the acquisition of pluripotency. This is supported by the observation that treatment of partially reprogrammed iPS cells with the DNA methyltransferase inhibitor 5-aza-cytidine was able to promote their transition into the fully reprogrammed pluripotent state [70]. Besides, inhibition of H3K27 methyltransferase Ezh2 by small molecule, GSK-126, reduced reprogramming efficiency [84], whereas inhibition of DOT1L by small molecule, EPZ004777, showed enhancement of reprogramming [83]. Various histone deacetylase inhibitors (HDACi) were also shown to improve reprogramming [96–100]. In combination with HDACi valproic acid, human iPS cells can be generated only with Oct4 and Sox2 with a comparable reprogramming efficiency by the four Yamanaka factors [101]. Interestingly, it was found that vitamin C can increase reprogramming efficiency by promoting the transition of pre-iPS cells to fully reprogrammed cells [102], potentially through acting as a cofactor of Kdm2b to induce H3K36me2/3 demethylation [78], activation of H3K9 demethylases (Kdm3a, 3b, 4c and 4d) to remove H3K9me3 [89], and promoting Tet-mediated DNA demethylation [103]. With the aid of small chemical molecules, the iPS cell reprogramming efficiency and duration could be further improved.

5. "Epigenetic memory" in reprogrammed cells

"Epigenetic memory" refers to the persistent expression of parental genes in the daughter cells through the inheritance of distinctive epigenetic marks. Consequently, the epigenetic profile of a parent cell is faithfully passed on to its daughter cells such that the gene expression pattern is memorized and maintained throughout cell generations. In the situation of reprogramming cell fate, the persistent somatic cell epigenetic signature and the expression of lineage genes in the reprogrammed cells is thus regarded as an example of epigenetic memory.

Even though it has been shown that prolonged *in vitro* culture of mammalian embryos can lead to aberrant expression of imprinted and non-imprinted genes owing to associated epigenetic 'errors' [104, 105]. It has been shown that many cloned embryos demonstrate different degrees of resemblance with donor cell gene expression patterns. The aberrant epigenetic pattern in cloned embryos is thought to be the result of persistence of the epigenetic memory of the donor cells. Indeed, the resemblance of DNA and histone modification patterns of NT embryos to those of the donor cell types supports this conclusion [45–47]. For example, both global and gene-specific patterns of DNA methylation in cloned bovine and mouse embryos were shown to be similar to those of their respective donor somatic cell types [28, 106, 107]. The phenomenon of epigenetic memory is also highlighted by the X inactivation pattern in cloned embryos. In normal fertilized embryos, the paternal X chromosome is preferentially silenced in the trophectoderm and extraembryonic endoderm lineages, whereas random X inactivation occurs in the inner cell mass. However, in NT embryos generated from female donor nuclei, the inactive X chromosome of the donor cell is preferentially chosen for inactivation in the trophectoderm, which is in contrast to the random X inactivation in the embryo proper, demonstrating a certain extent of memory of the inactive X chromosome status [48]. Random X inactivation in the placenta was found in deceased cloned bovine embryos, which suggests that the persistence of this inactive X chromosome memory in the placenta may be crucial for fetal survival [49].

The epigenetic memory in NT embryos that maintains the donor expression gene states can be explained by the inheritance of DNA methylation patterns. Heritability of DNA methylation is mediated by the activity of the maintenance DNA methyltransferase Dnmt1 which preferentially targets hemi-methylated DNA. Thus, Dnmt1 restores the parental methylation status on the newly synthesized daughter DNA strand, thereby maintaining a silent gene state after cell division [108]. Methylated donor genes, such as pluripotency genes, remain inactivated after SCNT, apparently owing to the persistent donor-specific methylation pattern in the cloned embryos, possibility mediated by the residual somatic form of Dnmt in the donor nucleus. DNA methylation therefore provides a plausible mechanism for the propagation of a silent memory state in SCNT [109]. On the other hand, an active gene memory of the donor differentiation state is also observed in NT embryos. For example, both the donor endoderm and neurectoderm markers, *Edd* and *Ncam*, were found to be aberrantly expressed in *Xenopus* NT embryos derived from the respective donor cell types [18]. This active gene memory was further demonstrated to be associated with the incorporation of a histone H3 variant H3.3 at the active gene loci in *Xenopus* NT embryos [110]. Histone variant H3.3 is enriched in the regulatory region of active genes and is preferentially marked by modifications associated with an active chromatin state, such as H3K4 methylation, H3K9 acetylation and H3K79 methylation [111, 112]. Experiments using a mutant form of H3.3 demonstrated that the K4 residue on H3.3 plays a key role in the inheritance of active epigenetic memory [110], proposing a model in which H3.3 K4 methylation creates an "active histone environment" for the recovery of active chromatin configuration in daughter cells after chromosomal replication [113].

Early studies of iPS cells demonstrated that the Nanog-iPS cells displayed not only a highly similar transcriptome to wild-type ES cells, but also an ES cell histone modification profile. Genome-wide comparison of histone modifications (H3K4 and H3K27 trimethylation) between ES cells, MEFs and MEF-derived iPS cells demonstrated that more than 94% of the ES- or MEF-signature genes in iPS cells have identical histone methylation marks as in ES cells. Only 0.7% of these signature genes retain the histone methylation status of the original MEFs [63]. However, other gene expression profile studies of iPS cells showed that a significant number of differentiation genes have a similar expression pattern to that in the somatic cell of origin, but not in ES cells [114–117]. This transcriptional memory in iPS cells was found to be correlated with biased differentiation towards the original cell lineage, and with less competence in differentiation to other unrelated somatic lineages [115, 118–121]. Importantly, the persistent expression of somatic genes in iPS cells was associated with the somatic DNA methylation pattern [117, 122–124], highlighting the crucial role of epigenetic regulation in the retention of memory. This is in fact similar to the observation that an incomplete removal of donor cell DNA methylation pattern was observed in some aberrantly developed NT embryos [125]. Strikingly, the addition of epigenetic modifying agents, such as DNA methyltransferase inhibitor 5-aza-2′-deoxycytidine (5-aza-dC), can enhance the iPS cell reprogramming efficiency [70] and improve the differentiation competency to other unrelated somatic lineages [120]. Interestingly, it has been demonstrated that continuous passaging of iPS cells abrogates somatic DNA methylation patterns [115], which suggests a passive replication-dependent mechanism in loss of the parental memory in iPS cells. Nevertheless, a study showed that the epigenetic memory in some iPS cell lines cannot be removed even after extended passages [124]. Apart from DNA methylation, microRNA expression pattern was also shown to have a role in the retention of somatic memory in iPS cells derived from hematopoietic progenitors [126]. However, it should be emphasized that other profiling studies of iPS cells failed to find the gene expression and epigenetic differences when compared to ES cells [127, 128]. It thus proposes that the "somatic memory" in iPS cells could be an artifact of incomplete reprogramming resulting in variation between iPS cell lines [129]. It is also possible that there are individual iPS cell lines expressing gene signatures owing to culture conditions and laboratory practices [130], similar to the scenario that some ES cell lines exhibit preferential differentiation towards specific lineages [131–133]. In summary, the epigenetic memory in iPS cells remains a contentious issue.

6. Conclusions and perspectives

Although the term "epigenetic landscape" was first introduced by Waddington in 1942 [1], our understanding of how the epigenome of a cell type is maintained and altered during differentiation is still far from complete. The reversal of the differentiated state of a cell has important implications for our understanding of normal development and for regenerative medicine. Epigenetic reprogramming provides heritable changes of cell identity, and thus is a key event for the complete and permanent conversion of cell fate (**Figure 2**). Although reprogramming of cell fate can be achieved by different strategies, the rate (reprogramming time) and efficiency (number of reprogrammed cells) are far from comparable to the natural event during fertilization/de-differentiation. Achieving a complete epigenetic reversion to generate reprogrammed cells or iPS cells with a comparable potency state of early embryos would imply that these cells can respond correctly to differentiation-promoting signals, and more importantly, decrease the tumorigenic potential owing to pre-disposing epimutations. Notably, the status of epigenetic memory in iPS cells can be regarded as a state of incomplete reprogramming. The biased differentiation owing to the persistent somatic epigenetic memory in iPS cells might be useful in efficient differentiation to the desired cell type of origin, which usually results in a heterogeneous cell population by using un-optimized differentiation protocols. On the contrary, it has been shown that *in vitro* culture condition can alter the epigenetic status of iPS cells [134]. With an optimized culture condition, a more homogeneous population of iPS cells can be obtained, which corresponds to the naïve state of pluripotency, and hence, further abrogate the somatic "epigenetic memory". A more recent approach in reprogramming involves the use of a combination of small chemical molecules and epigenetic modifying agents, without any ectopic expression of transcription factors [135, 136]. This approach seems to induce pluripotent reprogramming process different from the transcription factor-mediated approach. Therefore, unlocking the secrets

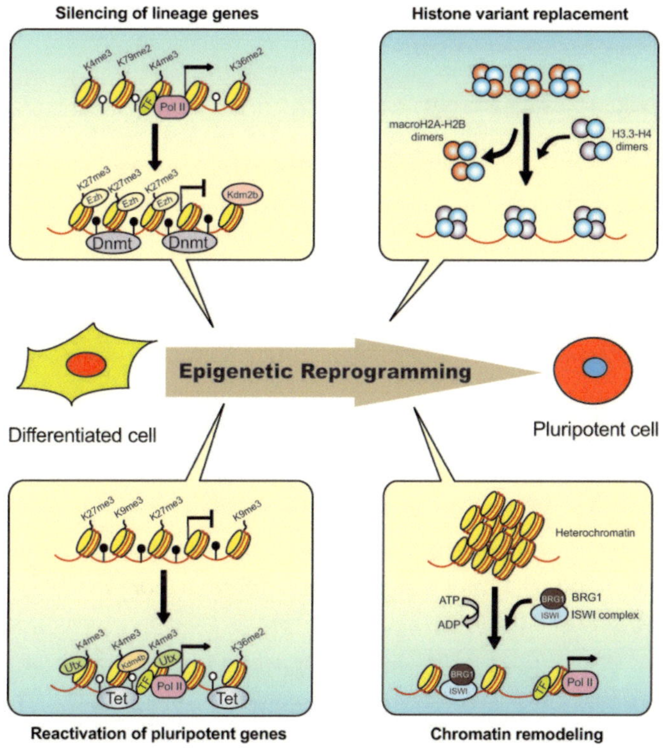

Figure 2. Mechanisms of epigenetic reprogramming. Epigenetic patterns in differentiated cells need to be reprogrammed to those of pluripotent cells, which results in silencing of differentiated genes and reactivation of pluripotency genes. Epigenetic reprogramming can be achieved by modulation of DNA methylation and histone modifications by various epigenetic modifying enzymes, such as Dnmt, Tet, Kdm2b, Kdm4b, Utx, Ezh2 (PcG), histone variant replacement, and chromatin remodeling enzymes. Other epigenetic mechanisms may also be involved. It is believed that a collaborative contribution of different epigenetic mechanisms is required for complete reversal of the differentiated cell state.

of epigenetic resetting mechanisms during cell differentiation can shed light on the development of more efficient and complete reprogramming approaches to further advance regenerative medicine.

Acknowledgements

This work was supported by the Research Grants Council of the Hong Kong Special Administrative Region (SAR), China (Grant no. HKU775510M and HKU774712M).

Conflict of interest

The authors declare that they have no conflict of interest.

Author details

Molly Pui Man Wong[1] and Ray Kit Ng[2*]

*Address all correspondence to: raykitng@hku.hk

1 School of Biomedical Sciences, Faculty of Medicine, The Chinese University of Hong Kong, Hong Kong SAR, China

2 School of Biomedical Sciences, Li Ka Shing Faculty of Medicine, The University of Hong Kong, Hong Kong SAR, China

References

[1] Waddington CH. The epigenotype. Endeavour. 1942;**1**:18-20

[2] Frid MG, Kale VA, Stenmark KR. Mature vascular endothelium can give rise to smooth muscle cells via endothelial-mesenchymal transdifferentiation: In vitro analysis. Circulation Research. 2002;**90**:1189-1196

[3] Tada M, Tada T, Lefebvre L, Barton SC, Surani MA. Embryonic germ cells induce epigenetic reprogramming of somatic nucleus in hybrid cells. The EMBO Journal. 1997;**16**:6510-6520

[4] Tada M, Takahama Y, Abe K, Nakatsuji N, Tada T. Nuclear reprogramming of somatic cells by in vitro hybridization with ES cells. Current Biology. 2001;**11**:1553-1558

[5] Tada M, Morizane A, Kimura H, Kawasaki H, Ainscough JF, Sasai Y, et al. Pluripotency of reprogrammed somatic genomes in embryonic stem hybrid cells. Developmental Dynamics. 2003;**227**:504-510

[6] Hakelien AM, Landsverk HB, Robl JM, Skalhegg BS, Collas P. Reprogramming fibroblasts to express T-cell functions using cell extracts. Nature Biotechnology. 2002;**20**:460-466

[7] Taranger CK, Noer A, Sorensen AL, Hakelien AM, Boquest AC, Collas P. Induction of dedifferentiation, genomewide transcriptional programming, and epigenetic reprogramming by extracts of carcinoma and embryonic stem cells. Molecular Biology of the Cell. 2005;**16**:5719-5735

[8] Hochedlinger K, Jaenisch R. Induced pluripotency and epigenetic reprogramming. Cold Spring Harbor Perspectives in Biology. 2015;**7**:1-24. DOI: 10.1101/cshperspect.a019448

[9] Briggs R, King TJ. Transplantation of living nuclei from blastula cells into enucleated frogs' eggs. Proceedings of the National Academy of Sciences of the United States of America. 1952;**38**:455-463

[10] Gurdon JB, Uehlinger V. "Fertile" intestine nuclei. Nature. 1966;**210**:1240-1241

[11] Wilmut I, Schnieke AE, McWhir J, Kind AJ, Campbell KH. Viable offspring derived from fetal and adult mammalian cells. Nature. 1997;**385**:810-813

[12] Loi P, Iuso D, Czernik M, Ogura A. A new, dynamic era for somatic cell nuclear transfer? Trends in Biotechnology. 2016;**34**:791-797. DOI: 10.1016/j.tibtech.2016.03.008

[13] Meissner A, Jaenisch R. Mammalian nuclear transfer. Developmental Dynamics. 2006;**235**: 2460-2469

[14] Yang X, Smith SL, Tian XC, Lewin HA, Renard JP, Wakayama T. Nuclear reprogramming of cloned embryos and its implications for therapeutic cloning. Nature Genetics. 2007;**39**:295-302

[15] Wilmut I, Bai Y, Taylor J. Somatic cell nuclear transfer: Origins, the present position and future opportunities. Philosophical Transactions of the Royal Society of London. Series B, Biological Sciences. 2015;**370**:20140366. DOI: 10.1098/rstb.2014.0366

[16] Boiani M, Eckardt S, Scholer HR, McLaughlin KJ. Oct4 distribution and level in mouse clones: Consequences for pluripotency. Genes & Development. 2002;**16**:1209-1219

[17] Bortvin A, Eggan K, Skaletsky H, Akutsu H, Berry DL, Yanagimachi R, et al. Incomplete reactivation of Oct4-related genes in mouse embryos cloned from somatic nuclei. Development. 2003;**130**:1673-1680

[18] Ng RK, Gurdon JB. Epigenetic memory of active gene transcription is inherited through somatic cell nuclear transfer. Proceedings of the National Academy of Sciences of the United States of America. 2005;**102**:1957-1962

[19] Inoue K, Kohda T, Lee J, Ogonuki N, Mochida K, Noguchi Y, et al. Faithful expression of imprinted genes in cloned mice. Science. 2002;**295**:297

[20] Mann MR, Chung YG, Nolen LD, Verona RI, Latham KE, Bartolomei MS. Disruption of imprinted gene methylation and expression in cloned preimplantation stage mouse embryos. Biology of Reproduction. 2003;**69**:902-914

[21] Liu JH, Yin S, Xiong B, Hou Y, Chen DY, Sun QY. Aberrant DNA methylation imprints in aborted bovine clones. Molecular Reproduction and Development. 2008;**75**:598-607

[22] Eggan K, Akutsu H, Loring J, Jackson-Grusby L, Klemm M, Rideout WM 3rd, et al. Hybrid vigor, fetal overgrowth, and viability of mice derived by nuclear cloning and tetraploid embryo complementation. Proceedings of the National Academy of Sciences of the United States of America. 2001;**98**:6209-6214

[23] Rideout WM 3rd, Eggan K, Jaenisch R. Nuclear cloning and epigenetic reprogramming of the genome. Science. 2001;**293**:1093-1098

[24] Shen CJ, Lin CC, Shen PC, Cheng WT, Chen HL, Chang TC, et al. Imprinted genes and satellite loci are differentially methylated in bovine somatic cell nuclear transfer clones. Cellular Reprogramming. 2013;**15**:413-424. DOI: 10.1089/cell.2013.0012

[25] Murphey P, Yamazaki Y, McMahan CA, Walter CA, Yanagimachi R, McCarrey JR. Epigenetic regulation of genetic integrity is reprogrammed during cloning. Proceedings of the National Academy of Sciences of the United States of America. 2009;**106**:4731-4735. DOI: 10.1073/pnas.0900687106

[26] Tamashiro KL, Wakayama T, Akutsu H, Yamazaki Y, Lachey JL, Wortman MD, et al. Cloned mice have an obese phenotype not transmitted to their offspring. Nature Medicine. 2002;**8**:262-267

[27] Chan MM, Smith ZD, Egli D, Regev A, Meissner A. Mouse ooplasm confers context-specific reprogramming capacity. Nature Genetics. 2012;**44**:978-980. DOI: 10.1038/ng.2382

[28] Dean W, Santos F, Reik W. Epigenetic reprogramming in early mammalian development and following somatic nuclear transfer. Seminars in Cell & Developmental Biology. 2003;**14**:93-100

[29] Niemann H, Carnwath JW, Herrmann D, Wieczorek G, Lemme E, Lucas-Hahn A, et al. DNA methylation patterns reflect epigenetic reprogramming in bovine embryos. Cellular Reprogramming. 2010;**12**:33-42. DOI: 10.1089/cell.2009.0063

[30] Simonsson S, Gurdon J. DNA demethylation is necessary for the epigenetic reprogramming of somatic cell nuclei. Nature Cell Biology. 2004;**6**:984-990

[31] Gu TP, Guo F, Yang H, Wu HP, Xu GF, Liu W, et al. The role of Tet3 DNA dioxygenase in epigenetic reprogramming by oocytes. Nature. 2011;**477**:606-610. DOI: 10.1038/nature10443

[32] Kikyo N, Wade PA, Guschin D, Ge H, Wolffe AP. Active remodeling of somatic nuclei in egg cytoplasm by the nucleosomal ATPase ISWI. Science. 2000;**289**:2360-2362

[33] Hansis C, Barreto G, Maltry N, Niehrs C. Nuclear reprogramming of human somatic cells by xenopus egg extract requires BRG1. Current Biology. 2004;**14**:1475-1480

[34] Wen D, Banaszynski LA, Rosenwaks Z, Allis CD, Rafii S. H3.3 replacement facilitates epigenetic reprogramming of donor nuclei in somatic cell nuclear transfer embryos. Nucleus. 2014;**5**:369-375. DOI: 10.4161/nucl.36231

[35] Pasque V, Gillich A, Garrett N, Gurdon JB. Histone variant macroH2A confers resistance to nuclear reprogramming. The EMBO Journal. 2011;**30**:2373-2387. DOI: 10.1038/emboj.2011.144

[36] Antony J, Oback F, Chamley LW, Oback B, Laible G. Transient JMJD2B-mediated reduction of H3K9me3 levels improves reprogramming of embryonic stem cells into cloned embryos. Molecular and Cellular Biology. 2013;**33**:974-983. DOI: 10.1128/MCB.01014-12

[37] Matoba S, Liu Y, Lu F, Iwabuchi KA, Shen L, Inoue A, et al. Embryonic development following somatic cell nuclear transfer impeded by persisting histone methylation. Cell. 2014;**159**:884-895. DOI: 10.1016/j.cell.2014.09.055

[38] Liu X, Wang Y, Gao Y, Su J, Zhang J, Xing X, et al. H3K9 demethylase KDM4E is an epigenetic regulator for bovine embryonic development and a defective factor for nuclear reprogramming. Development. 2018;**145**:dev158261. DOI: 10.1242/dev.158261

[39] Chung YG, Matoba S, Liu Y, Eum JH, Lu F, Jiang W, et al. Histone demethylase expression enhances human somatic cell nuclear transfer efficiency and promotes derivation of pluripotent stem cells. Cell Stem Cell. 2015;**17**:758-766. DOI: 10.1016/j.stem.2015.10.001

[40] Oback B, Wells DN. Donor cell differentiation, reprogramming, and cloning efficiency: Elusive or illusive correlation? Molecular Reproduction and Development. 2007;**74**:646-654. DOI: 10.1002/mrd.20654

[41] Hochedlinger K, Jaenisch R. Nuclear reprogramming and pluripotency. Nature. 2006; **441**:1061-1067

[42] Meshorer E, Misteli T. Chromatin in pluripotent embryonic stem cells and differentiation. Nature Reviews. Molecular Cell Biology. 2006;**7**:540-546

[43] Enright BP, Kubota C, Yang X, Tian XC. Epigenetic characteristics and development of embryos cloned from donor cells treated by trichostatin A or 5-aza-2'-deoxycytidine. Biology of Reproduction. 2003;**69**:896-901

[44] Enright BP, Sung LY, Chang CC, Yang X, Tian XC. Methylation and acetylation characteristics of cloned bovine embryos from donor cells treated with 5-aza-2'-deoxycytidine. Biology of Reproduction. 2005;**72**:944-948

[45] Kishigami S, Bui HT, Wakayama S, Tokunaga K, Van Thuan N, Hikichi T, et al. Successful mouse cloning of an outbred strain by trichostatin A treatment after somatic nuclear transfer. The Journal of Reproduction and Development. 2007;**53**:165-170

[46] Inoue K, Oikawa M, Kamimura S, Ogonuki N, Nakamura T, Nakano T, et al. Trichostatin A specifically improves the aberrant expression of transcription factor genes in embryos produced by somatic cell nuclear transfer. Scientific Reports. 2015;**5**:10127. DOI: 10.1038/srep10127

[47] Larsson LI, Bjerregaard B, Talts JF. Cell fusions in mammals. Histochemistry and Cell Biology. 2008;**129**:551-561

[48] Wang X, Willenbring H, Akkari Y, Torimaru Y, Foster M, Al-Dhalimy M, et al. Cell fusion is the principal source of bone-marrow-derived hepatocytes. Nature. 2003;**422**:897-901

[49] Vassilopoulos G, Wang PR, Russell DW. Transplanted bone marrow regenerates liver by cell fusion. Nature. 2003;**422**:901-904

[50] Cowan CA, Atienza J, Melton DA, Eggan K. Nuclear reprogramming of somatic cells after fusion with human embryonic stem cells. Science. 2005;**309**:1369-1373

[51] Malinowski AR, Fisher AG. Reprogramming of somatic cells towards pluripotency by cell fusion. Methods in Molecular Biology. 2016;**1480**:289-299. DOI: 10.1007/978-1-4939-6380-5_25

[52] Soza-Ried J, Fisher AG. Reprogramming somatic cells towards pluripotency by cellular fusion. Current Opinion in Genetics & Development. 2012;**22**:459-465. DOI: 10.1016/j.gde.2012.07.005

[53] Pereira CF, Terranova R, Ryan NK, Santos J, Morris KJ, Cui W, et al. Heterokaryon-based reprogramming of human B lymphocytes for pluripotency requires Oct4 but not Sox2. PLoS Genetics. 2008;**4**:e1000170. DOI: 10.1371/journal.pgen.1000170

[54] Bhutani N, Brady JJ, Damian M, Sacco A, Corbel SY, Blau HM. Reprogramming towards pluripotency requires AID-dependent DNA demethylation. Nature. 2010;**463**:1042-1047. DOI: 10.1038/nature08752

[55] Piccolo FM, Bagci H, Brown KE, Landeira D, Soza-Ried J, Feytout A, et al. Different roles for Tet1 and Tet2 proteins in reprogramming-mediated erasure of imprints induced by EGC fusion. Molecular Cell. 2013;**49**:1023-1033. DOI: 10.1016/j.molcel.2013.01.032

[56] Kimura H, Tada M, Nakatsuji N, Tada T. Histone code modifications on pluripotential nuclei of reprogrammed somatic cells. Molecular and Cellular Biology. 2004;**24**:5710-5720

[57] Pereira CF, Piccolo FM, Tsubouchi T, Sauer S, Ryan NK, Bruno L, et al. ESCs require PRC2 to direct the successful reprogramming of differentiated cells toward pluripotency. Cell Stem Cell. 2010;**6**:547-556. DOI: 10.1016/j.stem.2010.04.013

[58] Silva J, Chambers I, Pollard S, Smith A. Nanog promotes transfer of pluripotency after cell fusion. Nature. 2006;**441**:997-1001

[59] Theunissen TW, van Oosten AL, Castelo-Branco G, Hall J, Smith A, Silva JC. Nanog overcomes reprogramming barriers and induces pluripotency in minimal conditions. Current Biology. 2011;**21**:65-71. DOI: 10.1016/j.cub.2010.11.074

[60] Wong CC, Gaspar-Maia A, Ramalho-Santos M, Reijo Pera RA. High-efficiency stem cell fusion-mediated assay reveals Sall4 as an enhancer of reprogramming. PLoS One. 2008;**3**:e1955

[61] Lluis F, Pedone E, Pepe S, Cosma MP. Periodic activation of Wnt/beta-catenin signaling enhances somatic cell reprogramming mediated by cell fusion. Cell Stem Cell. 2008;**3**:493-507

[62] Takahashi K, Yamanaka S. Induction of pluripotent stem cells from mouse embryonic and adult fibroblast cultures by defined factors. Cell. 2006;**126**:663-676

[63] Wernig M, Meissner A, Foreman R, Brambrink T, Ku M, Hochedlinger K, et al. In vitro reprogramming of fibroblasts into a pluripotent ES-cell-like state. Nature. 2007;**448**:318-324

[64] Okita K, Ichisaka T, Yamanaka S. Generation of germline-competent induced pluripotent stem cells. Nature. 2007;**448**:313-317

[65] Yu J, Vodyanik MA, Smuga-Otto K, Antosiewicz-Bourget J, Frane JL, Tian S, et al. Induced pluripotent stem cell lines derived from human somatic cells. Science. 2007;**318**:1917-1920

[66] Takahashi K, Tanabe K, Ohnuki M, Narita M, Ichisaka T, Tomoda K, et al. Induction of pluripotent stem cells from adult human fibroblasts by defined factors. Cell. 2007;**131**:861-872

[67] Park IH, Zhao R, West JA, Yabuuchi A, Huo H, Ince TA, et al. Reprogramming of human somatic cells to pluripotency with defined factors. Nature. 2008;**451**:141-146

[68] Eminli S, Foudi A, Stadtfeld M, Maherali N, Ahfeldt T, Mostoslavsky G, et al. Differentiation stage determines potential of hematopoietic cells for reprogramming into induced pluripotent stem cells. Nature Genetics. 2009;**41**:968-976. DOI: 10.1038/ng.428

[69] Hanna J, Markoulaki S, Schorderet P, Carey BW, Beard C, Wernig M, et al. Direct reprogramming of terminally differentiated mature B lymphocytes to pluripotency. Cell. 2008;**133**:250-264

[70] Mikkelsen TS, Hanna J, Zhang X, Ku M, Wernig M, Schorderet P, et al. Dissecting direct reprogramming through integrative genomic analysis. Nature. 2008;**454**:49-55

[71] Koche RP, Smith ZD, Adli M, Gu H, Ku M, Gnirke A, et al. Reprogramming factor expression initiates widespread targeted chromatin remodeling. Cell Stem Cell. 2011;**8**:96-105. DOI: 10.1016/j.stem.2010.12.001

[72] Hussein SM, Puri MC, Tonge PD, Benevento M, Corso AJ, Clancy JL, et al. Genome-wide characterization of the routes to pluripotency. Nature. 2014;**516**:198-206. DOI: 10.1038/nature14046

[73] Stadtfeld M, Maherali N, Breault DT, Hochedlinger K. Defining molecular cornerstones during fibroblast to iPS cell reprogramming in mouse. Cell Stem Cell. 2008;**2**:230-240. DOI: 10.1016/j.stem.2008.02.001

[74] Sridharan R, Tchieu J, Mason MJ, Yachechko R, Kuoy E, Horvath S, et al. Role of the murine reprogramming factors in the induction of pluripotency. Cell. 2009;**136**:364-377. DOI: 10.1016/j.cell.2009.01.001

[75] Samavarchi-Tehrani P, Golipour A, David L, Sung HK, Beyer TA, Datti A, et al. Functional genomics reveals a BMP-driven mesenchymal-to-epithelial transition in the initiation of somatic cell reprogramming. Cell Stem Cell. 2010;**7**:64-77. DOI: 10.1016/j.stem.2010.04.015

[76] Li R, Liang J, Ni S, Zhou T, Qing X, Li H, et al. A mesenchymal-to-epithelial transition initiates and is required for the nuclear reprogramming of mouse fibroblasts. Cell Stem Cell. 2010;**7**:51-63. DOI: 10.1016/j.stem.2010.04.014

[77] Nguyen AT, Zhang Y. The diverse functions of Dot1 and H3K79 methylation. Genes & Development. 2011;**25**:1345-1358. DOI: 10.1101/gad.2057811

[78] Wang T, Chen K, Zeng X, Yang J, Wu Y, Shi X, et al. The histone demethylases Jhdm1a/1b enhance somatic cell reprogramming in a vitamin-C-dependent manner. Cell Stem Cell. 2011;**9**:575-587. DOI: 10.1016/j.stem.2011.10.005

[79] Polo JM, Anderssen E, Walsh RM, Schwarz BA, Nefzger CM, Lim SM, et al. A molecular roadmap of reprogramming somatic cells into iPS cells. Cell. 2012;**151**:1617-1632. DOI: 10.1016/j.cell.2012.11.039

[80] Pasque V, Tchieu J, Karnik R, Uyeda M, Sadhu Dimashkie A, Case D, et al. X chromosome reactivation dynamics reveal stages of reprogramming to pluripotency. Cell. 2014;**159**:1681-1697. DOI: 10.1016/j.cell.2014.11.040

[81] Kim JS, Choi HW, Arauzo-Bravo MJ, Scholer HR, Do JT. Reactivation of the inactive X chromosome and post-transcriptional reprogramming of Xist in iPSCs. Journal of Cell Science. 2015;**128**:81-87. DOI: 10.1242/jcs.154294

[82] Ang YS, Tsai SY, Lee DF, Monk J, Su J, Ratnakumar K, et al. Wdr5 mediates self-renewal and reprogramming via the embryonic stem cell core transcriptional network. Cell. 2011;**145**:183-197. DOI: 10.1016/j.cell.2011.03.003

[83] Onder TT, Kara N, Cherry A, Sinha AU, Zhu N, Bernt KM, et al. Chromatin-modifying enzymes as modulators of reprogramming. Nature. 2012;**483**:598-602. DOI: 10.1038/nature10953

[84] Rao RA, Dhele N, Cheemadan S, Ketkar A, Jayandharan GR, Palakodeti D, et al. Ezh2 mediated H3K27me3 activity facilitates somatic transition during human pluripotent reprogramming. Scientific Reports. 2015;**5**:8229. DOI: 10.1038/srep08229

[85] Mansour AA, Gafni O, Weinberger L, Zviran A, Ayyash M, Rais Y, et al. The H3K27 demethylase Utx regulates somatic and germ cell epigenetic reprogramming. Nature. 2012;**488**:409-413. DOI: 10.1038/nature11272

[86] Pasque V, Radzisheuskaya A, Gillich A, Halley-Stott RP, Panamarova M, Zernicka-Goetz M, et al. Histone variant macroH2A marks embryonic differentiation in vivo and acts as an epigenetic barrier to induced pluripotency. Journal of Cell Science. 2012;**125**:6094-6104. DOI: 10.1242/jcs.113019

[87] Gaspar-Maia A, Qadeer ZA, Hasson D, Ratnakumar K, Leu NA, Leroy G, et al. MacroH2A histone variants act as a barrier upon reprogramming towards pluripotency. Nature Communications. 2013;**4**:1565. DOI: 10.1038/ncomms2582

[88] Liang G, He J, Zhang Y. Kdm2b promotes induced pluripotent stem cell generation by facilitating gene activation early in reprogramming. Nature Cell Biology. 2012;**14**:457-466. DOI: 10.1038/ncb2483

[89] Chen J, Liu H, Liu J, Qi J, Wei B, Yang J, et al. H3K9 methylation is a barrier during somatic cell reprogramming into iPSCs. Nature Genetics. 2013;**45**:34-42. DOI: 10.1038/ng.2491

[90] Soufi A, Donahue G, Zaret KS. Facilitators and impediments of the pluripotency reprogramming factors' initial engagement with the genome. Cell. 2012;**151**:994-1004. DOI: 10.1016/j.cell.2012.09.045

[91] Sridharan R, Gonzales-Cope M, Chronis C, Bonora G, McKee R, Huang C, et al. Proteomic and genomic approaches reveal critical functions of H3K9 methylation and heterochromatin protein-1gamma in reprogramming to pluripotency. Nature Cell Biology. 2013;**15**:872-882. DOI: 10.1038/ncb2768

[92] Pawlak M, Jaenisch R. De novo DNA methylation by Dnmt3a and Dnmt3b is dispensable for nuclear reprogramming of somatic cells to a pluripotent state. Genes & Development. 2011;**25**:1035-1040. DOI: 10.1101/gad.2039011

[93] Gao Y, Chen J, Li K, Wu T, Huang B, Liu W, et al. Replacement of Oct4 by Tet1 during iPSC induction reveals an important role of DNA methylation and hydroxymethylation in reprogramming. Cell Stem Cell. 2013;**12**:453-469. DOI: 10.1016/j.stem.2013.02.005

[94] Doege CA, Inoue K, Yamashita T, Rhee DB, Travis S, Fujita R, et al. Early-stage epigenetic modification during somatic cell reprogramming by Parp1 and Tet2. Nature. 2012;**488**:652-655. DOI: 10.1038/nature11333

[95] Hu X, Zhang L, Mao SQ, Li Z, Chen J, Zhang RR, et al. Tet and TDG mediate DNA demethylation essential for mesenchymal-to-epithelial transition in somatic cell reprogramming. Cell Stem Cell. 2014;**14**:512-522. DOI: 10.1016/j.stem.2014.01.001

[96] Huangfu D, Maehr R, Guo W, Eijkelenboom A, Snitow M, Chen AE, et al. Induction of pluripotent stem cells by defined factors is greatly improved by small-molecule compounds. Nature Biotechnology. 2008;**26**:795-797. DOI: 10.1038/nbt1418

[97] Liang G, Taranova O, Xia K, Zhang Y. Butyrate promotes induced pluripotent stem cell generation. The Journal of Biological Chemistry. 2010;**285**:25516-25521. DOI: 10.1074/jbc.M110.142059

[98] Mali P, Chou BK, Yen J, Ye Z, Zou J, Dowey S, et al. Butyrate greatly enhances derivation of human induced pluripotent stem cells by promoting epigenetic remodeling and the expression of pluripotency-associated genes. Stem Cells. 2010;**28**:713-720. DOI: 10.1002/stem.402

[99] Zhang Z, Gao Y, Gordon A, Wang ZZ, Qian Z, Wu WS. Efficient generation of fully reprogrammed human iPS cells via polycistronic retroviral vector and a new cocktail of chemical compounds. PLoS One. 2011;**6**:e26592. DOI: 10.1371/journal.pone.0026592

[100] Zhang Z, Wu WS. Sodium butyrate promotes generation of human induced pluripotent stem cells through induction of the miR302/367 cluster. Stem Cells and Development. 2013;**22**:2268-2277. DOI: 10.1089/scd.2012.0650

[101] Huangfu D, Osafune K, Maehr R, Guo W, Eijkelenboom A, Chen S, et al. Induction of pluripotent stem cells from primary human fibroblasts with only Oct4 and Sox2. Nature Biotechnology. 2008;**26**:1269-1275

[102] Esteban MA, Wang T, Qin B, Yang J, Qin D, Cai J, et al. Vitamin C enhances the generation of mouse and human induced pluripotent stem cells. Cell Stem Cell. 2010;**6**:71-79. DOI: 10.1016/j.stem.2009.12.001

[103] Minor EA, Court BL, Young JI, Wang G. Ascorbate induces ten-eleven translocation (Tet) methylcytosine dioxygenase-mediated generation of 5-hydroxymethylcytosine. The Journal of Biological Chemistry. 2013;**288**:13669-13674. DOI: 10.1074/jbc.C113.464800

[104] Wrenzycki C, Herrmann D, Keskintepe L, Martins A Jr, Sirisathien S, Brackett B, et al. Effects of culture system and protein supplementation on mRNA expression in preimplantation bovine embryos. Human Reproduction. 2001;**16**:893-901

[105] Farin PW, Piedrahita JA, Farin CE. Errors in development of fetuses and placentas from in vitro-produced bovine embryos. Theriogenology. 2006;**65**:178-191. DOI: 10.1016/j.theriogenology.2005.09.022

[106] Kang YK, Koo DB, Park JS, Choi YH, Chung AS, Lee KK, et al. Aberrant methylation of donor genome in cloned bovine embryos. Nature Genetics. 2001;**28**:173-177

[107] Dean W, Santos F, Stojkovic M, Zakhartchenko V, Walter J, Wolf E, et al. Conservation of methylation reprogramming in mammalian development: Aberrant reprogramming in cloned embryos. Proceedings of the National Academy of Sciences of the United States of America. 2001;**98**:13734-13738

[108] Pradhan S, Esteve PO. Mammalian DNA (cytosine-5) methyltransferases and their expression. Clinical Immunology. 2003;**109**:6-16

[109] Ng RK, Gurdon JB. Maintenance of epigenetic memory in cloned embryos. Cell Cycle. 2005;**4**:760-763

[110] Ng RK, Gurdon JB. Epigenetic memory of an active gene state depends on histone H3.3 incorporation into chromatin in the absence of transcription. Nature Cell Biology. 2008;**10**:102-109

[111] McKittrick E, Gafken PR, Ahmad K, Henikoff S. Histone H3.3 is enriched in covalent modifications associated with active chromatin. Proceedings of the National Academy of Sciences of the United States of America. 2004;**101**:1525-1530

[112] Hake SB, Garcia BA, Duncan EM, Kauer M, Dellaire G, Shabanowitz J, et al. Expression patterns and post-translational modifications associated with mammalian histone H3 variants. The Journal of Biological Chemistry. 2006;**281**:559-568

[113] Ng RK, Gurdon JB. Epigenetic inheritance of cell differentiation status. Cell Cycle. 2008;**7**:1173-1177

[114] Chin MH, Mason MJ, Xie W, Volinia S, Singer M, Peterson C, et al. Induced pluripotent stem cells and embryonic stem cells are distinguished by gene expression signatures. Cell Stem Cell. 2009;**5**:111-123. DOI: 10.1016/j.stem.2009.06.008

[115] Polo JM, Liu S, Figueroa ME, Kulalert W, Eminli S, Tan KY, et al. Cell type of origin influences the molecular and functional properties of mouse induced pluripotent stem cells. Nature Biotechnology. 2010;**28**:848-855. DOI: 10.1038/nbt.1667

[116] Marchetto MC, Yeo GW, Kainohana O, Marsala M, Gage FH, Muotri AR. Transcriptional signature and memory retention of human-induced pluripotent stem cells. PLoS One. 2009;**4**:e7076. DOI: 10.1371/journal.pone.0007076

[117] Ohi Y, Qin H, Hong C, Blouin L, Polo JM, Guo T, et al. Incomplete DNA methylation underlies a transcriptional memory of somatic cells in human iPS cells. Nature Cell Biology. 2011;**13**:541-549. DOI: 10.1038/ncb2239

[118] Bar-Nur O, Russ HA, Efrat S, Benvenisty N. Epigenetic memory and preferential lineage-specific differentiation in induced pluripotent stem cells derived from human pancreatic islet beta cells. Cell Stem Cell. 2011;**9**:17-23. DOI: 10.1016/j.stem.2011.06.007

[119] Sanchez-Freire V, Lee AS, Hu S, Abilez OJ, Liang P, Lan F, et al. Effect of human donor cell source on differentiation and function of cardiac induced pluripotent stem cells. Journal of the American College of Cardiology. 2014;**64**:436-448. DOI: 10.1016/j.jacc.2014.04.056

[120] Kim K, Doi A, Wen B, Ng K, Zhao R, Cahan P, et al. Epigenetic memory in induced pluripotent stem cells. Nature. 2010;**467**:285-290. DOI: 10.1038/nature09342

[121] Feng Q, Lu SJ, Klimanskaya I, Gomes I, Kim D, Chung Y, et al. Hemangioblastic derivatives from human induced pluripotent stem cells exhibit limited expansion and early senescence. Stem Cells. 2010;**28**:704-712. DOI: 10.1002/stem.321

[122] Lister R, Pelizzola M, Kida YS, Hawkins RD, Nery JR, Hon G, et al. Hotspots of aberrant epigenomic reprogramming in human induced pluripotent stem cells. Nature. 2011;**471**:68-73. DOI: 10.1038/nature09798

[123] Ruiz S, Diep D, Gore A, Panopoulos AD, Montserrat N, Plongthongkum N, et al. Identification of a specific reprogramming-associated epigenetic signature in human induced pluripotent stem cells. Proceedings of the National Academy of Sciences of the United States of America. 2012;**109**:16196-16201. DOI: 10.1073/pnas.1202352109

[124] Kim K, Zhao R, Doi A, Ng K, Unternaehrer J, Cahan P, et al. Donor cell type can influence the epigenome and differentiation potential of human induced pluripotent stem cells. Nature Biotechnology. 2011;**29**:1117-1119. DOI: 10.1038/nbt.2052

[125] Kawasumi M, Unno Y, Matsuoka T, Nishiwaki M, Anzai M, Amano T, et al. Abnormal DNA methylation of the Oct-4 enhancer region in cloned mouse embryos. Molecular Reproduction and Development. 2009;**76**:342-350. DOI: 10.1002/mrd.20966

[126] Vitaloni M, Pulecio J, Bilic J, Kuebler B, Laricchia-Robbio L, Izpisua Belmonte JC. MicroRNAs contribute to induced pluripotent stem cell somatic donor memory. The Journal of Biological Chemistry. 2014;**289**:2084-2098. DOI: 10.1074/jbc.M113.538702

[127] Guenther MG, Frampton GM, Soldner F, Hockemeyer D, Mitalipova M, Jaenisch R, et al. Chromatin structure and gene expression programs of human embryonic and induced pluripotent stem cells. Cell Stem Cell. 2010;**7**:249-257. DOI: 10.1016/j.stem.2010.06.015

[128] Bock C, Kiskinis E, Verstappen G, Gu H, Boulting G, Smith ZD, et al. Reference maps of human ES and iPS cell variation enable high-throughput characterization of pluripotent cell lines. Cell. 2011;**144**:439-452. DOI: 10.1016/j.cell.2010.12.032

[129] Sullivan GJ, Bai Y, Fletcher J, Wilmut I. Induced pluripotent stem cells: Epigenetic memories and practical implications. Molecular Human Reproduction. 2010;**16**:880-885. DOI: 10.1093/molehr/gaq091

[130] Newman AM, Cooper JB. Lab-specific gene expression signatures in pluripotent stem cells. Cell Stem Cell. 2010;**7**:258-262. DOI: 10.1016/j.stem.2010.06.016

[131] Allegrucci C, Young LE. Differences between human embryonic stem cell lines. Human Reproduction Update. 2007;**13**:103-120. DOI: 10.1093/humupd/dml041

[132] Osafune K, Caron L, Borowiak M, Martinez RJ, Fitz-Gerald CS, Sato Y, et al. Marked differences in differentiation propensity among human embryonic stem cell lines. Nature Biotechnology. 2008;**26**:313-315. DOI: 10.1038/nbt1383

[133] Sharova LV, Sharov AA, Piao Y, Shaik N, Sullivan T, Stewart CL, et al. Global gene expression profiling reveals similarities and differences among mouse pluripotent stem cells of different origins and strains. Developmental Biology. 2007;**307**:446-459. DOI: 10.1016/j.ydbio.2007.05.004

[134] Stadtfeld M, Apostolou E, Ferrari F, Choi J, Walsh RM, Chen T, et al. Ascorbic acid prevents loss of Dlk1-Dio3 imprinting and facilitates generation of all-iPS cell mice from terminally differentiated B cells. Nature Genetics. 2012;**44**:398-405, S1-2. DOI: 10.1038/ng.1110

[135] Hou P, Li Y, Zhang X, Liu C, Guan J, Li H, et al. Pluripotent stem cells induced from mouse somatic cells by small-molecule compounds. Science. 2013;**341**:651-654. DOI: 10.1126/science.1239278

[136] Zhao Y, Zhao T, Guan J, Zhang X, Fu Y, Ye J, et al. A XEN-like state bridges somatic cells to pluripotency during chemical reprogramming. Cell. 2015;**163**:1678-1691. DOI: 10.1016/j.cell.2015.11.017

Chapter 9

Cytosine Modifications and Distinct Functions of TET1 on Tumorigenesis

Cuili Ma, Pengfei Ji, Nina Xie and Yujing Li

Additional information is available at the end of the chapter

http://dx.doi.org/10.5772/intechopen.83709

Abstract

Vast emerging evidences are linking the base modifications and gene expression involved in essential metabolic pathways. Among the base modification markers extensively studied, 5-methylcytosine (5mC) and its oxidative derivatives (5-hydroxymethylcytosine (5-hmC), 5-formylcytosine (5-fC), and 5-carboxylcytosine (5-caC)) dynamically occur in DNA and RNA and have been acknowledged as the important epigenetic markers involved in regulation of cellular biological processes. The modification of C has been characterized biochemically, molecularly, and phenotypically, including elucidation of its methyltransferase complexes (writer), demethylases (eraser), 10-11 translocation proteins (TETs), and direct interaction proteins (readers). The levels and the landscapes of these epigenetic markers in the epitranscriptomes and epigenomes are precisely and dynamically regulated by the fine-tuned coordination of the writers and erasers in accordance with stages of the growth, development, and reproduction as naturally programmed during the life span. In mammalian genome, the TET family is consisted of three members, including TET1, TET2, and TET3. The link between aberrant modifications and diseases, such as cancers, neurodegenerative disorders, and heart diseases, has been appreciated. This review article will highlight the research advances in the writers and erasers for the modifications of cytosine in genome, as well as the dual function of TET1 in tumorigenesis as a tumor suppressor and a promoter. Additionally, the future research directions are addressed.

Keywords: methylation, demethylation, 5-methylcytosine (5-mC), 5-hydroxymethylcytosine (5-hmC), DNA methyltransferases (DNMTs), DNA demethylase, 10-11 translocation protein (TET), tumorigenesis, suppressor, promoter, 5-mC binding protein, 5-hmC binding protein

1. Introduction

Epigenetics is defined as the investigation on gene expression alterations heritable to next generations caused by nongenetic but heritable cellular memory other than DNA sequence variations [1]. The epigenetic memories including dynamic base modifications (DNA methylation/demethylation), histone modifications, chromatin architecture, and noncoding RNAs maintain all the biological processes in the programmed tracks. Any aberrant alterations could lead to development of abnormality and initiation of diseases such as neurological disorders and cancers as reviewed in [2–8]. A micro-event in base modification could lead to strong "earthquake" in the signaling pathways and the consequent alteration of organism phenotypes, even diseases. The most extensively studied modifications are methylation and demethylation of 5-cytosine (5-C).

DNA base modifications such as methylation of 5-mC [9–14] and 5-hydroxymethylcytosine (5-hmC) [15–21] have been acknowledged as the best characterized epigenetic markers in mammalian brains [20, 22–24] and ES cells [25–27], essentially regulating chromatin structure and consequently gene expression with the potential mechanisms. This review article mainly focuses on the recent advances in methylation/demethylation modifications of 5-C in mammalian genomes, including methylation/demethylation machineries, methyltransferase complexes (writers) and demethylase complexes (erasers), as well as the distinct functions of TET1 in the regulation of tumorigenesis.

2. Cytosine modifications

To maintain the normal life process, any base modification must be dynamically and tightly regulated in accordance with stages of the growth, development, and reproduction, including modification generation by methyltransferase complexes (writers), removal by demethyltransferases (erasers), as well as the preferential binding protein components (readers), to get the related epigenetic markers into the biochemical effects.

2.1. Methyltransferases of cytosine methylation

DNA methylation, particularly the most abundant CpG methylation marker 5-mC, is an essential modification of DNA in the mammalian genome, typically linked with gene silencing and involved in gene regulation, development, genome defense, and disease. A family of DNA methyltransferases named (DNMTs) is responsible for the addition of methyl groups to the 5-position of the carbon, including DNMT1, DNMT2, DNMT3a, DNMT3b, and DNMT3L. The five members are structurally and functionally distinct. The three methyltransferase enzymes DNMT1, DNMT3a, and DNMT3b serve as writers for the de novo CpG methylation pattern during embryogenesis [28, 29], while DNMT1 could confer the maintenance of parent DNA methylation patterns to the new daughter strand DNA during DNA replication [30].

Traditionally, DNMT1 was regarded as the maintenance methyltransferase copying methylation marks of hemimethylated DNA to the newly synthesized daughter strand during DNA replication, making the enzyme indispensable for dividing progenitor cells [29, 39, 40]. This is supported by the finding that DNMT1 has higher affinity to hemimethylated DNA [41, 42] and that gene knockout of Dnmt1 in the central nervous system leads to lethal in mice [43]. While Dnmt1 deletion in all dividing somatic cells is also lethal [43–47], mouse embryonic stem cells are viable, despite the resulting global loss of DNA methylation [48]. Notably, human embryonic stem cells (ESCs) also displayed a global demethylation upon Dnmt1 deletion [49].

However, in accordance with the special requirement, the DNMT1 and DNMT3A are functionally correlated. For example, in the adult brain, both methyltransferases could carry out cytosine methylation in the promoter and gene body regions, leading to transcription repression [31].

While DNMT1 is believed to function mainly for the maintenance of established patterns of DNA methylation in normal living cells, in the diseased cells such as cancer cells, DNMT1 alone is not sufficient to maintain the programmed normal gene hypermethylation. As such, the collaboration of DNMT1 and DNMT3b is indispensable for the maintenance function.

Dnmt3l is believed to function as a stimulator of the Dnmt3A and Dnmt3B, and has related function with DNMT2 [32, 36–38].

Sirt1 regulates DNA methylation and differentiation potential of embryonic stem cells by antagonizing Dnmt3l. DNMT2, a tRNA methyltransferase and the most conserved member of the DNMTs methylates tRNAs to protect them from ribonuclease digestion. More importantly, DNMT2 is functionally related to the sperm small RNA (sncRNAs) mediated essentially in writing the "paternal epigenetic signature" to sperm RNA [32]. The mechanism is that the DNMT2-conferred m5C in sncRNAs regulates the secondary structure and biological properties of sncRNAs, suggesting that sperm RNA modifications could serve as one of the carriers for paternally imprinted epigenetic memories [33].

Coordination of the DNA methylation by DNMTs as well as histone modifications contributes to the regulation of cell death through development, aging, and disease [34, 35].

2.2. Demethylation and demethylases

The dynamic DNA methylation/demethylation is tightly regulated during the whole life span. DNA demethylation, the removal of a methyl group, is not just a reverse process of methylation, but rather very complicated metabolic pathways indispensable for reactivation of genes and directly involved in pathogenesis of diseases such as cancers and neurological disorders. Either passive, active, or combination of both, leads to demethylation of DNA. The passive mechanism renders the automatic demethylation in a way that dilution and gradual loss of methylation in the newly synthesized DNA strands during successive replication rounds. In contrast, the active demethylation is believed to be the most important mechanism for active DNA demethylation via 5-mC oxidation catalyzed by the 10-11 translocation proteins (TETs) in alpha-ketoglutarate (a-KG) and Fe(II) dependent manner [22]. In addition to TETs, several other enzymes are acknowledged to be involved in the active mechanisms for demethylation, such as activation-induced cytidine deaminase (AID) [50], TET [51, 52], and thymine DNA glycosylase (TDG) [53–55].

5-hmC is generated by oxidation of 5-mC by TET, and the 5-hmC faces several fates once it is generated. First, the 5-hmC could be directly converted to regular cytosine through mechanisms involving the base excision repair pathway. Second, stepwise, a small percentage (~10%) of the 5-hmC is converted to 5-formylcytosine (5-fC) and 5-carboxylcytosine (5-caC), respectively [56, 57]. The 5-fC and 5-caC are finally converted into regular cytosine [58] with the help of Thymine-DNA glycosylase (TDG). Finally, in some tissues such as stem cells and adult neuron cells, high 5-hmC levels could be detected particularly in transcribed regions adjacent to the promoter and enhancers, positively correlating with gene expression. The low turnover rates of 5-hmC in some tissues suggest that besides serving as an intermediate of active demethylation, the stable accumulation of the 5-hmC forms a dynamic 5-hmC landscape to serve as special epigenetic markers, potentially altering the local chromatin structures via recruiting or repelling some special protein components with high affinity to or low even repellent to 5-hmC-harboring DNA [59, 60]. For example, 5-hmC loss has become a hall marker for cancer cells [61–66]. In addition, the TET members are acknowledged as the tumor suppressors as Tet gene mutations or deletions have been identified in some tumor tissues [67].

In mammalian genome, the TET family is consisted of three members, including TET1, TET2, and TET3. While all three TET members could function as hydroxylases for conversion of 5-mC to 5-hmC and further stepwise from 5-hmC to 5-fC and 5-fC to 5-caC, their functions involved in diverse biological pathways are in the development stage and specifically in tissue-dependent manners [25, 68].

2.2.1. TET1 and regulation of its target gene expression

Highly expressed in ESCs, PGCs, and inner cell mass of blastocyst, TET1 protein has been proven to be mainly responsible for the initial oxidation of 5-mC to 5-hmC, and to establish the paradoxically dual distinct epigenetic patterns in transcriptional activation and repression in accordance with life processes of growth and development. Alternative splicing mechanism leads to several TET1 isoforms, including the full-length canonical and the short transcripts [69–73]. TET1 expression is regulated by very complicated factors including the reprogramming factors such as Oct3/4, Nanog, and Myc [68, 70] in early embryos, ESCs and PGCs [69], the transcription factors in the differentiated cells, and STAT3/STAT5 in acute myeloid leukemia (AML) [74].

The full length of TET1 protein is believed to have multiple functions in regulation of gene expression. In general, TET1 catalyzes the oxidation of 5-mC to 5-hmC, which serves as an epigenetic marker and intermediate for active demethylation, leading to transcription activation. The more emerging evidence has supported the TET1 conferred transcription activation and repression of its direct target genes [75–77] at the transcriptional level. At the molecular level, the interaction between TET1 and SIN3a facilitates transcription activation of their target genes at the transcription level. More importantly, the interaction has been detected between TET1/TET2 and E26 transformation-specific or E-twenty-six (ETS) family, one of the largest transcription factor families. For example, ETS variant 2 (ETV2), an ETS family transcription factor, interacts with TET1/TET2 to recruit the demethylases to the Robo4 promoter for demethylation-mediated transcription activation during endothelial differentiation. More recently, the Methyl-CpG-binding domain (MBD) protein, such as MBD1, through its CXXC domain recruits TET1 other than TET2 and TET3 to the heterochromatin for oxidation of 5-mC to 5-hmC, whereas the resulting 5-hmC releases the MBD1 from the binding sites by affinity-based displacement [78].

On the other hand, TET1 also confers transcription repression of its target genes. It is accepted that the TET1-mediated transcription repression does not require the catalytic activity of the TET1 in conversion of 5-mC to 5-hmC, but rather the interaction between TET1 and some other protein components that contain repressor complexes [79]. Several mechanisms for TET1-mediated transcription repression have been proposed. First of all, TET1 binds a large number of polycomb target genes and interacts with SIN3A, the core component of the SIN3A co-repressor complex, leading to the transcription repression of their co-target genes via the SIN3A conferred histone deacetylation [76, 80].

The second mechanism of the TET1 conferred transcription repression is involved in TET1 interaction with recruitment of MBD repression complexes such as MBD3 [78, 81] at least in ES cells. The evidence of the mechanism includes the co-localization of TET1 and MBD3 in ESCs, higher affinity to 5-hmC than 5-mC, and association of the MBD3 knockdown with reduced level of 5-hmC as well as the enhanced expression of the 5-hmC-modified genes.

Several other mechanisms that TET1 represses the transcription have been also uncovered. It is convinced that TET1 is involved in the repression of polycomb-targeted regulator genes in accordance with the development stage by recruiting polycomb repressive complex 2 (PRC2) to the CpG-rich promoters of these genes [82]. Further study indicated requirement of the catalytic activity in oxidation of 5-mC to 5-hmC for the PRC repressive complex-mediated repression, evidenced by the fact that the PRC2 was co-localized with 5-hmC [80], while TET1 recruits the EZH2 DNMT-containing PRC complex targeting H3K27 methylation.

During the early stages of epiblast differentiation, repression of TET1 target genes was conferred by the interaction between TET1 and the JMJD8 and enhancement of the JMJD8 demethylase transcriptional repressor expression [83], but does not require the TET1 oxidation activity. Although TET1, TET2, and TET3 are all expressed in gonadotrope-precursor cells, the TET1 expression was dramatically decreased in the differentiated cells. Differentiation with according increase in the expression of the luteinizing hormone gene (Lhb). The short isoform of TET1 with deletion of the N-terminal CXXC-domain binds the H3K27me2/3 enriched region located at the upstream promoter of the Lhb gene, downregulating its expression and leading to differentiation deficiency [73].

3. Distinct functions of TET1 on tumorigenesis

3.1. *Tet1* functions as an oncogene in some cancers

Initially, given the mutations and the deletions as predominant variation of TET proteins, particularly TET1, in human cancer genomes, it was accepted that TET1 functions as a tumor suppressor [61, 65, 66]. Indeed, TET1 and TET3 bear the predominant mutations in some tumors including colorectal cancer, melanoma, and cutaneous squamous cell carcinoma [88–90]. However, emerging evidences are connecting the TET1 overexpression and tumorigenesis as well, most likely attributed to activation of cancer-specific oncogenic pathways mediated by TET1 conferred hypomethylation [72, 84] (**Figures 1** and **2**).

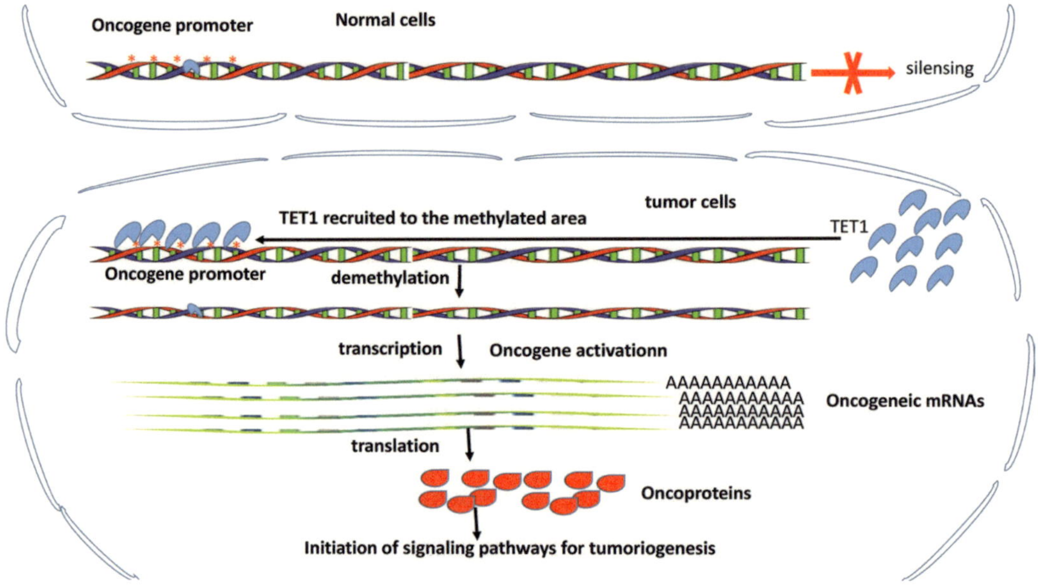

Figure 1. TET1 functions as a tumor promoter by activation of the oncogenes via demethylation of the methylated promoter regions in the oncogenes. (A) In normal cells, the promoter regions of the oncogenes are usually methylated and therefore silenced. (B) However, in some cells, TET1 is highly expressed and recruited by its interaction partners to the methylated promoter regions of the oncogenes, leading to demethylation and the activation of the oncogenes. Consequently, the oncoproteins initiate tumorigenesis.

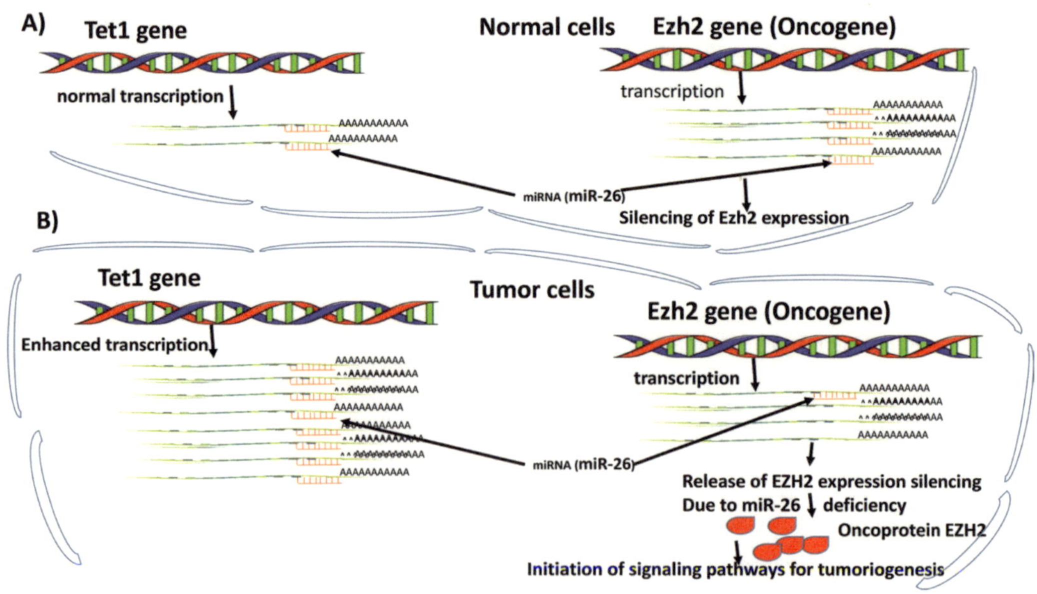

Figure 2. Sequestration of miR-26 by its target 3′UTRs of Tet1 leads to miR-26 deficiency to target its target Ezh2, an oncogene. (A) In normal cells, due to low level of Tet1 expression, majority of the miR-26 targets to Ezh2 leads to sufficient silencing of the oncogene. (B) In some cancer cells, dramatically enhanced transcription of Tet1 sequestrates the miR-26, conferring the miR-26 deficiency to target its Ezh2 target 3′UTRs. Consequently, miR-26 deficiency to the Ezh2 releases the miRNA-mediated expression repression of the oncogenes, conferring the initiation of tumorigenesis.

3.1.1. TET1 demethylation associated activation of the members in the oncogenic pathways

TET1 overexpression accounts for about 40% of patients with triple-negative breast cancer (TNBC) that belongs to the most hypomethylated cancers observed, leading to about 10% hypomethylation of the queried CGI and activation of oncogenic pathways including PI3K, EGFR, and PDGF. Thus, TET1 seems functioning as a potential oncogene and could serve as a target for intervention therapy [84]. This phenomenon was observed not only in NTBC, but also in MLL-rearranged leukemia where TET1 is believed to activate the downstream oncogenic pathways by its demethylase activity, serving as an oncogene [86]. Additionally, via DNA hypomethylation, TET1 was demonstrated to regulate the expression of MUC4, one member of the mucin (MUC) family and an essential factor for carcinogenesis and tumor invasion in lung neoplasms, functioning as the potential oncogene [86, 87].

TET1 functions as an important oncoprotein in acute myeloid leukemia (AML) as evidenced by the high level expression of TET1 in AML, indicating that efficient inhibition of TET1 expression could serve as a powerful strategy for AML therapy. Drug screening led to identification of two compounds NSC-370284 and its structure analogue UC-514321, which repress TET1 transcription by targeting directly to target STAT3/5, TET1 transcriptional activators, suggesting the potential of the compounds targeting the STAT/TET1 for efficient therapy of AML [74].

Full length TET1 (TET1FL) has a CXXC domain that binds to unmethylated CpG islands (CGIs), allowing TET1 to protect CGIs from aberrant methylation and limiting its ability to regulate genes outside of CGIs. An isoform of TET1 (TET1ALT) without CXXC domain but still with catalytic domain is repressed in ES cells while it is activated in embryonic and adult tissues in contrast to TET1FL's expression in ESCs and repression in adult tissues. TET1ALT aberrant activation is detected in breast cancer, uterine and ovarian cancer, and glioblastoma, leading to worse overall survival in these types of cancers. As for the pathogenesis mediated by the TET1ALT isoform, a predominantly activated isoform of TET1 in cancer cells does not protect from CGI methylation but likely mediates dynamic site-specific demethylation outside of CGIs.

3.1.2. Hypoxia induced promotion of TET1 expression

Enhanced expression of TET1 by hypoxia induction has been reported to upregulate cancer cell migration, invasion, and proliferation via the HIF1α signaling pathway in JEG3 cells [100], suggesting the oncogenic function of TET1 under hypoxia condition.

3.1.3. Overexpression of Tet1 mRNA 3′UTRs leads to sequestration of miRNAs, which target the oncogenic transcripts as well, leading to miRNA deficiency to target the oncogenic transcript

Transcription levels of TETs were significantly elevated while the protein levels were not in gastric cancer (GC) tissues compared to the adjacent normal tissues, suggesting the essential role(s) of the endogenous TET transcripts in gastric carcinogenesis and prognosis. Further study showed that overexpression of 5′UTRs, CDs, and 3′UTRs contributed to varied effects in a way that overexpression of TET 3′UTRS promoted GC growth and proliferation. Given that miR-26 targets 3′UTRs of both TET1 and EZH2 mRNAs, overexpression of TET members mRNA sequestrates miR-26 competitively and leads to release of the miR-26 mediated repression of EZH2. Thus, activation of EZH2 expression facilitates gastric carcinogenesis and progression [87] (**Figure 2**).

3.2. TET1 serves as a tumor suppressor

The pathogenic contributions the TET members made in various human cancers by functioning as tumor suppressors or promoters have been proven to be versatile. The hypermethylation-based transcriptional silencing of TET1 is frequently detected in non-Hodgkin B cell lymphoma (B-NHL), suggesting TET1 as a tumor suppressor of hematopoietic malignancy [91]. Similarly, TET1 is downregulated upon NF-κB activation in multiple cancers including basal-like breast cancer (BLBC), melanoma, lung, and thyroid cancers, demonstrating that TET1 is the tumor suppressor that relies on involvement of the immune system [92].

3.2.1. TET1 methylation-mediated activation of tumor suppressor genes

It is acknowledged that 5hmC depletion initiates carcinogenesis caused by either TET1 expression repression or aberrant localization. Significantly lower 5-hmC and TET1 expression level and subcellular mislocalization in gastric cancer tissues demonstrate the crucial role of TET1 as a cancer repressor [97] (**Figure 3**).

In the tested epithelial ovarian cancer (EOC), undetected TET1 expression suggests that the consequence of TET1 repression induces the tumorigenesis, in accordance with the inhibition

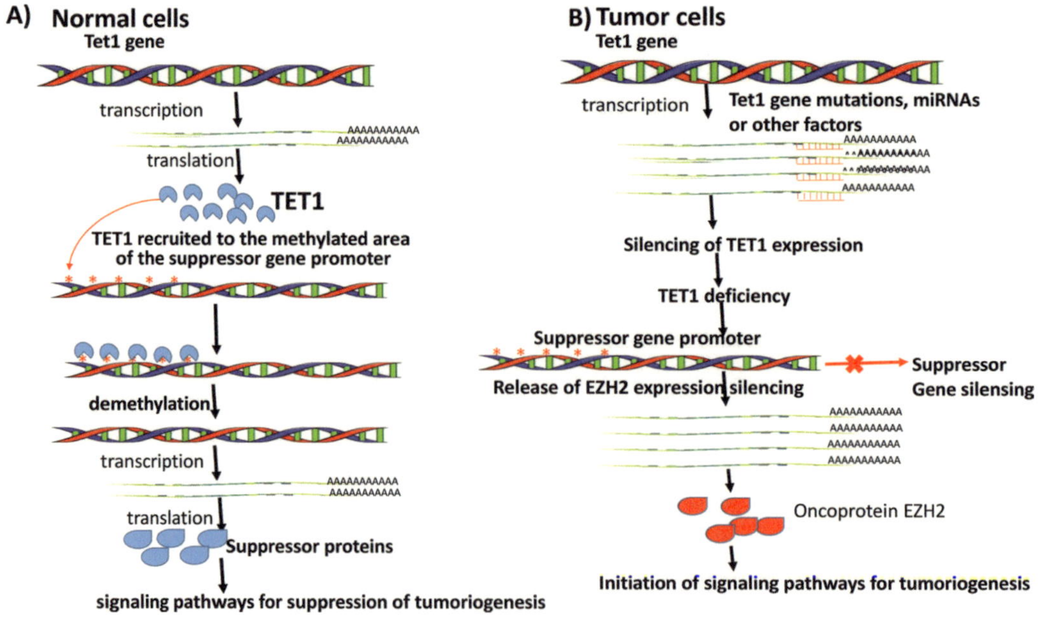

Figure 3. TET1 functions as a tumor suppressor. (A) In normal cells, TET1 expression is maintained at a regular level. The TET1 is recruited to the methylated promoter area of the tumor suppressor genes for demethylation so that the tumor suppressor could be expressed at the normal levels. (B) Due to mutation of the Tet1 genes or some other factors, Tet1 expression silenced, leading to silencing of the suppressor genes and activation of the oncogenes such as Ezh2. Expression of the oncoproteins initiates the tumorigenesis or enhances cancer cell growth and metastasis.

of colony formation, cell migration, and invasion by ectopic expression of TET1 in SKOV3 and OVCAR3 cells. The potential mechanism is the TET1 conferred demethylation and the consequent activation of the expression of two key proteins SFRP2 and DKK1 in the canonical Wnt/β-catenin signaling pathway, associated with inhibition of EMT and metastasis [101].

TET1 is identified as a key tumor suppressor player in ovarian cancer cell lines as well by demethylating a CpG site within the Ras association domain family member 5 (RASSF5) promoter to enhance expression of the RASSF5, leading to the growth inhibition of ovarian cancer cells [102].

More evidences show that EGFR-mediated TET1 repression induces silencing of tumor suppressors in cancer cells such as lung adenocarcinomas and glioblastomas. If only the oncogenic EGFR expression is inhibited, TET1 could bind to promoters of the tumor suppressors to activate their expression via DNA demethylation. TET1 overexpression inhibits lung and glioblastoma tumor growth, and vice versa, in agreement with the significant decrease in TET1 expression or TET1 cytoplasmic localization in the majority of lung cancer samples. Thus, it is plausible to speculate that TET1 may serve as the therapeutic target for oncogenic EGFR-induced lung cancers and glioblastomas [93]. However, Lai et al. could not draw the same conclusion in human NSCLC patient samples. They did not detect the EGFR-mediated TET1 silencing, but rather observed the significant elevation of the TET1 expression levels in patient samples with EGFR mutations, suggesting the inconclusiveness in EGFR-mediated TET1 silencing among the cellular and animal models and human lung cancer patients [94].

Eicosapentaenoic acid (EPA), one of the major polyunsaturated fatty acids, could enhance the formation of PPARγ-RXRα-TET1 to recruit TET1 to a hypermethylated CpG island on the p21 gene for rapid demethylation and consequent expression of p21Waf1/Cip1, leading to inhibition of cancer cell-cycle progression in hepatocarcinoma cells. This suggests the bridge requirement for TET1 exerting the anti-tumor function and potential of EPA for solid tumor therapy such as live cancer [97].

3.2.2. TET1 silencing and loss of 5-hmC induces initiation of tumors

Loss of 5-hydroxymethylcytosine (5 hmC) caused by TET1 dysfunction could induce tumor initiation and enhance malignancy by promoting cancer cell growth, migration, and invasion in DLD1 colon cancer cells mediated by EZH2 [96]. With loss of TET1, EZH2 repression is released, but H3K27 demethylase UTX-1 expression is repressed, enhancing histone H3K27 tri-methylation and consequently repressing the target gene E-cadherin (DH1). Accordingly, even at the condition of TET1 deficiency, either the H3K27 demethylase UTX-1 overexpression or EZH2 depletion both could enhance H3K27 demethylation at CDH1 promoter, thereby impeding EMT and tumor invasion. Likewise, either EZH2 overexpression or UTX-1 depletion both could promote EMT and tumor metastasis in DLD1 cells. Thus, these results elucidate regulation interplay among TET1, E-cadherin, and EZH2 and indicate the critical mediator role the EZH2 plays in the E-cadherin repression and tumor progression [95].

3.2.3. miRNA-mediated repression of TET1 expression

Some miRNAs are identified to be involved in regulation of cancer progression or repression, and one of the mechanisms refers to the oncogenic miRNA-mediated TET1 repression and the consequent loss of 5-hmC. Indeed, miR-21-5p has been confirmed to target Tet1 in colorectal cancer (CRC), serving as a biomarker for diagnostics and prognostics in CRC [98]. Similarly, miR-4284 directly targeting Tet1 mRNA downregulates TET1 levels of both mRNA and protein in human gastric cancer SGC-7901 cells, and thus serves as an oncogenic marker [99], suggesting that miR-4284 could provide a potential target for gastric cancer therapy.

3.2.4. Dual function of miRNA by interaction with Tet mRNA 3′UTRs

Some miRNAs are reported to function as both a suppressor and a promoter in some cancers such as miR29b in breast cancer (BC) cells by regulation of BC cell proliferation, metastasis, and epithelial-mesenchymal transition (EMT). Significantly decreased expression of miR-29b in BC samples and cell lines suggests the role of TET1 as a BC suppressor. However, miR-29b overexpression promotes cell proliferation, colony formation, migration, and EMT, indicating that miR-29b functions as a BC promoter [103]. In vitro assay TET1 has been identified as one of the miR-29b targets, and it turns out that overexpression of miR29b leads to TET1 downregulation-mediated promotion of proliferation, colony formation, invasion, and EMT in GC cells such as MDA-MB-231 and MCF-7. Further study showed that the TET1-mediated suppression of the BC attributed to TET1 conferred disruption of ZEB2 expression by binding to the promoter of ZEB2. While the miR-29b/TET1/ZEB2 pathway offers understanding for the mechanism of miR-29b and TET1-mediated BC promotion, the suppression mechanism for TET1 remains to be elusive in GC [104].

4. Concluding remarks

In the past decades, particularly recent years, significant achievements have been made in epigenetic study particularly 5-mC and its derivatives such as 5-hmC, 5-fC, and 5-caC, in understanding the generation, dynamic alteration, machinery, distribution, and biological functions and connection between the modifications and the pathogenesis of diseases such as neurological disorders and cancers. However, a large number of unknown epigenetic events related to pathogenesis of many diseases particularly cancers remain to be elusive. Although the individual members of the methyltransferase complexes (writer) for cytosine modifications have been characterized, their coordination in conducting the methylation in response to tumorigenesis has not yet been comprehensively investigated. Similarly, the functional study on the TET proteins (the erasers for methylation) stays only at the conversion of 5-mC to 5-hmC, identification of the targeting miRNAs, and identification of serving as tumor suppressors or promoters by several known mechanisms. However, it is logical to speculate that as such huge protein molecules, TET proteins may have much more unidentified functions. Further study on the unknown functions will provide essential information for dissecting the cancer pathogenesis. First, only limited information is available for the physical interaction components of the TETs; identification of the TET interaction proteins may help us better understand how and where the TETs are recruited to function as demethylase to maintain the dynamic balance of 5-mC/5-hmC and the chromatin remodeling. Then, identification of other functions of TETs other than demethylase will be of importance. Given that the 5-hmC is not so much serving an intermediate of demethylation as the important dynamic 5-hmC landscape, it is essential to investigate how the epigenetic information stored in the landscape is transformed into the biological effect. To this end, for identification of the readers of the 5-hmC modification, the specific 5-hmC binding proteins might be the prerequisite. A better understanding the functions of methyltransferase complex for cytosine methylation, TETs for demethylation of 5-mC and interaction protein components as well as the other known functions, and the specific readers of the 5-hmC marker could identify some epigenetic components for therapeutic targets for treatments of cancers and other diseases such as neurological disorders.

It has been reported that Tet1 alternative splicing forms have distinct functions [74]. However, the information regarding the Tet1 alternative splicing is still limited. Further alternative splicing study may identify more unknown functions conferred by the different isoforms which may bear the potential for therapeutic targets.

Additionally, the chemical biology approach based on further identification of small molecule compounds that target the 5-mC/5-hmC machineries or the signaling pathways in which 5-mC/5-hmC involved could help explore therapeutic targets for some stubborn diseases such as cancers and neurological diseases.

Author details

Cuili Ma[1], Pengfei Ji[2], Nina Xie[3] and Yujing Li[4*]

*Address all correspondence to: yli29@emory.edu

1 Department of Internal Medicine, Qingdao Agricultural University Associated Hospital, Qingdao, Shandong, PR China

2 Department of Nutrition and Food Science, Texas A&M University, College Station, TX, USA

3 Department of Neurology of Xiangya Hospital, Central South University, Changsha, Hunan, China

4 Department of Human Genetics, Emory University School of Medicine, Atlanta, GA, USA

References

[1] Dupont C, Armant DR, Brenner CA. Epigenetics: Definition, mechanisms and clinical perspective. Seminars in Reproductive Medicine. 2009;**27**(5):351-357. DOI: 10.1055/s-0029-1237423

[2] Soubry A. Epigenetics as a driver of developmental origins of health and disease: Did we forget the Fathers? BioEssays. 2018;**40**(1):1700113. DOI: 10.1002/bies.201700113

[3] Joshita S, Umemura T, Tanaka E, Ota M. Genetics and epigenetics in the pathogenesis of primary biliary cholangitis. Clinical Journal of Gastroenterology. 2018;**11**(1):11-18. DOI: 10.1007/s12328-017-0799-z

[4] Lockwood LE, Youssef NA. Systematic review of epigenetic effects of pharmacological agents for bipolar disorders. Brain Sciences. 2017;**7**(11):E154. DOI: 10.3390/brainsci7110154

[5] Uchida S, Yamagata H, Seki T, Watanabe Y. Epigenetic mechanisms of major depression: Targeting neuronal plasticity. Psychiatry and Clinical Neurosciences. 2018;**72**(4):212-227. DOI: 10.1111/pcn.12621

[6] Cheng Z, Zheng L, Almeida FA. Epigenetic reprogramming in metabolic disorders: Nutritional factors and beyond. The Journal of Nutritional Biochemistry. 2018;**54**:1-10. DOI: 10.1016/j.jnutbio.2017.10.004

[7] Flotho C, Sommer S, Lübbert M. DNA-hypomethylating agents as epigenetic therapy before and after allogeneic hematopoietic stem cell transplantation in myelodysplastic syndromes and juvenile myelomonocytic leukemia. Seminars in Cancer Biology. 2018;**51**: 68-79. DOI: 10.1016/j.semcancer.2017.10.011

[8] Neal M, Richardson JR. Epigenetic regulation of astrocyte function in neuroinflammation and neurodegeneration. Biochimica et Biophysica Acta. 2018;**1864**(2):432-443. DOI: 10.1016/j.bbadis.2017.11.004

[9] Ginder GD, Williams DC Jr. Readers of DNA methylation, the MBD family as potential therapeutic targets. Pharmacology & Therapeutics. 2018;**184**:98-111. DOI: 10.1016/j.pharmthera.2017.11.002

[10] Fransquet PD, Lacaze P, Saffery R, McNeil J, Woods R, Ryan J. Blood DNA methylation as a potential biomarker of dementia: A systematic review. Alzheimer's & Dementia. 2018;**14**(1):81-103. DOI: 10.1016/j.jalz.2017.10.002

[11] Antoniani C, Romano O, Miccio A. Concise review: Epigenetic regulation of hematopoiesis: Biological insights and therapeutic applications. Stem Cells Translational Medicine. 2017;**6**(12):2106-2114. DOI: 10.1002/sctm.17-0192

[12] He X, Ou C, Xiao Y, Han Q, Li H, Zhou S. LncRNAs: key players and novel insights into diabetes mellitus. Oncotarget. 2017;**8**(41):71325-71341. DOI: 10.18632/oncotarget.19921

[13] Christopher MA, Kyle SM, Katz DJ. Neuroepigenetic mechanisms in disease. Epigenetics & Chromatin. 2017;**10**(1):47. DOI: 10.1186/s13072-017-0150-4

[14] Schmauss C. The roles of class I histone deacetylases (HDACs) in memory, learning, and executive cognitive functions: A review. Neuroscience and Biobehavioral Reviews. 2017;**83**:63-71. DOI: 10.1016/j.neubiorev.2017.10.004

[15] Wyatt GR, Cohen SS. The bases of the nucleic acids of some bacterial and animal viruses: The occurrence of 5-hydroxymethylcytosine. The Biochemical Journal. 1953;**55**(5):774-782

[16] Cohen SS, Weed LL. Some precursors of the 5-hydroxymethylcytosine of T6r+ bacteriophage. The Journal of Biological Chemistry. 1954;**209**(2):789-794

[17] Anisymova NI, Gabrilovich IM, Soshina NV, Cherenkevich SN. 5-Hydroxymethylcytosine-containing Klebsiella bacteriophage. Biochimica et Biophysica Acta. 1969;**190**(1):225-227

[18] Wiberg JS. Amber mutants of bacteriophage T4 defective in deoxycytidine diphosphatase and deoxycytidine triphosphatase. On the role of 5-hydroxymethylcytosine in bacteriophage deoxyribonucleic acid. The Journal of Biological Chemistry. 1967;**242**(24):5824-5829

[19] Wyatt GR, Cohen SS. A new pyrimidine base from bacteriophage nucleic acids. Nature. 1952;**170**(4338):1072-1073

[20] Bird AP, Wolffe AP. Methylation-induced repression—belts, braces, and chromatin. Cell. 1999;**99**(5):451-454

[21] Klose RJ, Bird AP. Genomic DNA methylation: The mark and its mediators. Trends in Biochemical Sciences. 2006;**31**(2):89-97

[22] Tahiliani M, Koh KP, Shen Y, Pastor WA, Bandukwala H, Brudno Y, et al. Conversion of 5-methylcytosine to 5-hydroxymethylcytosine in mammalian DNA by MLL partner TET1. Science. 2009;**324**(5929):930-935. DOI: 10.1126/science.1170116

[23] Guo JU, Szulwach KE, Su Y, Li Y, Yao B, Xu Z, et al. Genome-wide antagonism between 5-hydroxymethylcytosine and DNA methylation in the adult mouse brain. Frontiers in Biology. 2014;**9**(1):66-74

[24] Bernstein AI, Lin Y, Street RC, Lin L, Dai Q, Yu L, et al. 5-Hydroxymethylation-associated epigenetic modifiers of Alzheimer's disease modulate Tau-induced neurotoxicity. Human Molecular Genetics. 2016;**25**(12):2437-2450

[25] Ito S, D'Alessio AC, Taranova OV, Hong K, Sowers LC, Zhang Y. Role of Tet proteins in 5mC to 5hmC conversion, ES-cell self-renewal and inner cell mass specification. Nature. 2010;**466**(7310):1129-1133. DOI: 10.1038/nature09303

[26] Wu H, D'Alessio AC, Ito S, Wang Z, Cui K, Zhao K, et al. Genome-wide analysis of 5-hydroxymethylcytosine distribution reveals its dual function in transcriptional regulation in mouse embryonic stem cells. Genes & Development. 2011;**25**(7):679-684. DOI: 10.1101/gad.2036011

[27] Shih AH, Abdel-Wahab O, Patel JP, Levine RL. The role of mutations in epigenetic regulators in myeloid malignancies. Nature Reviews. Cancer. 2012;**12**(9):599-612. DOI: 10.1038/nrc3343

[28] Han L, Witmer PD, Casey E, Valle D, Sukumar S. DNA methylation regulates MicroRNA expression. Cancer Biology & Therapy. 2007;**6**(8):1284-1288

[29] Hermann A, Goyal R, Jeltsch A. The Dnmt1 DNA-(cytosine-C5)-methyltransferase methylates DNA processively with high preference for hemimethylated target sites. The Journal of Biological Chemistry. 2004;**279**(46):48350-48359

[30] Hashimshony T, Zhang J, Keshet I, Bustin M, Cedar H. The role of DNA methylation in setting up chromatin structure during development. Nature Genetics. 2003;**34**(2):187-192

[31] Jang HS, Shin WJ, Lee JE, Do JT. CpG and Non-CpG Methylation in Epigenetic Gene Regulation and Brain Function. Genes (Basel). 2017;**8**(6):E148. DOI: 10.3390/genes8060148

[32] Heo J, Lim J, Lee S, Jeong J, Kang H, Kim Y, et al. Sirt1 regulates DNA methylation and differentiation potential of embryonic stem cells by antagonizing Dnmt3l. Cell Reports. 2017;**18**(8):1930-1945. DOI: 10.1016/j.celrep.2017.01.074

[33] Zhang Y, Zhang X, Shi J, Tuorto F, Li X, Liu Y, et al. Dnmt2 mediates intergenerational transmission of paternally acquired metabolic disorders through sperm small non-coding RNAs. Nature Cell Biology. 2018;**20**(5):535-540. DOI: 10.1038/s41556-018-0087-2

[34] Fagiolini M, Jensen CL, Champagne FA. Epigenetic influences on brain development and plasticity. Current Opinion in Neurobiology. 2009;**19**(2):207-212. DOI: 10.1016/j.conb.2009.05.009

[35] Akbarian S, Beeri MS, Haroutunian V. Epigenetic determinants of healthy and diseased brain aging and cognition. JAMA Neurology. 2013;**70**(6):711-718. DOI: 10.1001/jamaneurol.2013.1459

[36] Kareta MS, Botello ZM, Ennis JJ, Chou C, Chédin F. Reconstitution and mechanism of the stimulation of de novo methylation by human DNMT3L. The Journal of Biological Chemistry. 2006;**281**(36):25893-25902

[37] Niles KM, Yeh JR, Chan D, Landry M, Nagano MC, Trasler JM. Haploinsufficiency of the paternal-effect gene Dnmt3L results in transient DNA hypomethylation in progenitor cells of the male germline. Human Reproduction. 2013;**28**(2):519-530. DOI: 10.1093/humrep/des395

[38] Tajima S, Suetake I, Takeshita K, Nakagawa A, Kimura H. Domain structure of the Dnmt1, Dnmt3a, and Dnmt3b DNA methyltransferases. Advances in Experimental Medicine and Biology. 2016;**945**:63-86

[39] Hermann A, Gowher H, Jeltsch A. Biochemistry and biology of mammalian DNA methyltransferases. Cellular and Molecular Life Sciences. 2004;**61**(19-20):2571-2587

[40] Hirasawa R, Chiba H, Kaneda M, Tajima S, Li E, Jaenisch R, et al. Maternal and zygotic Dnmt1 are necessary and sufficient for the maintenance of DNA methylation imprints during preimplantation development. Genes & Development. 2008;**22**(12):1607-1616. DOI: 10.1101/gad.1667008

[41] Bashtrykov P, Ragozin S, Jeltsch A. Mechanistic details of the DNA recognition by the Dnmt1 DNA methyltransferase. FEBS Letters. 2012;**586**(13):1821-1823. DOI: 10.1016/j.febslet.2012.05.026

[42] Bashtrykov P, Jankevicius G, Smarandache A, Jurkowska RZ, Ragozin S, Jeltsch A. Specificity of Dnmt1 for methylation of hemimethylated CpG sites resides in its catalytic domain. Chemistry & Biology. 2012;**19**(5):572-578. DOI: 10.1016/j.chembiol.2012.03.010

[43] Fan G, Beard C, Chen RZ, Csankovszki G, Sun Y, Siniaia M, et al. DNA hypomethylation perturbs the function and survival of CNS neurons in postnatal animals. The Journal of Neuroscience. 2001;**21**(3):788-797

[44] Li E, Bestor TH, Jaenisch R. Targeted mutation of the DNA methyltransferase gene results in embryonic lethality. Cell. 1992;**69**(6):915-926

[45] Jackson-Grusby L, Beard C, Possemato R, Tudor M, Fambrough D, Csankovszki G, et al. Loss of genomic methylation causes p53-dependent apoptosis and epigenetic deregulation. Nature Genetics. 2001;**27**(1):31-39

[46] Trowbridge JJ, Snow JW, Kim J, Orkin SH. DNA methyltransferase 1 is essential for and uniquely regulates hematopoietic stem and progenitor cells. Cell Stem Cell. 2009;**5**(4): 442-449. DOI: 10.1016/j.stem.2009.08.016

[47] Sen GL, Reuter JA, Webster DE, Zhu L, Khavari PA. DNMT1 maintains progenitor function in self-renewing somatic tissue. Nature. 2010;**463**(7280):563-567. DOI: 10.1038/nature08683

[48] Tsumura A, Hayakawa T, Kumaki Y, Takebayashi S, Sakaue M, Matsuoka C, et al. Maintenance of self-renewal ability of mouse embryonic stem cells in the absence of DNA methyltransferases Dnmt1, Dnmt3a and Dnmt3b. Genes to Cells. 2006;**11**(7):805-814

[49] Liao J, Karnik R, Gu H, Ziller MJ, Clement K, Tsankov AM, et al. Targeted disruption of DNMT1, DNMT3A and DNMT3B in human embryonic stem cells. Nature Genetics. 2015;**47**(5):469-478. DOI: 10.1038/ng.3258

[50] Bhutani N, Brady JJ, Damian M, Sacco A, Corbel SY, Blau HM. Reprogramming towards pluripotency requires AID-dependent DNA demethylation. Nature. 2010;**463**(7284): 1042-1047. DOI: 10.1038/nature08752

[51] Costa Y, Ding J, Theunissen TW, Faiola F, Hore TA, Shliaha PV, et al. NANOG-dependent function of TET1 and TET2 in establishment of pluripotency. Nature. 2013;**495**(7441): 370-374. DOI: 10.1038/nature11925

[52] Pastor WA, Aravind L, Rao A. TETonic shift: biological roles of TET proteins in DNA demethylation and transcription. Nature Reviews. Molecular Cell Biology. 2013;**14**(6):341-356. DOI: 10.1038/nrm3589

[53] Gao Y, Chen J, Li K, Wu T, Huang B, Liu W, et al. Replacement of Oct4 by Tet1 during iPSC induction reveals an important role of DNA methylation and hydroxymethylation in reprogramming. Cell Stem Cell. 2013;**12**(4):453-469. DOI: 10.1016/j.stem.2013.02.005

[54] Chen J, Guo L, Zhang L, Wu H, Yang J, Liu H, et al. Vitamin C modulates TET1 function during somatic cell reprogramming. Nature Genetics. 2013;**45**(12):1504-1509. DOI: 10.1038/ng.2807

[55] Bagci H, Fisher AG. DNA demethylation in pluripotency and reprogramming: The role of tet proteins and cell division. Cell Stem Cell. 2013;**13**(3):265-269. DOI: 10.1016/j.stem.2013.08.005

[56] He YF, Li BZ, Li Z, Liu P, Wang Y, Tang Q, et al. Tet-mediated formation of 5-carboxylcytosine and its excision by TDG in mammalian DNA. Science. 2011;**333**(6047):1303-1307. DOI: 10.1126/science.1210944

[57] Ito S, Shen L, Dai Q, Wu SC, Collins LB, Swenberg JA, et al. Tet proteins can convert 5-methylcytosine to 5-formylcytosine and 5-carboxylcytosine. Science. 2011;**333**(6047): 1300-1303. DOI: 10.1126/science.1210597

[58] Song CX, Szulwach KE, Dai Q, Fu Y, Mao SQ, Lin L, et al. Genome-wide profiling of 5-formylcytosine reveals its roles in epigenetic priming. Cell. 2013;**153**(3):678-691. DOI: 10.1016/j.cell.2013.04.001

[59] Hashimoto H, Liu Y, Upadhyay AK, Chang Y, Howerton SB, Vertino PM, et al. Recognition and potential mechanisms for replication and erasure of cytosine hydroxymethylation. Nucleic Acids Research. 2012;**40**(11):4841-4849. DOI: 10.1093/nar/gks155

[60] Mellén M, Ayata P, Dewell S, Kriaucionis S, Heintz N. MeCP2 binds to 5hmC enriched within active genes and accessible chromatin in the nervous system. Cell. 2012;**151**(7): 1417-1430. DOI: 10.1016/j.cell.2012.11.022

[61] Yang H, Liu Y, Bai F, Zhang JY, Ma SH, Liu J, et al. Tumor development is associated with decrease of TET gene expression and 5-methylcytosine hydroxylation. Oncogene. 2013;**32**(5):663-669. DOI: 10.1038/onc.2012.67

[62] Moran-Crusio K, Reavie L, Shih A, Abdel-Wahab O, Ndiaye-Lobry D, Lobry C, et al. Tet2 loss leads to increased hematopoietic stem cell self-renewal and myeloid transformation. Cancer Cell. 2011;**20**(1):11-24. DOI: 10.1016/j.ccr.2011.06.001

[63] Ko M, Bandukwala HS, An J, Lamperti ED, Thompson EC, Hastie R, et al. Ten-Eleven-Translocation 2 (TET2) negatively regulates homeostasis and differentiation of hematopoietic stem cells in mice. Proceedings of the National Academy of Sciences of the United States of America. 2011;**108**(35):14566-14571. DOI: 10.1073/pnas.1112317108

[64] An J, González-Avalos E, Chawla A, Jeong M, López-Moyado IF, Li W, et al. Acute loss of TET function results in aggressive myeloid cancer in mice. Nature Communications. 2015;**6**:10071. DOI: 10.1038/ncomms10071

[65] Haffner MC, Chaux A, Meeker AK, Esopi DM, Gerber J, Pellakuru LG, et al. Global 5-hydroxymethylcytosine content is significantly reduced in tissue stem/progenitor cell compartments and in human cancers. Oncotarget. 2011;**2**(8):627-637

[66] Kudo Y, Tateishi K, Yamamoto K, Yamamoto S, Asaoka Y, Ijichi H, et al. Loss of 5-hydroxymethylcytosine is accompanied with malignant cellular transformation. Cancer Science. 2012;**103**(4):670-676. DOI: 10.1111/j.1349-7006.2012.02213.x

[67] Cimmino L, Aifantis I. Alternative roles for oxidized mCs and TETs. Current Opinion in Genetics & Development. 2017;**42**:1-7. DOI: 10.1016/j.gde.2016.11.003

[68] Koh KP, Yabuuchi A, Rao S, Huang Y, Cunniff K, Nardone J, et al. Tet1 and Tet2 regulate 5-hydroxymethylcytosine production and cell lineage specification in mouse embryonic stem cells. Cell Stem Cell. 2011;**8**(2):200-213. DOI: 10.1016/j.stem.2011.01.008

[69] Zhang W, Xia W, Wang Q, Towers AJ, Chen J, Gao R, et al. Isoform switch of TET1 regulates DNA demethylation and mouse development. Molecular Cell. 2016;**64**(6): 1062-1073. DOI: 10.1016/j.molcel.2016.10.030

[70] Neri F, Dettori D, Incarnato D, Krepelova A, Rapelli S, Maldotti M, et al. TET1 is a tumour suppressor that inhibits colon cancer growth by derepressing inhibitors of the WNT pathway. Oncogene. 2015;**34**(32):4168-4176. DOI: 10.1038/onc.2014.356

[71] Sohni A, Bartoccetti M, Khoueiry R, Spans L, Vande Velde J, De Troyer L, et al. Dynamic switching of active promoter and enhancer domains regulates Tet1 and Tet2 expression during cell state transitions between pluripotency and differentiation. Molecular and Cellular Biology. 2015;**35**(6):1026-1042. DOI: 10.1128/MCB.01172-14

[72] Good CR, Madzo J, Patel B, Maegawa S, Engel N, Jelinek J, et al. A novel isoform of TET1 that lacks a CXXC domain is overexpressed in cancer. Nucleic Acids Research. 2017;**45**(14):8269-8281. DOI: 10.1093/nar/gkx435

[73] Yosefzon Y, David C, Tsukerman A, Pnueli L, Qiao S, Boehm U, et al. An epigenetic switch repressing Tet1 in gonadotropes activates the reproductive axis. Proceedings of the National Academy of Sciences of the United States of America. 2017;**114**(38): 10131-10136. DOI: 10.1073/pnas.1704393114

[74] Jiang X, Hu C, Ferchen K, Nie J, Cui X, Chen CH, et al. Targeted inhibition of STAT/TET1 axis as a therapeutic strategy for acute myeloid leukemia. Nature Communications. 2017;**8**(1):2099. DOI: 10.1038/s41467-017-02290-w

[75] Wu H, Zhang Y. Tet1 and 5-hydroxymethylation: a genome-wide view in mouse embryonic stem cells. Cell Cycle. 2011;**10**(15):2428-2436

[76] Williams K, Christensen J, Pedersen MT, Johansen JV, Cloos PA, Rappsilber J, et al. TET1 and hydroxymethylcytosine in transcription and DNA methylation fidelity. Nature. 2011;**473**(7347):343-348. DOI: 10.1038/nature10066

[77] Xu Y, Wu F, Tan L, Kong L, Xiong L, Deng J, et al. Genome-wide regulation of 5hmC, 5mC, and gene expression by Tet1 hydroxylase in mouse embryonic stem cells. Molecular Cell. 2011;**42**(4):451-464. DOI: 10.1016/j.molcel.2011.04.005

[78] Zhang P, Rausch C, Hastert FD, Boneva B, Filatova A, Patil SJ, et al. Methyl-CpG binding domain protein 1 regulates localization and activity of Tet1 in a CXXC3 domain-dependent manner. Nucleic Acids Research. 2017;**45**(12):7118-7136. DOI: 10.1093/nar/gkx281

[79] Melamed P, Yosefzon Y, David C, Tsukerman A, Pnueli L. Tet enzymes, variants, and differential effects on function. Frontiers in Cell and Development Biology. 2018;**6**:22. DOI: 10.3389/fcell.2018.00022

[80] Neri F, Incarnato D, Krepelova A, Rapelli S, Pagnani A, Zecchina R, et al. Genome-wide analysis identifies a functional association of Tet1 and Polycomb repressive complex 2 in mouse embryonic stem cells. Genome Biology. 2013;**14**(8):R91. DOI: 10.1186/gb-2013-14-8-r91

[81] Yildirim O, Li R, Hung JH, Chen PB, Dong X, Ee LS, et al. Mbd3/NURD complex regulates expression of 5-hydroxymethylcytosine marked genes in embryonic stem cells. Cell. 2011;**147**(7):1498-1510. DOI: 10.1016/j.cell.2011.11.054

[82] Wu H, D'Alessio AC, Ito S, Xia K, Wang Z, Cui K, et al. Dual functions of Tet1 in transcriptional regulation in mouse embryonic stem cells. Nature. 2011;**473**(7347):389-393. DOI: 10.1038/nature09934

[83] Khoueiry R, Sohni A, Thienpont B, Luo X, Velde JV, Bartoccetti M, et al. Lineage-specific functions of TET1 in the postimplantation mouse embryo. Nature Genetics. 2017;**49**(7):1061-1072. DOI: 10.1038/ng.3868

[84] Good CR, Panjarian S, Kelly AD, Madzo J, Patel B, Jelinek J, et al. TET1-mediated hypomethylation activates oncogenic signaling in triple-negative breast cancer. Cancer Research. 2018;**78**(15):4126-4137. DOI: 10.1158/0008-5472.CAN-17-2082

[85] Huang H, Jiang X, Li Z, Li Y, Song CX, He C, et al. TET1 plays an essential oncogenic role in MLL-rearranged leukemia. Proceedings of the National Academy of Sciences of the United States of America. 2013;**110**(29):11994-11999. DOI: 10.1073/pnas.1310656110

[86] Yokoyama S, Higashi M, Tsutsumida H, Wakimoto J, Hamada T, Wiest E, et al. TET1-mediated DNA hypomethylation regulates the expression of MUC4 in lung cancer. Genes & Cancer. 2017;**8**(3-4):517-527. DOI: 10.18632/genesandcancer.139

[87] Deng M, Zhang R, He Z, Qiu Q, Lu X, Yin J, et al. TET-Mediated Sequestration of miR-26 Drives EZH2 Expression and Gastric Carcinogenesis. Cancer Research. 2017;**77**(22): 6069-6082. DOI: 10.1158/0008-5472.CAN-16-2964

[88] Li YY, Hanna GJ, Laga AC, Haddad RI, Lorch JH, Hammerman PS. Genomic analysis of metastatic cutaneous squamous cell carcinoma. Clinical Cancer Research. 2015;**21**(6): 1447-1456. DOI: 10.1158/1078-0432.CCR-14-1773

[89] Seshagiri S, Stawiski EW, Durinck S, Modrusan Z, Storm EE, Conboy CB, et al. Recurrent R-spondin fusions in colon cancer. Nature. 2012;**488**(7413):660-664. DOI: 10.1038/nature11282

[90] Hodis E, Watson IR, Kryukov GV, Arold ST, Imielinski M, Theurillat JP, et al. A landscape of driver mutations in melanoma. Cell. 2012;**150**(2):251-263. DOI: 10.1016/j.cell.2012.06.024

[91] Cimmino L, Dawlaty MM, Ndiaye-Lobry D, Yap YS, Bakogianni S, Yu Y, et al. TET1 is a tumor suppressor of hematopoietic malignancy. Nature Immunology. 2015;**16**(6): 653-662. DOI: 10.1038/ni.3148

[92] Collignon E, Canale A, Al Wardi C, Bizet M, Calonne E, Dedeurwaerder S, et al. Immunity drives TET1 regulation in cancer through NF-κB. Science Advances. 2018;**4**(6):eaap7309. DOI: 10.1126/sciadv.aap7309

[93] Forloni M, Gupta R, Nagarajan A, Sun LS, Dong Y, Pirazzoli V, et al. Oncogenic EGFR represses the TET1 DNA demethylase to induce silencing of tumor suppressors in cancer cells. Cell Reports. 2016;**16**(2):457-471. DOI: 10.1016/j.celrep.2016.05.087

[94] Lai JI, Lai YC, Chen YC, Wang NK, Pan JN, Wang WS, et al. Clinical analysis of NSCLC patients reveals lack of association between EGFR mutation and TET1 downregulation. Cancer Gene Therapy. 2017;**24**(9):373-380. DOI: 10.1038/cgt.2017.26

[95] Zhou Z, Zhang HS, Liu Y, Zhang ZG, Du GY, Li H, et al. Loss of TET1 facilitates DLD1 colon cancer cell migration via H3K27me3-mediated down-regulation of E-cadherin. Journal of Cellular Physiology. 2018;**233**(2):1359-1369. DOI: 10.1002/jcp.26012

[96] Wang KC, Kang CH, Tsai CY, Chou NH, Tu YT, Li GC, et al. Ten-eleven translocation 1 dysfunction reduces 5-hydroxymethylcytosine expression levels in gastric cancer cells. Oncology Letters. 2018;**15**(1):278-284. DOI: 10.3892/ol.2017.7264

[97] Ceccarelli V, Valentini V, Ronchetti S, Cannarile L, Billi M, Riccardi C, et al. Eicosapentaenoic acid induces DNA demethylation in carcinoma cells through a TET1-dependent mechanism. The FASEB Journal. 2018:fj201800245R. DOI: 10.1096/fj.201800245R

[98] Cheng YW, Chou CJ, Yang PM. Ten-eleven translocation 1 (TET1) gene is a potential target of miR-21-5p in human colorectal cancer. Surgical Oncology. 2018;**27**(1):76-81. DOI: 10.1016/j.suronc.2017.12.004

[99] Li Y, Shen Z, Jiang H, Lai Z, Wang Z, Jiang K, et al. MicroRNA-4284 promotes gastric cancer tumorigenicity by targeting ten-eleven translocation 1. Molecular Medicine Reports. 2018;**17**(5):6569-6575. DOI: 10.3892/mmr.2018.8671

[100] Zhu J, Wang K, Li T, Chen J, Xie D, Chang X, et al. Hypoxia-induced TET1 facilitates trophoblast cell migration and invasion through HIF1α signaling pathway. Scientific Reports. 2017;**7**(1):8077. DOI: 10.1038/s41598-017-07560-7

[101] Duan H, Yan Z, Chen W, Wu Y, Han J, Guo H, et al. TET1 inhibits EMT of ovarian cancer cells through activating Wnt/β-catenin signaling inhibitors DKK1 and SFRP2. Gynecologic Oncology. 2017;**147**(2):408-417. DOI: 10.1016/j.ygyno.2017.08.010

[102] Li BT, Yu C, Xu Y, Liu SB, Fan HY, Pan WW. TET1 inhibits cell proliferation by inducing RASSF5 expression. Oncotarget. 2017;**8**(49):86395-86409. DOI: 10.18632/oncotarget.21189

[103] Wang H, An X, Yu H, Zhang S, Tang B, Zhang X, et al. MiR-29b/TET1/ZEB2 signaling axis regulates metastatic properties and epithelial-mesenchymal transition in breast cancer cells. Oncotarget. 2017;**8**(60):102119-102133. DOI: 10.18632/oncotarget.22183

[104] Tsai KW, Li GC, Chen CH, Yeh MH, Huang JS, Tseng HH, et al. Reduction of global 5-hydroxymethylcytosine is a poor prognostic factor in breast cancer patients, especially for an ER/PR-negative subtype. Breast Cancer Research and Treatment. 2015;**153**(1):219-234. DOI: 10.1007/s10549-015-3525-x

Role of *COX-2* Promoter Methylation and *Helicobacter pylori* Infection in Impaired Gastric Ulcer Healing

Hiroshi Yasuda, Yoshiyuki Watanabe,
Ritsuko Oikawa and Fumio Itoh

Additional information is available at the end of the chapter

http://dx.doi.org/10.5772/intechopen.79973

Abstract

Helicobacter pylori (*H. pylori*) infection causes aberrant DNA methylation of various genes in the gastric mucosa. Cyclooxygenases (COX) play a critical role in peptic ulcer development and healing. Human *COX-2* has CpG islands (CGIs) in its promoter region, suggesting a possible epigenetic regulation. Here, we evaluated *COX-2* promoter methylation in the gastric mucosa of patients with various gastric diseases and found that *COX-2* methylation levels in the gastric mucosa were significantly increased in patients with *H. pylori* infection. We further investigated the roles of COX-2 during the healing of acetic acid-induced gastric ulcers in *H. pylori*-infected Mongolian gerbils (MGs). While *COX-2* mRNA expression levels on the edges of acetic acid-induced gastric ulcers were significantly increased after ulcer induction in MGs in the absence of *H. pylori*, no such induction was observed in *H. pylori*-infected gastric mucosa. Cloning of the MG *COX-2* gene revealed abundant CGIs in the promoter region. *COX-2* mRNA expression in MG-derived gastric carcinoma MGC2 cells was significantly increased by addition of the demethylating agent 5-Aza-dC. Additionally, *COX-2* methylation levels were higher in *H. pylori*-infected MG gastric mucosa than in control mucosa. These results indicated that epigenetic inhibition of *COX-2* mRNA induced by *H. pylori* impairs gastric ulcer healing.

Keywords: COX-2, gastric ulcer healing, *Helicobacter pylori*, methylation, Mongolian gerbils, MGC2 cells

1. Introduction

Helicobacter pylori (*H. pylori*) is a Gram-negative bacterium that selectively colonizes the gastric epithelium of humans and is the leading cause of peptic ulcers [1]. Although the majority of individuals infected with *H. pylori* remain asymptomatic throughout their life, essentially all infected individuals develop chronic inflammation. Patients with antral-predominant gastritis are predisposed to duodenal ulcers, while patients with corpus-predominant gastritis and multifocal atrophy are more likely to have gastric ulcers. Eradication of *H. pylori* drastically lowers the recurrence of *H. pylori*-associated peptic ulcers. In addition, the observed delayed ulcer healing has been reported to involve *H. pylori*-induced inflammation, increased apoptosis of epithelial cells at the ulcer margin, overexpression of inflammatory cytokines, and reduced gastric microcirculation [2]. Cyclooxygenase (COX) is a membrane-bound glycoprotein that functions as the rate-limiting enzyme in prostaglandin (PG) synthesis. PGs increase the resistance of the gastric mucosa to injury by downregulating inflammatory responses. Two major COX isoforms have been identified, COX-1, which is constitutively expressed and considered a housekeeping enzyme, and COX-2, whose mRNA levels rise rapidly in response to inflammatory and mitogenic stimuli. There is a known synergism between *H. pylori* infection and non-steroidal anti-inflammatory drug (NSAID) use in the development of peptic ulcers and ulcer bleeding [3]. These findings indicate possible interactions between *H. pylori* infection, COX activity, and ulcerogenesis. Aberrant methylation of 5′-CpG islands (CGIs) has been implicated in the transcriptional silencing of a wide range of genes involved in various diseases, such as cancer. Human *COX-2* has CG-rich CGIs in its promoter region, which suggests epigenetic regulation. Here, we report and discuss our recent results on the epigenetic regulation of *COX-2* activity in *H. pylori*-infected gastric mucosa of humans and Mongolian gerbils (MGs), and the possible relationship between *COX-2* methylation and delayed gastric ulcer healing.

2. COX and the gastric mucosal barrier

In 1971, Vane and colleagues first demonstrated that aspirin and other NSAIDs inhibited the synthesis of PGs by blocking COX activity [4]. COX plays pivotal roles in the gastric mucosal barrier [5, 6]. COX catalyzes the conversion of arachidonic acid to the common precursor prostanoids, prostaglandin (PG) H_2, and PGG_2. The major PGs produced by the human and rodent gastric mucosa are PGE_2 and PGI_2, with lesser amounts of PGF_2 and PGD_2. Each of these metabolites binds to a specific G protein-coupled receptor to trigger intracellular responses. PGs have been shown to accelerate ulcer healing in experimental models and humans [7]. COX exists in two isoforms commonly referred to as COX-1 and COX-2. COX-1 is constitutively expressed in various tissues. In the stomach, prostanoids synthesized via the COX-1 pathway are responsible for cytoprotection of the gastric mucosa and the production of thromboxane by platelets. Although COX-2 is generally expressed at very low levels in healthy tissues, including in the stomach, it is expressed at particularly high levels at sites of inflammation.

It was originally thought that only COX-1 was involved in the gastric mucosal defense system; however, several clinical trials have suggested that a COX-2 selective inhibitor produces lesser, but some, gastrointestinal toxicity compared to traditional NSAIDs [8, 9]. In accordance with this finding, animal studies have suggested that both COX-1 and COX-2 are necessary for gastric mucosal healing, and COX-1 inhibition alone, which can be induced pharmacologically by specific inhibitors or genetically by gene targeting [10], does not cause gastric mucosal injury. It has been shown that a combination of selective COX-1 and COX-2 inhibitors is required to cause hemorrhagic erosion of the gastric mucosa, which is comparable to that observed with indomethacin [11]. Prostanoids produced by COX-2, especially PGE_2, enhance cell proliferation. The beneficial effects of PGE_2 on gastric ulcer healing in rodents appear to be mediated via the EP4 receptor [12]. In addition, COX-2-derived PG stimulates vascular endothelial growth factor (VEGF) release from gastric fibroblasts, which is an important contributor to ulcer healing [13, 14], likely via stimulation of new blood vessel growth. The increase in COX-2 immunoreactivity that is observed in monocytes, macrophages, fibroblasts, and endothelial cells at the ulcer margin is closely correlated, both temporally and spatially, with the increase in cell proliferation [15]. COX-2 appears to represent a second line of defense that is activated during ulcer healing to compensate for the temporary loss of COX-1 in the mucosa adjacent to the ulcer and assists COX-1 in protecting gastric mucosal integrity. The healing-impairment effect of NSAIDs is also observed with selective COX-2 inhibitors [16].

3. *H. pylori* infection and COX-2 in gastric mucosa

The pathophysiological roles of COX-2 in *H. pylori*-infected gastric mucosa are intriguing. *H. pylori* has been implicated as an inducer of COX-2 in the stomach [17–19]. In addition, COX-2 expression is elevated within *H. pylori*-induced gastritis and malignant lesions [20], and *H. pylori* induced COX-2 expression and enhanced PGE_2 production in a human gastric carcinoma cell line. Both *H. pylori* infection and NSAID use independently and significantly increase the risk of peptic ulcers and ulcer bleeding. While COX-2 is necessary for gastric mucosal healing [21], *H. pylori* infection is the leading cause of gastric ulceration. To understand the bimodal effects of *H. pylori* infection on COX-2 induction during ulcer healing, we explored the effects of epigenetic regulation and *H. pylori* infection on the induction of COX-2 *in vivo* and *in vitro*.

H. pylori infection causes aberrant DNA methylation of various genes in the gastric mucosa, including *COX-2* [22–25]. Human *COX-2* has CGIs in its promoter region. Thus, we compared *COX-2* promoter methylation levels in the gastric mucosa in *H. pylori*-positive and *H. pylori*-negative cases [26]. As mentioned above, in qualitative experiments, *COX-2* gene promoter methylation levels were significantly higher in *H. pylori*-positive cases than in *H. pylori*-negative cases (**Figure 1**). *COX-2* promoter methylation levels were significantly lower in patients with *H. pylori* eradication than in those with *H. pylori* infection. We then investigated the effects of *COX-2* promoter methylation on *COX-2* mRNA expression *in vitro* using the human gastric adenocarcinoma cell line Kato III, in which the *COX-2* promoter is densely methylated [22]. *COX-2* mRNA expression was not observed in these cells, despite the addition of the protein kinase C stimulator α-phorbol 12,13-dibutyrate (PDBu). However, *COX-2* expression was observed after the addition of the demethylating agent 5-Aza-dC, and expression was enhanced by adding PDBu (**Figure 2**) [26]. These results indicate that *H. pylori* infection causes reversible *COX-2* promoter methylation in the gastric mucosa, and that *COX-2* mRNA expression is regulated through an epigenetic mechanism.

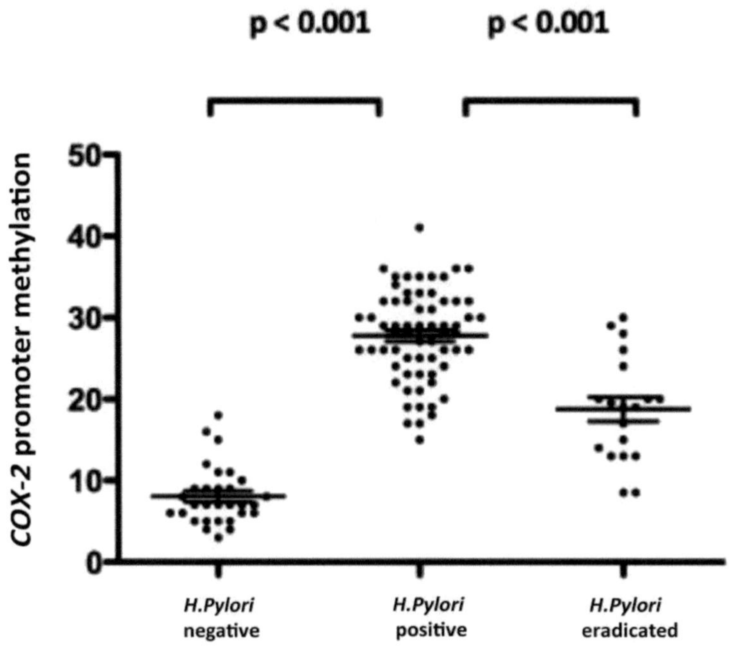

Figure 1. *COX-2* DNA methylation levels in patients with or without *H. pylori* infection, and patients previously with successfully eradicated *H. pylori* infection. The figure is modified from Ref. [26].

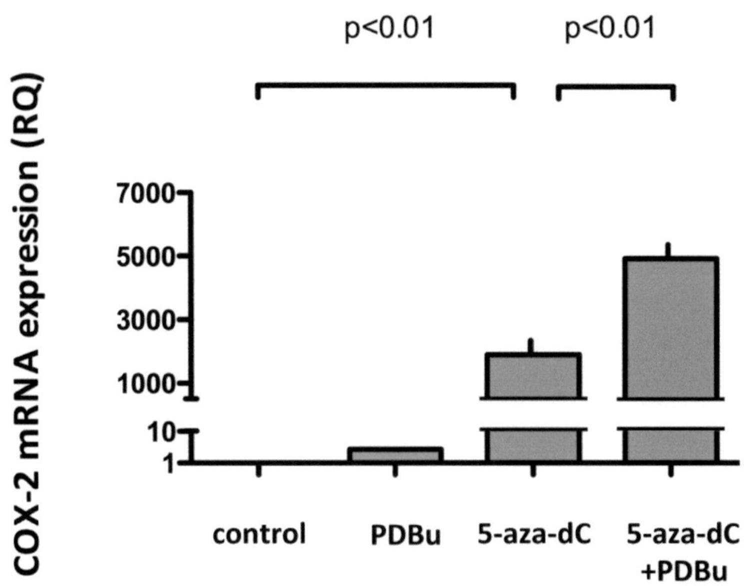

Figure 2. Effects of a PKC stimulator (α-phorbol 12,13-dibutyrate; PDBu) on COX-2 mRNA expression with or without 5-aza-dC in the human gastric adenocarcinoma cell line KATO-III. KATO-III cells were treated with vehicle (1 μmol/L) with or without 5-Aza-dC for 5 days. The figure is modified from Ref. [26].

4. Acetic acid-induced gastric ulcer healing and *COX* mRNA levels in Mongolian gerbils with or without *H. pylori* infection

H. pylori infection in humans is best modeled in Mongolian gerbils (MGs), and chronic infection with *H. pylori* induces inflammatory cell infiltration in the gastric mucosa in MGs (**Figure 3A**). It was previously shown that *H. pylori* infection significantly delayed acetic acid-induced ulcer healing in mice and MGs [27–29]. *H. pylori* infection induces aberrant DNA methylation of several CGIs in MGs [30]. As described above, COX-2-derived PGs are important for gastric ulcer healing, and the *COX-2* promoter is densely methylated in the human gastric mucosa in the presence of *H. pylori* infection. To investigate the roles of *COX-2*

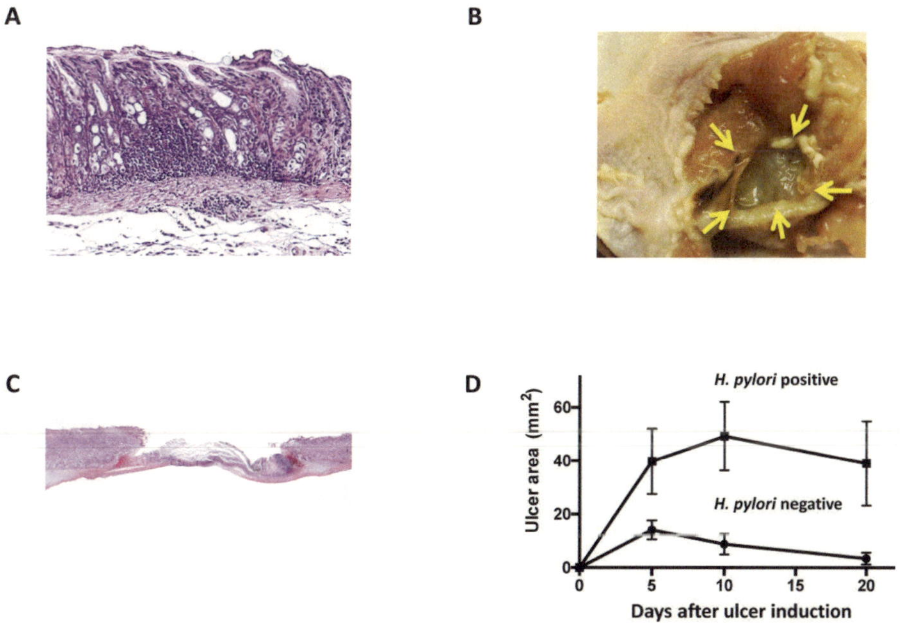

Figure 3. A. Microscopic features of *H. pylori*-infected gastric mucosa of Mongolian gerbils (MGs). Sections were stained with hematoxylin and eosin. B. Gross appearance 10 days after gastric ulcer induction of *H. pylori*-infected MGs. C. Microphotograph 10 days after gastric ulcer induction in *H. pylori*-infected MGs. D. Serial changes in acetic acid-induced gastric ulcer areas in MGs with or without *H. pylori* infection. *H. pylori* (ATCC43504; American Type Culture Collection, Rockville, MD) was grown in Brucella broth (Becton Dickinson, Cockeysville, MD) containing 10% v/v horse serum for 40 h at 37°C under microaerobic conditions (15% CO_2) and high humidity with shaking (150 rpm). Male MGs (MGs/Sea) were purchased from Kyudo (Saga, Japan). At 11 weeks of age, *H. pylori* (0.8 mL samples of Brucella broth containing 1.0×10^9 colony-forming units) was delivered intragastrically using an oral catheter after fasting for 24 h. Gastric ulcers were induced experimentally in MGs according to the method described by Wang et al. [40]. Briefly, after anesthetization with ketalar, the abdomen was opened through a midline incision, and 50 μL of 25% acetic acid was injected in the subserosa of the anterior wall of the stomach. The MGs were killed at5, 10, and 20 days after ulcer induction, and the stomach was dissected and removed. The maximum and minimum diameters of the ulcers were measured, and the ulcer area, which was approximately elliptical, was calculated and was compared between MGs with and without *H. pylori* infection. Values are the mean ± SE. The animal care committee of St. Marianna University approved the experimental design, and the animals were cared for in accordance with institutional guidelines.

methylation and *H. pylori* infection in gastric ulcer healing, *COX* mRNA levels in samples prepared from acetic acid-induced gastric ulcers in MGs were examined. Then, the effects of *COX-2* methylation on *COX-2* mRNA expression were also investigated *in vitro* using an *H. pylori*-infected MG stomach-derived cell line.

Gastric ulcers were produced by injecting 25% of acetic acid (0.03 mL) into the submucosal layer of the gastric wall of the antral-oxyntic border in MGs 48 weeks after inoculation with an *H. pylori* suspension in Brucella broth. The MGs were killed at 5, 10, and 20 days after ulcer induction, and the stomachs were dissected and removed. The maximum and minimum diameters of the ulcers were measured, and the area of each ulcer, which was approximately elliptical, was calculated and compared between MGs with and without *H. pylori* infection. The ulcer area was largest on day 5, and then gradually decreased. In accordance with previous reports [29], the gastric ulcer area was larger in *H. pylori*-infected MGs than in uninfected MGs (**Figure 3B–D**). While *COX-2* mRNA expression at the ulcer edge was increased 5 days after acetic acid injection in MGs without *H. pylori* infection, as was reported in rats and mice [31], no increases in *COX-2* mRNA levels were observed in *H. pylori*-infected MGs. In contrast, gastric ulceration was not associated with a change in *COX-1* mRNA levels in MGs with or without *H. pylori* infection (**Figure 4A,B**). Thus, *H. pylori* infection caused delayed ulcer healing and impaired COX-2 induction in MG stomachs.

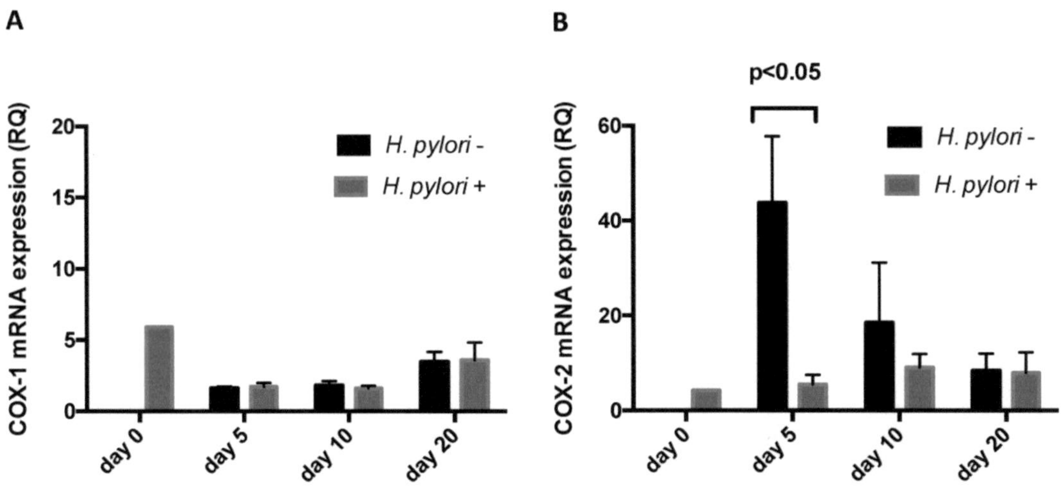

Figure 4. *COX* mRNA levels during healing of an acetic acid-induced gastric ulcer in MGs with or without *H. pylori* infection. *COX* mRNA levels in acetic-acid-induced gastric ulcers in MGs were measured by real-time PCR. First-strand cDNA was prepared by reverse transcription of 5 μg of total RNA using superscript III reverse transcriptase (Applied Biosystems, Carlsbad, CA, USA). Real-time quantitative reverse transcription PCR was carried out using TaqMan Gene Expression Assays with a 7500 real-time PCR system (Applied Biosystems) according to the manufacturer's instructions. Primers for *COX-1*, *COX-2*, and *β-actin* were designed based on their cDNA sequences (GenBank accession nos. AB 044783, AB044784, and AB040445, respectively) and according to previous reports [33]. SDS2.1 software (Applied Biosystems) was used to perform the comparative Δ-Ct analysis. Glyceraldehyde-3-phosphate dehydrogenase was used as an endogenous control. Values are the mean ± SE. β-Actin left: gctacagcttcaccaccaca, right: ccatctcttgctcgaagtcc, 93 bp. COX-1 left: gtggctatttcctgcagctc, right: agtgggtgccagtggtagag, 112 bp. COX-2 left: tgggcgtgaaaggaaataag, right: ggggatcagggatgaacttt, 87 bp.

5. Effects of *COX-2* methylation on its mRNA expression in MGs *in vitro*

As discussed above, *COX-2* mRNA expression is regulated by an epigenetic mechanism in KATO-III human gastric carcinoma cells in which *COX-2* is densely methylated [22]. To investigate the role of methylation in *COX-2* mRNA expression in *H. pylori*-infected gastric mucosa of MGs, we treated MGC2 cells with 5-aza-dC, a methyltransferase inhibitor, or trichostatin A (TSA), a histone deacetylase inhibitor. MGC2 is an adenocarcinoma cell line established from the gastric cancer tissue of a *H. pylori*-infected MG [32]. *COX-2* mRNA expression levels in these cells were restored after the addition of 5-aza-dC. In contrast, treatment with TSA did not induce *COX-2* mRNA expression (**Figure 5**). These results indicated that *COX-2* mRNA expression in MGs is regulated via both transcriptional and epigenetic mechanisms. Histone acetylation was not involved in silencing *COX-2* expression in these cells.

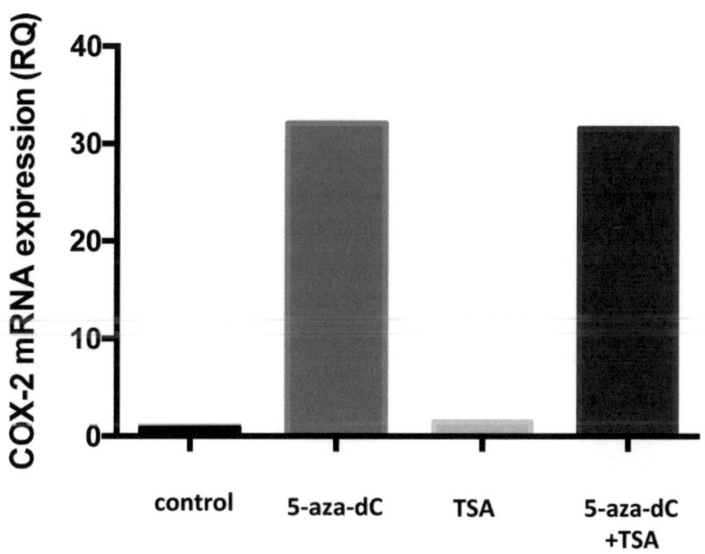

Figure 5. Effects of 5-aza-dC, a methyltransferase inhibitor, or trichostatin A (TSA), a histone deacetylase inhibitor, on *COX-2* mRNA expression in MG gastric adenocarcinoma MGC2 cells. The MGC2 cells [32] were a generous gift from Dr. Tatematsu and were maintained in RPMI-1640 medium supplemented with 10% fetal bovine serum (FBS; Gibco, Grand Island, NY, USA) plus serum expander MITO (0.1%, Collaborative Biomedical Products Bedford, MA, USA) on a type I collagen-coated dish (Asahi Techno Glass, Japan). All the cultures were incubated at 37°C with 95% air and 5% CO_2. *COX-2* mRNA expression was measured by real-time PCR in the cell lines grown in the presence of vehicle with or without 5-Aza-dC (1 μmol/L) or TSA for 5 days. Values shown are the mean ± SE.

6. Cloning the MG *COX-2* promoter region

As mentioned above, human *COX-2* has abundant CGIs in the promoter region. To examine whether CGIs are present in the *COX-2* promoter region of MGs, we performed genomic

PCR with primers designed using as previously reported MG cDNA (Accession #: AB177842) [33]. The PCR product has an intron sequence in the promoter region. Then, gene walking was performed using a primer in this intron. The promoter region of the MG *COX-2* gene was obtained (**Figure 6**), and the sequence contains more CGIs than the mouse genome, and

```
PTGS2 human promoter
CCCCCTCTGCTCCCAAATTGGGGCAGCTTCCTGGGTTTCCGATTTTCTCATTTCCGTG
GGTAAAAAACCCTGCCCCCACCGGGCTTACGCAATTTTTTTAAGGGGAGAGGAGGGAA
AAATTTGTGGGGGGTACGAAAAGGCGGAAAGAAACAGTCATTTCGTCACATGGGCTTG
GTTTTCAGTCTTATAAAAAGGAAGGTTCTCTCGGTTAGCGACCAATTGTCATACGACT
TGCAGTGAGCGTCAGGAGCACGTCCAGGAACTCCTCAGCAGCGCCTCCTTCAGCTCCA
CAGCCAGACGCCCTCAGACAGCAAAGCCTACCCCCGCGCCGCGCCCTGCCCGCCGCTG
CGATGCTCGCCCGCGCCCTGCTGCTGTGCGCGGTCCTGGCGCTCAGCCATACAGGTGA
GTACCTGGCGCCGCGCACCGGGGACTCCGGTTCCACGCACCCGGGCAGAGTTTCCGCT
CTGACCTCCTGGGTCTATCCCAGTACTCCGACTTCTCT

PTGS2 mongolian gerbil promoter
CCACCTGGGGCCGCCGAGCTTTCCCGGCTTTCCTGGTCTCCCGATGCGTGGGTCGGGC
CTTGCACAACTCGTTGACTGAGATGGATGGAGCTGTTGGGGGTTGGGGGAAAGACCGG
GCGGAAAGGGAGCGTCATCCCACCAACGTCACGTTGTGTGGGTGAGGGGAACCTCACA
GACTTTTTAAACGCAAGGTTCTCCCCCGTTAGCAGTCAGTTGTCAAACTGCGAGCGAA
GAGCTTCAGGAATATCAGGAGTCTGCCCAGGAAGACCTCAGCGCTGCGTCCCGCCAGC
TCCGCCGCCACCGCCAACGCCACCGCCGCCGCTACCGCCGCCGCTACCGCCGCCGCC
ACCTCCGCGATGCTAGTCCGAGCTGCGCTGCTCTGCGTCGCCCTGGCGCTCTGCCGTG
CAGGTGAGTGACCCCACGCCCCAAGGCGCCCGCCGCGCTCCGCGGGATCACACTCTCC
GCCGTGCTGCTTACTGCGCAGGGGCGAAGACTCCGGACCACCTGAGTCCGCGGAGACT
GC

PTGS2 mouse promoter
CCACCTGGGCAGCCAAAGGGCAGCTTCCCGGCTTCCTTCGTCTCTCATTTGCGTGGGT
AAAAGCCTGCCGCTGCGGTTCTTGCGCAACTCACTGAAGCAGAGAGGGGGAAAAGTTG
GTGGGGGTTGGGGAAAGCCTAAGCGGAAAGACAGAGTCACCACTACGTCACGTGGAGT
CCGCTTTACAGACTTAAAAGCAAGGTTCTCCCCATTAGCAGCCAGTTGTCAAACTGCG
AGCTAAGAGCTTCAGGAGTCAGTCAGGACTCTGCTCACGAAGGAACTCAGCACTGCAT
CCTGCCAGCTCCACCGCCACCACTACTGCCACCTCCGCTGCCACCTCTGCGATGCTCT
TCCGAGCTGTGCTGCTCTGCGCTGCCCTGGGGCTCAGCCAGGCAGGTGAGTTGCGCCC
CGAGGCGCCCTGGGACTCCTCAGGCTCAGACCCCCTTTCTAGCAGTGTGGGCCTGGCT
GTGTATTGGCAAAGAGCCTGGACTGCTTA
```

Figure 6. Human, MG, and mouse *COX-2* promoter regions. Genome walking to isolate genomic, the *COX-2* promoter region of MGs, was performed by using the Straight Walk Kit [41] according to the manufacturer's instructions (Bex, Tokyo, Japan). Amplified fragments were cloned into the pSTBlue-1 vector (Novagen, Madison, WI, USA) and sequenced. More abundant CGIs were observed in the MG genome when compared to the mouse genome, and the number was comparable to human genome.

is comparable to that of the human genome. Pyrosequencing showed several methylated CGIs (>15%) in the MGC2 cells. Treatment with 5-aza-dC decreased the methylation levels in these cells (**Figure 7**). As was observed in the human stomach, *COX-2* methylation levels were increased in *H. pylori*-infected gastric mucosa of MGs when compared to the levels in the control mucosa (**Figure 8**).

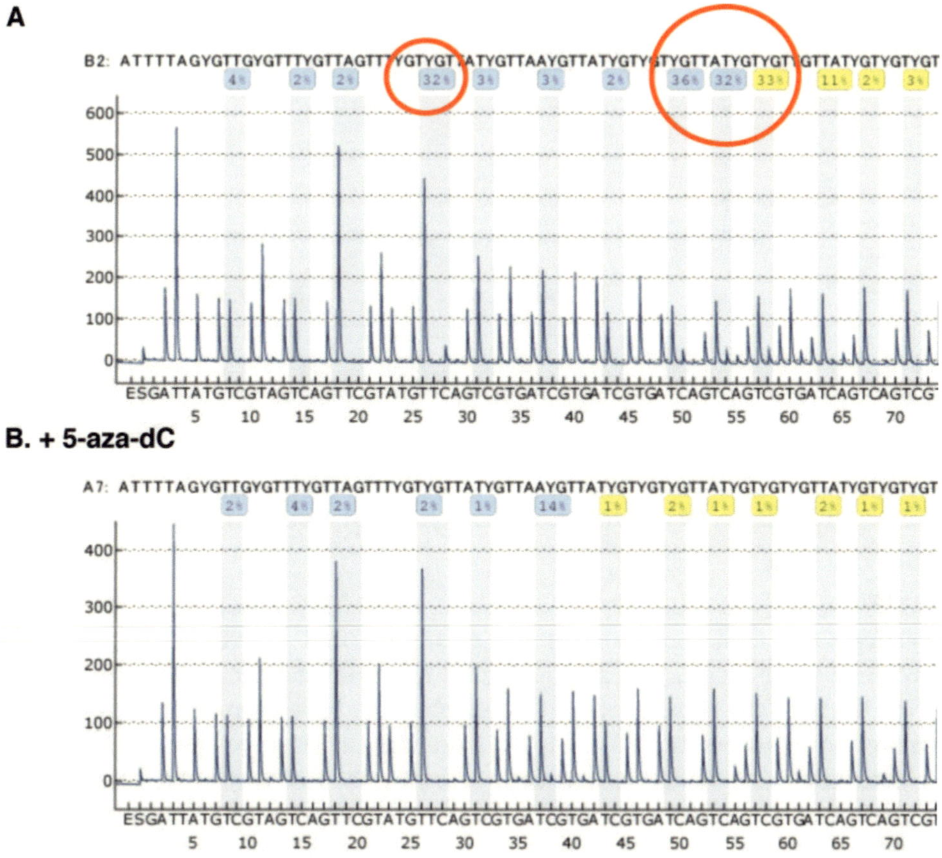

Figure 7. Effect of 5-aza-dC treatment on *COX-2* methylation levels in MGC2 cells. Cells were treated with vehicle (A) with 5-aza-dC (B, 1 μmol/L) for 5 days, and then methylation levels of *COX-2* were analyzed by quantitative bisulfite-pyrosequencing methods as previously reported [42]. Bisulfite treatment of gDNA was performed with the EpiTect bisulfite kit (QIAGEN, Valencia, CA, USA) according to the manufacturer's protocol. Bisulfite-treated DNA (1 μL) was used as a template in subsequent PCR experiments. For most assays, we used touchdown PCR. All PCR assays included a denaturation step at 95°C for 30 s, followed by annealing at various temperatures for 30 s, and extension at 72°C for 30 s. After PCR, the biotinylated strand was captured on streptavidin-coated beads (Amersham Bioscience, Uppsala, Sweden) and incubated with sequencing primers. Pyrosequencing was performed using PSQ HS 96 Gold single-nucleotide polymorphism reagents on a PyroMark Q24 pyrosequencing machine (QIAGEN). The protocol for pyrosequencing was described in detail previously [42]. Pyrosequencing quantitatively measures the methylation status of CpG sites in a target region. Adjacent sites usually show highly concordant methylation. Therefore, the mean percent methylation in detected sites can be used as a representative value for each gene promoter. Cases with the methylation density > 15% were regarded as methylation positive. Forward primer: tgggtgaggggaattttataga, reverse primer: aaaccctaaccatccttacaa; and sequencing primer: aggagtttgtttaggaag.

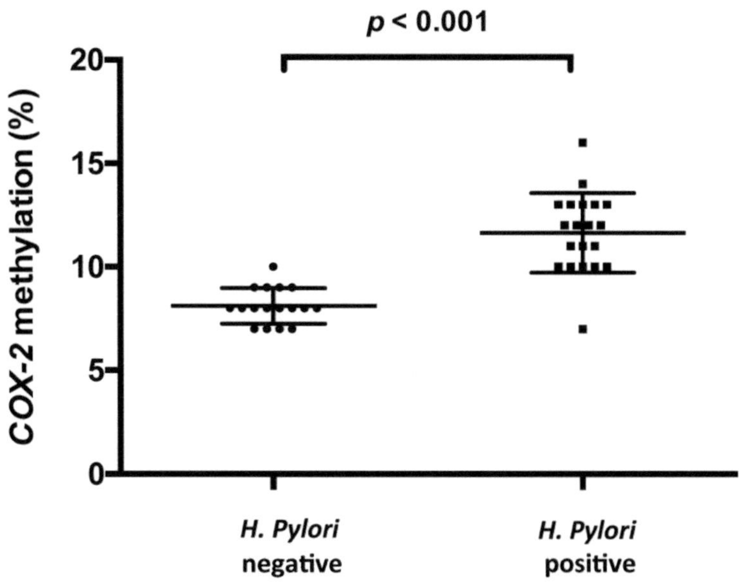

Figure 8. DNA methylation levels at *COX-2* in the MG gastric mucosa in the presence or absences of *H. pylori* infection. Methylation levels at *COX-2* in the MG gastric mucosa were analyzed by quantitative bisulfite pyrosequencing.

7. Role of *COX-2* methylation and *H. pylori* infection in gastric mucosal healing

Nonselective COX inhibitors damage the gastrointestinal mucosa, and gastrointestinal injury represents the most significant side effect of chronic nonsteroidal anti-inflammatory drug (NSAID) use [34]. Thus, selective COX-2 inhibitors have been developed as ideal anti-inflammatory drugs, devoid of GI toxicity, and clinical trials have suggested that selective COX-2 inhibitors produce less gastrointestinal injury than conventional NSAIDs [8, 9]. In support of this, animal studies have also suggested that, in contrast to the initial concept, the importance of COX-2 in the repair of gastric mucosal damage has been recognized [31, 35]. In rat gastric mucosa, markedly elevated levels of *COX-2* mRNA were observed after induction of damage by ischemia-reperfusion [35, 36]. COX-2 mRNA and protein expression increase during repair of gastric mucosal lesions, and selective COX-2 inhibitors delay mucosal healing in mice [31]. PGs derived from ulceration-induced COX-2 at the ulcer margin enhanced epithelial cell proliferation and increased the expression of growth factors, including hepatocyte growth factor, epidermal growth factor, transforming growth factor-α, and VEGF [16].

The release of inflammatory cytokines and recruitment of inflammatory cells have been considered the potential factors for delayed ulcer healing in *H. pylori* infection [2, 28]. Chronic inflammation, including *H. pylori* infection, is known to cause aberrant DNA methylation [37, 38]. Here, we focused on *COX-2* methylation in *H. pylori*-infected gastric mucosa. Human *COX-2* has CGIs in the promoter region, and methylation levels in this region are increased in

H. pylori-infected gastric mucosa. In addition, *COX-2* promoter methylation levels in patients with successfully eradicated *H. pylori* infection are lower than the levels in *H. pylori*-positive cases. As mentioned above, experiments using rodents have shown that *COX-2* mRNA expression is enhanced during ulcer healing. Thus, we investigated the influence of *H. pylori* infection on COX-2 expression during stomach ulcer healing in MGs. Acetic acid-induced ulcer healing was delayed in *H. pylori*-infected MG stomachs when compared to the healing of ulcers without *H. pylori* infection [29]. The *COX-2* promoter region of MGs also has abundant CGIs, comparable to human *COX-2*. Thus, MGs are good models for investigating the role of *COX-2* methylation in gastric mucosal healing. While *COX-2* mRNA expression at the ulcer edge was increased 5 days after acetic acid injection in MG stomachs without *H. pylori* infection, as was observed in the mouse and rat models, such increases in *COX-2* mRNA expression were not observed in *H. pylori*-infected MG gastric mucosa. However, a demethylating agent restored COX-2 mRNA expression in both human and MG gastric carcinoma cell lines in which COX-2 is densely methylated. Accordingly, we demonstrated *in vitro* and *in vivo* that *COX-2* mRNA expression is regulated through an epigenetic mechanism in human and MG gastric mucosa. Aberrant DNA methylation has been extensively investigated in the context of the pathogenesis of various cancers, including stomach cancer. In addition, the involvement of epigenetic modifications has been reported in the pathogenesis of various chronic diseases, such as essential hypertension and cardiovascular disease [39]. Taken together, the epigenetic inhibition of COX-2 expression due to chronic inflammation induced by *H. pylori* infection seems to interfere with ulcer healing and increase the vulnerability of the gastric mucosa. The decrease in *COX-2* methylation levels after *H. pylori* eradication may restore gastric mucosal defense.

8. Conclusions

H. pylori infection, the leading cause of peptic ulcer disease, induces sustained inflammation in the gastric mucosa. This chronic inflammation causes aberrant DNA methylation in various genes. The *COX-2* promoter regions in both MGs and humans contain abundant CGIs. COX plays critical roles in peptic ulcer development and healing, and both COX-1 and COX-2 are necessary for the gastric mucosal defense system. COX-1 is a housekeeping enzyme, which maintains microcirculation and mucous production. During ulcer healing, COX-2 is expressed at the margin of an ulcer, and COX-2-derived PGs induce various growth factors to promote mucosal healing and angiogenesis. Ulcer healing is delayed in *H. pylori* infected-MG stomachs, which is the best model of *H. pylori* infection in humans. We showed that *COX-2* mRNA induction during ulcer healing was impaired in *H. pylori*-infected-MG stomachs. The *COX-2* promoter region is methylated in both *H. pylori*-infected human and MG gastric mucosa. These CGIs are also methylated in human Kato III cells and MGC2 gastric carcinoma cells. *COX-2* mRNA expression in these cells is restored by treatment with a demethylating agent. In *H. pylori*-infected gastric mucosa, *COX-2* promoter methylation appears to be involved in the impaired *COX-2* mRNA induction typically observed during ulcer healing, which leads to delayed ulcer healing.

Acknowledgements

This work was supported in part by a grant-in-aid for scientific research from the Ministry of Education, Science, and Culture of Japan to HY.

Conflict of interest

There is no conflict of interest to disclose.

Author details

Hiroshi Yasuda*, Yoshiyuki Watanabe, Ritsuko Oikawa and Fumio Itoh

*Address all correspondence to: hyasuda@marianna-u.ac.jp

Division of Gastroenterology and Hepatology, St. Marianna University School of Medicine, Kawasaki, Japan

References

[1] Marshall BJ, Warren JR. Unidentified curved bacilli in the stomach of patients with gastritis and peptic ulceration. Lancet. 1984;**1**:1311-1315

[2] Li H, Kalies I, Mellgard B, Helander HF. A rat model of chronic *Helicobacter pylori* infection. Studies of epithelial cell turnover and gastric ulcer healing. Scandinavian Journal of Gastroenterology. 1998;**33**:370-378

[3] Huang JQ, Sridhar S, Hunt RH. Role of *Helicobacter pylori* infection and non-steroidal anti-inflammatory drugs in peptic-ulcer disease: A meta-analysis. Lancet. 2002;**359**: 14-22. DOI: 10.1016/S0140-6736(02)07273-2

[4] Vane JR. Inhibition of prostaglandin synthesis as a mechanism of action for aspirin-like drugs. Nature: New Biology. 1971;**231**:232-235

[5] Hinz B, Brune K. Cyclooxygenase-2 10 year later. The Journal of Pharmacology and Experimental Therapeutics. 2002;**300**:367-375

[6] Gudis K, Sakamoto C. The role of cyclooxygenase in gastric mucosal protection. Digestive Diseases and Sciences. 2005;**50**(Suppl 1):S16-S23. DOI: 10.1007/s10620-005-2802-7

[7] Sontag SJ, Schnell TG, Budiman-Mak E, Adelman K, Fleischmann R, Cohen S, Roth SH, Ipe D, Schwartz KE. Healing of NSAID-induced gastric ulcers with a synthetic prostaglandin analog (enprostil). The American Journal of Gastroenterology. 1994;**89**:1014-1020

[8] Bombardier C, Laine L, Reicin A, Shapiro D, Burgos-Vargas R, Davis B, Day R, Ferraz MB, Hawkey CJ, Hochberg MC, Kvien TK, Schnitzer TJ, VIGOR Study Group. Comparison

of upper gastrointestinal toxicity of rofecoxib and naproxen in patients with rheumatoid arthritis. The New England Journal of Medicine. 2000;**343**:1520-1528. DOI: 10.1056/NEJM200011233432103

[9] Chan FK, Hung LC, Suen BY, Wu JC, Lee KC, Leung VK, Hui AJ, To KF, Leung WK, Wong VW, Chung SC, Sung JJ. Celecoxib versus diclofenac and omeprazole in reducing the risk of recurrent ulcer bleeding in patients with arthritis. The New England Journal of Medicine. 2002;**347**:2104-2110. DOI: 10.1056/NEJMoa021907

[10] Langenbach R, Morham SG, Tiano HF, Loftin CD, Ghanayem BI, Chulada PC, Mahler JF, Lee CA, Goulding EH, Kluckman KD, Kim HS, Smithies O. Prostaglandin synthase 1 gene disruption in mice reduces arachidonic acid-induced inflammation and indomethacin-induced gastric ulceration. Cell. 1995;**83**:483-492

[11] Wallace JL, McKnight W, Reuter BK, Vergnolle N. NSAID induced gastric damage in rats: Requirement for inhibition of both cyclooxygenase 1 and 2. Gastroenterology. 2000;**119**:706-714

[12] Hatazawa R, Tanaka A, Tanigami M, Amagase K, Kato S, Ashida Y, Takeuchi K. Cyclooxygenase-2/prostaglandin E2 accelerates the healing of gastric ulcers via EP4 receptors. American Journal of Physiology. Gastrointestinal and Liver Physiology. 2007;**293**:G788-G797. DOI: 10.1152/ajpgi.00131.2007

[13] Miura S, Tatsuguchi A, Wada K, Takeyama H, Shinji Y, Hiratsuka T, Futagami S, Miyake K, Gudis K, Mizokami Y, Matsuoka T, Sakamoto C. Cyclooxygenase-2-regulated vascular endothelial growth factor release in gastric fibroblasts. American Journal of Physiology. Gastrointestinal and Liver Physiology. 2004;**287**:G444-G451. DOI: 10.1152/ajpgi.00537.2003

[14] Wallace JL, Dicay M, McKnight W, Dudar GK. Platelets accelerate gastric ulcer healing through presentation of vascular endothelial growth factor. British Journal of Pharmacology. 2006;**148**:274-278. DOI: 10.1038/sj.bjp.0706722

[15] Schmassmann A, Peskar BM, Stettler C, et al. Effects of inhibition of prostaglandin endoperoxide synthase-2 in chronic gastrointestinal ulcer models in rats. British Journal of Pharmacology. 1998;**123**:795-804

[16] Halter F, Tarnawski AS, Schmassmann A, Peskar BM. Cyclooxygenase 2-implications on maintenance of gastric mucosal integrity and ulcer healing: Controversial issues and perspectives. Gut. 2001;**49**:443-453

[17] Romano M, Vittorio R, Memoli A, Tuccilli C, Di Popolo A, Sommi P, Acquaviva AM, CDV B, Bruni CB, Zarrilli R. *Helicobacter pylori* up-regulates cyclooxygenase-2 mRNA expression and prostaglandin E2 synthesis in MKN 28 gastric mucosal cells in vitro. The Journal of Biological Chemistry. 1998;**273**:28560-28563

[18] Fu S, Ramanujam KS, Wong A, Wong A, Fantry GT, Drachenberg CB, James SP, Meltzer SJ, Wilson KT. Increased expression and cellular localization of inducible nitric oxide synthase and cyclooxygenase 2 in *Helicobacter pylori* gastritis. Gastroenterologia. 1999;**116**:1319-1329

[19] McCarthy CJ, Crofford LJ, Greenson J, Scheiman JM. Cyclooxygenase-2 expression in gastric antral mucosa before and after eradication of *Helicobacter pylori* infection. The American Journal of Gastroenterology. 1999;**94**:1218-1223

[20] Yasuda H, Yamada M, Endo Y, Inoue K, Yoshiba M. Elevated cyclooxygenase-2 expression in patients with early gastric cancer in the gastric pylorus. Journal of Gastroenterology. 2005;**40**:690-697. DOI: 10.1007/s00535-005-1612-1

[21] Wallace JL. Prostaglandins, NSAIDs, and gastric mucosal protection: Why doesn't the stomach digest itself? Physiological Reviews. 2008;**88**:1547-1565. DOI: 10.1152/physrev.00004.2008

[22] Kikuchi T, Itoh F, Toyota M, Suzuki H, Yamamoto H, Fujita M, Hosokawa M, Imai K. Aberrant methylation and histone deacetylation of cyclooxygenase 2 in gastric cancer. International Journal of Cancer. 2002;**97**:272-277

[23] Chan AO, Peng JZ, Lam SK, Lai KC, Yuen MF, Cheung HKL, Kwong YL, Rashid A, Chan CK, Wong BC. Promoter methylation of E-cadherin gene in gastric mucosa associated with *Helicobacter pylori* infection and in gastric cancer. Gut. 2003;**52**:502-506

[24] Maekita T, Nakazawa K, Mihara M, Nakajima T, Yanaoka K, Iguchi M, Arii K, Kaneda A, Tsukamoto T, Tatematsu M, Tamura G, Saito D, Sugimura T, Ichinose M, Ushijima T. High levels of aberrant DNA methylation in *Helicobacter pylori*-infected gastric mucosae and its possible association with gastric cancer risk. Clinical Cancer Research. 2006;**12**:989-995. DOI: 10.1158/1078-0432.CCR-05-2096

[25] Perri F, Cotugno R, Piepoli A, Merla A, Quitadamo M, Gentile A, Pilotto A, Annese V, Andriulli A. Aberrant DNA methylation in non-neoplastic gastric mucosa of *H. Pylori* infected patients and effect of eradication. The American Journal of Gastroenterology. 2007;**102**:1361-1371. DOI: 10.1111/j.1572-0241.2007.01284.x

[26] Michikawa Y, Yasuda H, Watanabe Y, Oikawa R, Ohishi Y, Maehata T, Itoh F. COX-2 gene promoter methylation in patients infected with *Helicobacter pylori*. Clinical Medicine Insights: Gastroenterology. 2013;**6**:13-19. DOI: 10.4137/CGast.S11917

[27] Makino M, Koga T, Ito K, Kawada H, Tabata K. Delayed healing of chronic gastric ulcer after *Helicobacter pylori* infection in mice. The Journal of Pharmacy and Pharmacology. 1998;**50**:943-994

[28] Konturek PC, Brzozowski T, Konturek SJ, Stachura J, Karczewska E, Panjdo R, Ghiara P, Hahn EG. Mouse model of *Helicobacter pylori* infection: Studies of gastric function and ulcer healing. Alimentary Pharmacology & Therapeutics. 1999;**13**:333-346

[29] Keto Y, Ebata M, Tomita K, Okabe S. Influence of *Helicobacter pylori* infection on healing and relapse of acetic acid ulcers in Mongolian gerbils. Digestive Diseases and Sciences. 2002;**47**:837-849

[30] Niwa T, Tsukamoto T, Toyoda T, Mori M, Tanaka H, Maekita T, Ichinose M, Tatematsu M, Ushijima T. Inflammatory processes triggered by *Helicobacter pylori* infection cause aberrant DNA methylation in gastric epithelial cells. Cancer Research. 2010;**70**:1430-1440. DOI: 10.1158/0008-5472.CAN-09-2755

[31] Mizuno H, Sakamoto C, Matsuda K, Wada K, Uchida T, Noguchi H, Akamatsu T, Kasuga M. Induction of cyclooxygenase 2 in gastric mucosal lesions and its inhibition by the specific antagonist delays healing in mice. Gastroenterology. 1997;**112**:387-397

[32] Nozaki K, Tanaka H, Ikehara Y, Cao X, Nakanishi H, Azuma T, Yamazaki S, Yamaoka Y, Shimizu N, Mafune K, Kaminishi M, Tatematsu M. *Helicobacter pylori*-dependent NF-kappa B activation in newly established Mongolian gerbil gastric cancer cell lines. Cancer Science. 2005;**96**:170-175. DOI: 10.1111/j.1349-7006.2005.00030.x

[33] Matsubara S, Shibata H, Takahashi M, Ishikawa F, Yokokura T, Sugimura T, Wakabayashi K. Cloning of Mongolian gerbil cDNAs encoding inflammatory proteins, and their expression in glandular stomach during *H. pylori* infection. Cancer Science. 2004;**95**:798-802

[34] Wallace JL. Nonsteroidal anti-inflammatory drugs and gastroenteropathy: The second hundred years. Gastroenterology. 1997;**112**:1000-1016

[35] Maricic N, Ehrlich K, Gretzer B, Schuligoi R, Respondek M, Peskar BM. Selective cyclo-oxygenase-2 inhibitors aggravate ischemia-reperfusion injury in the rat stomach. British Journal of Pharmacology. 1999;**128**:1659-1666

[36] Kishimoto Y, Wada K, Nakamoto K, Ashida K, Kamisaki Y, Kawasaki H, Itoh T. Quantitative analysis of cyclooxygenase-2 gene expression on acute gastric injury induced by ischemia-reperfusion in rats. Life Sciences. 1997;**60**:PL127-PL133

[37] Chiba T, Marusawa H, Ushijima T. Inflammation-associated cancer development in digestive organs: Mechanisms and roles for genetic and epigenetic modulation. Gastroenterology. 2012;**143**:550-563. DOI: 10.1053/j.gastro.2012.07.009

[38] Asada K, Nakajima T, Shimazu T, Yamamichi N, Maekita T, Yokoi C, Oda I, Ando T, Yoshida T, Nanjo S, Fujishiro M, Gotoda T, Ichinose M, Ushijima T. Demonstration of the usefulness of epigenetic cancer risk prediction by a multicentre prospective cohort study. Gut. 2015;**64**:388-396. DOI: 10.1136/gutjnl-2014-307094

[39] Fan R, Wang WJ, Zhong QL, Duan SW, Xu XT, Hao LM, Zhao J, Zhang LN. Aberrant methylation of the GCK gene body is associated with the risk of essential hypertension. Molecular Medicine Reports. 2015;**12**:2390-2394. DOI: 10.3892/mmr.2015.3631

[40] Wang JY, Yamasaki S, Takeuchi K, Okabe S. Delayed healing of acetic acid-induced gastric ulcers in rats by indomethacin. Gastroenterology. 1989;**96**:393-402

[41] Tsuchiya T, Kameya N, Nakamura I. Straight walk: A modified method of ligation-mediated genome walking for plant species with large genomes. Analytical Biochemistry. 2009;**388**:158-160. DOI: 10.1016/j.ab.2009.02.002

[42] Watanabe Y, Kim H-S, Castoro RJ, Chung W, Estecio MRH, Kondo K, Guo Y, Ahmed SS, Toyota M, Itoh F, Suk KT, Cho M-Y, Shen L, Jelinek J, Issa J-JJ. Sensitive and specific detection of early gastric cancer with DNA methylation analysis of gastric washes. Gastroenterologia. 2009;**136**:2149-2158. DOI: 10.1053/j.gastro.2009.02.085

Chapter 11

Epigenetic Regulation of Hepatitis B Virus Replication

In Young Moon and Jin-Wook Kim

Additional information is available at the end of the chapter

http://dx.doi.org/10.5772/intechopen.81711

Abstract

Hepatitis B virus (HBV) is the most important cause of chronic viral hepatitis worldwide. The genome of HBV is 3.2 kb partially double-stranded DNA, which is translocated to the nuclei of infected hepatocytes and converted to complete double-stranded DNA, aka covalently closed circular DNA (cccDNA). Typical course of chronic HBV infection results in inactive carrier state with clearance of viral particles in the bloodstream. However, the cccDNA can be detected in the hepatocytes from inactive carriers by sensitive methods. It has been increasingly known that epigenetic mechanisms contribute to the control of HBV replication in the inactive stage of HBV infection. Histone modification and DNA methylation have been identified in the HBV cccDNA, leading to modification of transcriptional activity. The understanding of epigenetic control of transcription will shed light on the development of new therapeutic strategy against HBV cccDNA.

Keywords: hepatitis B virus, covalently closed circular DNA, histone modification, inactive carrier

1. Introduction

Hepatitis B virus (HBV) is the most important cause of chronic viral hepatitis worldwide. About 240–350 million people are infected with HBV globally [1]. Development of potent nucleos(t)ide analogs (Nas) has revolutionized the treatment of HBV, but current treatment cannot eradicate the DNA genome of HBV, i.e., covalently closed circular DNA (cccDNA) from the nuclei of infected hepatocytes. Since there have been no innate clearing mechanisms identified for foreign double-stranded DNA in mammalian cells, theoretically HBV cccDNA in the liver stem cells will dilute out but persist indefinitely in some portion of the hepatocytes [2]. Therefore, prolonged use of current NA therapy is recommended without interruption, which is very costly.

It has been known that transcriptional activity of HBV cccDNA varies according to the stage of natural history of chronic hepatitis B (CHB) [3, 4]. Interestingly, many patients with chronic hepatitis B are free from circulating HBV during the natural course despite the presence of HBV cccDNA in the infected nuclei [5]. These findings raise the possibility that replication of HBV is regulated at the transcriptional level. Genetic changes, i.e., DNA mutation, are an attractive explanation for the variable transcriptional activity since the reverse transcriptase activity of HBV is error-prone. However, no universal mutations have been identified associated with transcriptional suppression [6]. Consequently, epigenetic control has been proposed as the mechanism of these variable transcriptional activities in CHB patients [7], and this article covers the current knowledge of epigenetic mechanisms contributing to the transcriptional control of HBV replication.

2. Organization of HBV cccDNA and its transcriptional control

Hepatitis B virus (HBV) is a partially double-stranded circular DNA virus [8]. The viral DNA goes into nuclei of infected hepatocytes where it is converted to cccDNA [8]. The cccDNA is a viral minichromosome, which takes the form of "beads-on-a-string" conformation of nucleosomal packaging [9], analogous to DNA packaging by mammalian nucleosome. HBV core protein and X protein along with histone H3 and H4 are components of HBV minichromosome [9–11]. A variety of cellular transcription factors bind HBV cccDNA, which in turn control transcriptional activity of HBV promoters: the preC/pregenomic, S1, S2, and X promoters [12]. The core promoter initiates transcription of preC and pregenomic RNA, the template for the viral genome by reverse transcription. Ubiquitous transcription factors such as specificity protein 1 (SP1), nuclear factor kappa B (NF-κβ), activator protein 1 (AP-1), and liver-enriched transcription factors such as hepatocyte nuclear factor 3 (HNF3), CAAT enhancer-binding protein (C/EBP), and several nuclear receptors such as hepatocyte nuclear factor 4 (HNF4), peroxisome proliferator-activated receptors (PPAR) and retinoid X receptors (RXRα), farnesoid acid receptor (FXR), small heterodimer partner (SHP), and testicular orphan receptor 4 (TR4) can bind core promoter [13, 14].

3. Epigenetic control of HBV transcription: histone modification

As described above, ultrastructure of HBV cccDNA simulates that of mammalian nucleosome, suggesting the possibility of histone molecules as a main factor for transcriptional control [9]. Indeed, Pollicino et al. demonstrated the feasibility of HBV-associated histone modification as a major transcriptional regulator of HBV [11]. Genome-wise search for posttranslational modification (PTM) of HBV-infected liver cell lines has revealed that active marks of transcription such as H3K4me3, H3K27ac, and H3K122ac are abundant in active chromatin, especially in the core promoter region [19]. Interestingly, however, the repressive marks of transcription, i.e., H3K27me3 and H3K9me2, are depleted in the HBV cccDNA, suggesting that modified

Histone change	Modifier	HBV response and mechanisms	References
H3K4me3	MLL3	**Activation** (H3K4me modification of the NTCP promoter by MLL3 may facilitate HBV infection in vivo.)	[19, 30]
	Zinc finger and homeoboxes (ZHX2)	**Repression (**ZHX2 inhibited trimethylation of H3K4. Overexpression of ZHX2 also decreases the acetylation levels of H3K27 and H3K122.)	[26]
H3K27ac	EZH2	**Activation** (Knockdown of EZH2 resulted in upregulation of HBeAg and ABsAg, indicating a repressive function of EZH2 on HBV gene expression.)	[28, 31]
H3K122ac	P300/CBP	**Activation**	[19, 32, 33]
AcH3/AcH4 H3K4me3	HBx	**Activation**	[20–22, 24]
H4R3me2s	PRMT5	**Repression** (PRMT5-mediated histone H4 dimethyl Arg3 symmetric (H4R3me2s) represses cccDNA transcription.)	[28]
H3K9me3	SETDB1	**Repression** (upon HBV infection, cellular mechanisms involving SETDB1-mediated H3K9me3 and HP1 induce silencing of HBV cccDNA transcription through modulation of chromatin structure.)	[24]
	SIRT3	**Repression** (SIRT3 is a novel host factor epigenetically restricting HBV cccDNA transcription by acting cooperatively with histone methyltransferase.)	[25]
H3K27me3	Suz12	**Repression** (downregulation of Suz12 and Znf198 enhances HBV replication.)	[34]
H3K79me	KDM2B	**Repression** (KDM2B as an H3K79 demethylase and link its function to transcriptional repression via SIRT1-mediated chromatin silencing.)	[27]
etc	BCP mutation	**Repression** (BCP mutations decrease viral replication capacity possibly by modulating the acetylation and deacetylation of cccDNA-bound histones.)	[29]
	IFNα	**Repression**	[19, 35, 36]

Table 1. Histone modification affecting HBV transcription.

histones regulate HBV transcription mainly in favor of active replication. Pol2 enrichment is co-localized at the H3K27me-enriched transcription start site of precore/pregenomic area, and treatment with interferon alpha reduces the active PTMs, also suggesting that active marks of histone modification contribute to the transcriptional activity of HBV.

Much is unknown regarding the effects of histone modification on the binding of these transcription factors. Hepatitis X protein (HBx) is the most studied modulator of HBV-bound histone [20, 21]. HBx is bound to the cccDNA and enhance transcription by increasing histone acetylation and recruiting cellular coactivators p300, CBP, and PCAF [22], by inhibiting protein arginine methyltransferase 1 and reducing H4 methylation [23]. HBx also increases histone acetylation and H3K4me3 and decreases HP1 binding and

H3K9me3 on the cccDNA [24]. Other host transcription factors, mainly suppressors, that act via epigenetic control of HBV include SIRT3 [25], zinc finger and homeoboxes 2 (ZHX2) [26], KDM2B [27], protein arginine methyltransferase 5 (PRMT5) [28], and SETDB1 [24]. Interestingly, mutations in the basal core promoter are also reported to be associated with histone modification [29]. The effect of histone modification on HBV replication is summarized in **Table 1**.

4. Epigenetic control of HBV transcription: cccDNA methylation

Isolation and testing of HBV cccDNA shows methylation not only of HBV DNA integrated in host chromosome [39, 40] but also of HBV cccDNA. Methylation is speculated to affect replicative activity of HBV [38], and this hypothesis was confirmed in the hepatoma cell lines [41] and human liver tissue [15, 16]. HBV cccDNA has three CpG islands which harbor methylation in human liver and hepatoma cell lines [37, 38]. Methylation of HBV in the CpG islands of cccDNA is associated with suppressed transcriptional activity of HBV [15, 16] (for recent reviews, see [17, 18]). Especially, methylation of CpG island II is associated with reduced HBV replicability [42].

Although DNA methyltransferases, i.e., DNMT1, DNMT3a, and DNMT3b, are expressed in normal tissues [45], the level of expression is higher in HCC [46], which may explain the increased levels of methylation in hepatocellular carcinoma (HCC) compared to noncancerous tissues [43, 44]. In addition, since the hepatic expression of DNMT3a and DNMT3b, the de novo methylators, increases with age [47], methylation of HBV may also increase with age [42], which might explain suppressed replicative activity of HBV in the later stage of natural history. Degree of methylation also depends on HBeAg positivity [15, 42] and degree of hepatic fibrosis [42].

The mechanism of HBV DNA methylation is still unknown (**Table 2**). From the specific patterns of methylation in the HBV genome [42], it can be speculated that some kind of molecular chaperon (s) may guide the de novo methylation enzymes to the specific target sequence. HBx may play a role, because it recruits DNMT3A to the regulatory promoters [48]. Small RNAs may be a plausible candidate, as suggested by our in vitro study in which short hairpin RNA induced methylation of the target site in HBV [49].

HBV area	Modifier	HBV replication	References
CpG2	HBc	**Increased** (The relative abundances of HBc binding to CpG island 2 were associated with the binding of CREB binding protein (CBP) and with hypomethylation in CpG island 2 of HBV cccDNA minichromosomes.)	[50]
CpG3	PEG-IFN	**Decreased** (PEG-IFN treatment significantly increased methylation of HBV cccDNA in CpG island III.)	[51]

Table 2. Mechanisms of methylation in HBV cccDNA.

5. Therapeutic implications and future perspectives of epigenetics in chronic hepatitis B

NAs, the most commonly used modality of CHB therapy, is costly without definite duration. Interferon induces sustained virologic response with finite duration, but the response rate is suboptimal. The realistic goal of CHB therapy is to render the patients to the clinical situation similar to inactive carrier stage, i.e., normal alanine aminotransferase levels with low or negative serum HBV DNA levels. Since epigenetic silencing may contribute to the suppressive HBV replication status of inactive carrier stage, it would be theoretically feasible and clinically useful to induce epigenetic suppression of HBV replication simulating natural inactive stage of disease. Further studies will be needed to elucidate the mechanisms and long-term consequences of epigenetic suppression of HBV replication.

6. Conclusions

Epigenetic modification is an important mechanism of host-viral interaction in the transcriptional control of HBV. Current treatment strategy focuses on the inactivation/elimination of HBV cccDNA [17, 52], and knowledge on the epigenetic control is prerequisite for the novel development of HBV cure in the foreseeable future.

Acknowledgements

This work is supported by a National Research Foundation of Korea (NRF) grant to J-W Kim that was funded by the Korean Government (2017R1D1A1B03031483).

Author details

In Young Moon and Jin-Wook Kim*

*Address all correspondence to: kimjinw@snu.ac.kr

Seoul National University College of Medicine, Seongnam, South Korea

References

[1] MacLachlan JH, Cowie BC. Hepatitis B virus epidemiology. Cold Spring Harbor Perspectives in Medicine. 2015;**5**:a021410

[2] Li M, Sohn JA, Seeger C. Distribution of hepatitis B virus nuclear DNA. Journal of Virology. 2018;**92**:e01391-e01317

[3] Laras A, Koskinas J, Dimou E, Kostamena A, Hadziyannis SJ. Intrahepatic levels and replicative activity of covalently closed circular hepatitis B virus DNA in chronically infected patients. Hepatology. 2006;**44**:694-702

[4] Volz T, Lutgehetmann M, Wachtler P, Jacob A, Quaas A, Murray JM, et al. Impaired intrahepatic hepatitis B virus productivity contributes to low viremia in most HBeAg-negative patients. Gastroenterology. 2007;**133**:843-852

[5] Werle-Lapostolle B, Bowden S, Locarnini S, Wursthorn K, Petersen J, Lau G, et al. Persistence of cccDNA during the natural history of chronic hepatitis B and decline during adefovir dipivoxil therapy. Gastroenterology. 2004;**126**:1750-1758

[6] Jammeh S, Tavner F, Watson R, Thomas HC, Karayiannis P. Effect of basal core promoter and pre-core mutations on hepatitis B virus replication. The Journal of General Virology. 2008;**89**:901-909

[7] Levrero M, Pollicino T, Petersen J, Belloni L, Raimondo G, Dandri M. Control of cccDNA function in hepatitis B virus infection. Journal of Hepatology. 2009;**51**:581-592

[8] Seeger C, Mason WS. Molecular biology of hepatitis B virus infection. Virology. 2015; **479-480**:672-686

[9] Bock CT, Schwinn S, Locarnini S, Fyfe J, Manns MP, Trautwein C, et al. Structural organization of the hepatitis B virus minichromosome. Journal of Molecular Biology. 2001;**307**:183-196

[10] Lucifora J, Arzberger S, Durantel D, Belloni L, Strubin M, Levrero M, et al. Hepatitis B virus X protein is essential to initiate and maintain virus replication after infection. Journal of Hepatology. 2011;**55**:996-1003

[11] Pollicino T, Belloni L, Raffa G, Pediconi N, Squadrito G, Raimondo G, et al. Hepatitis B virus replication is regulated by the acetylation status of hepatitis B virus cccDNA-bound H3 and H4 histones. Gastroenterology. 2006;**130**:823-837

[12] Moolla N, Kew M, Arbuthnot P. Regulatory elements of hepatitis B virus transcription. Journal of Viral Hepatitis. 2002;**9**:323-331

[13] Quasdorff M, Protzer U. Control of hepatitis B virus at the level of transcription. Journal of Viral Hepatitis. 2010;**17**:527-536

[14] Hensel KO, Rendon JC, Navas MC, Rots MG, Postberg J. Virus-host interplay in hepatitis B virus infection and epigenetic treatment strategies. The FEBS Journal. 2017;**284**:3550-3572

[15] Guo Y, Li Y, Mu S, Zhang J, Yan Z. Evidence that methylation of hepatitis B virus covalently closed circular DNA in liver tissues of patients with chronic hepatitis B modulates HBV replication. Journal of Medical Virology. 2009;**81**:1177-1183

[16] Kim JW, Lee SH, Park YS, Hwang JH, Jeong SH, Kim N, et al. Replicative activity of hepatitis B virus is negatively associated with methylation of covalently closed circular DNA in advanced hepatitis B virus infection. Intervirology. 2011;**54**:316-325

[17] Hong X, Kim ES, Guo H. Epigenetic regulation of hepatitis B virus covalently closed circular DNA: Implications for epigenetic therapy against chronic hepatitis B. Hepatology. 2017;**66**:2066-2077

[18] Koumbi L, Karayiannis P. The epigenetic control of hepatitis B virus modulates the outcome of infection. Frontiers in Microbiology. 2015;**6**:1491

[19] Tropberger P, Mercier A, Robinson M, Zhong W, Ganem DE, Holdorf M. Mapping of histone modifications in episomal HBV cccDNA uncovers an unusual chromatin organization amenable to epigenetic manipulation. Proceedings of the National Academy of Sciences of the United States of America. 2015;**112**:E5715-E5724

[20] Luo L, Chen S, Gong Q, Luo N, Lei Y, Guo J, et al. Hepatitis B virus X protein modulates remodelling of minichromosomes related to hepatitis B virus replication in HepG2 cells. International Journal of Molecular Medicine. 2013;**31**:197-204

[21] Protzer U. Hepatitis: Epigenetic control of HBV by HBx protein—Releasing the break? Nature Reviews Gastroenterology & Hepatology. 2015;**12**:558-559

[22] Belloni L, Pollicino T, De Nicola F, Guerrieri F, Raffa G, Fanciulli M, et al. Nuclear HBx binds the HBV minichromosome and modifies the epigenetic regulation of cccDNA function. Proceedings of the National Academy of Sciences of the United States of America. 2009;**106**:19975-19979

[23] Benhenda S, Ducroux A, Riviere L, Sobhian B, Ward MD, Dion S, et al. Methyltransferase PRMT1 is a binding partner of HBx and a negative regulator of hepatitis B virus transcription. Journal of Virology. 2013;**87**:4360-4371

[24] Riviere L, Gerossier L, Ducroux A, Dion S, Deng Q, Michel ML, et al. HBx relieves chromatin-mediated transcriptional repression of hepatitis B viral cccDNA involving SETDB1 histone methyltransferase. Journal of Hepatology. 2015;**63**:1093-1102

[25] Ren JH, Hu JL, Cheng ST, Yu HB, Wong VKW, Law BYK, et al. SIRT3 restricts HBV transcription and replication via epigenetic regulation of cccDNA involving SUV39H1 and SETD1A histone methyltransferases. Hepatology. 2018;**68**(4):1260-1276

[26] Xu L, Wu Z, Tan S, Wang Z, Lin Q, Li X, et al. Tumor suppressor ZHX2 restricts hepatitis B virus replication via epigenetic and non-epigenetic manners. Antiviral Research. 2018;**153**:114-123

[27] Kang JY, Kim JY, Kim KB, Park JW, Cho H, Hahm JY, et al. KDM2B is a histone H3K79 demethylase and induces transcriptional repression via sirtuin-1-mediated chromatin silencing. The FASEB Journal. 2018;**32**(10):5737-5750, fj201800242R

[28] Zhang W, Chen J, Wu M, Zhang X, Zhang M, Yue L, et al. PRMT5 restricts hepatitis B virus replication through epigenetic repression of covalently closed circular DNA transcription and interference with pregenomic RNA encapsidation. Hepatology. 2017;**66**:398-415

[29] Koumbi L, Pollicino T, Raimondo G, Stampoulis D, Khakoo S, Karayiannis P. Hepatitis B virus basal core promoter mutations show lower replication fitness associated with cccDNA acetylation status. Virus Research. 2016;**220**:150-160

[30] Ananthanarayanan M, Li Y, Surapureddi S, Balasubramaniyan N, Ahn J, Goldstein JA, et al. Histone H3K4 trimethylation by MLL3 as part of ASCOM complex is critical for NR activation of bile acid transporter genes and is downregulated in cholestasis. American Journal of Physiology. Gastrointestinal and Liver Physiology. 2011;**300**:G771-G781

[31] Allis CD, Jenuwein T. The molecular hallmarks of epigenetic control. Nature Reviews. Genetics. 2016;**17**:487-500

[32] Tropberger P, Pott S, Keller C, Kamieniarz-Gdula K, Caron M, Richter F, et al. Regulation of transcription through acetylation of H3K122 on the lateral surface of the histone octamer. Cell. 2013;**152**:859-872

[33] Pradeepa MM, Grimes GR, Kumar Y, Olley G, Taylor GC, Schneider R, et al. Histone H3 globular domain acetylation identifies a new class of enhancers. Nature Genetics. 2016;**48**:681-686

[34] Andrisani OM. Deregulation of epigenetic mechanisms by the hepatitis B virus X protein in hepatocarcinogenesis. Viruses. 2013;**5**:858-872

[35] Belloni L, Allweiss L, Guerrieri F, Pediconi N, Volz T, Pollicino T, et al. IFN-alpha inhibits HBV transcription and replication in cell culture and in humanized mice by targeting the epigenetic regulation of the nuclear cccDNA minichromosome. The Journal of Clinical Investigation. 2012;**122**:529-537

[36] Liu F, Campagna M, Qi Y, Zhao X, Guo F, Xu C, et al. Alpha-interferon suppresses hepadnavirus transcription by altering epigenetic modification of cccDNA minichromosomes. PLoS Pathogens. 2013;**9**:e1003613

[37] Vivekanandan P, Thomas D, Torbenson M. Hepatitis B viral DNA is methylated in liver tissues. Journal of Viral Hepatitis. 2008;**15**:103-107

[38] Vivekanandan P, Kannangai R, Ray SC, Thomas DL, Torbenson M. Comprehensive genetic and epigenetic analysis of occult hepatitis B from liver tissue samples. Clinical Infectious Diseases. 2008;**46**:1227-1236

[39] Yamamura K, Tsurimoto T, Ebihara T, Kamino K, Fujiyama A, Ochiya T, et al. Methylation of hepatitis B virus DNA and liver-specific suppression of RNA production in transgenic mouse. Japanese Journal of Cancer Research. 1987;**78**:681-688

[40] Watanabe Y, Yamamoto H, Oikawa R, Toyota M, Yamamoto M, Kokudo N, et al. DNA methylation at hepatitis B viral integrants is associated with methylation at flanking human genomic sequences. Genome Research. 2015;**25**:328-337

[41] Vivekanandan P, Thomas D, Torbenson M. Methylation regulates hepatitis B viral protein expression. The Journal of Infectious Diseases. 2009;**199**:1286-1291

[42] Zhang Y, Mao R, Yan R, Cai D, Zhang Y, Zhu H, et al. Transcription of hepatitis B virus covalently closed circular DNA is regulated by CpG methylation during chronic infection. PLoS One. 2014;**9**:e110442

[43] Jain S, Chang TT, Chen S, Boldbaatar B, Clemens A, Lin SY, et al. Comprehensive DNA methylation analysis of hepatitis B virus genome in infected liver tissues. Scientific Reports. 2015;**5**:10478

[44] Kaur P, Paliwal A, Durantel D, Hainaut P, Scoazec JY, Zoulim F, et al. DNA methylation of hepatitis B virus (HBV) genome associated with the development of hepatocellular carcinoma and occult HBV infection. The Journal of Infectious Diseases. 2010;**202**:700-704

[45] Robertson KD, Uzvolgyi E, Liang G, Talmadge C, Sumegi J, Gonzales FA, et al. The human DNA methyltransferases (DNMTs) 1, 3a and 3b: Coordinate mRNA expression in normal tissues and overexpression in tumors. Nucleic Acids Research. 1999;**27**:2291-2298

[46] Oh BK, Kim H, Park HJ, Shim YH, Choi J, Park C, et al. DNA methyltransferase expression and DNA methylation in human hepatocellular carcinoma and their clinicopathological correlation. International Journal of Molecular Medicine. 2007;**20**:65-73

[47] Xiao Y, Word B, Starlard-Davenport A, Haefele A, Lyn-Cook BD, Hammons G. Age and gender affect DNMT3a and DNMT3b expression in human liver. Cell Biology and Toxicology. 2008;**24**:265-272

[48] Zheng DL, Zhang L, Cheng N, Xu X, Deng Q, Teng XM, et al. Epigenetic modification induced by hepatitis B virus X protein via interaction with de novo DNA methyltransferase DNMT3A. Journal of Hepatology. 2009;**50**:377-387

[49] Park HK, Min BY, Kim NY, Jang ES, Shin CM, Park YS, et al. Short hairpin RNA induces methylation of hepatitis B virus covalently closed circular DNA in human hepatoma cells. Biochemical and Biophysical Research Communications. 2013;**436**:152-155

[50] Guo YH, Li YN, Zhao JR, Zhang J, Yan Z. HBc binds to the CpG islands of HBV cccDNA and promotes an epigenetic permissive state. Epigenetics. 2011;**6**:720-726

[51] Uchida T, Imamura M, Hayes CN, Hiraga N, Kan H, Tsuge M, et al. Persistent loss of hepatitis B virus markers in serum without cellular immunity by combination of peginterferon and entecavir therapy in humanized mice. Antimicrobial Agents and Chemotherapy. 2017;**61**:e00725-e00717

[52] Seeger C. Control of viral transcripts as a concept for future HBV therapies. Current Opinion in Virology. 2018;**30**:18-23

Chapter 12

Part 1: The PIWI-piRNA Pathway Is an Immune-Like Surveillance Process That Controls Genome Integrity by Silencing Transposable Elements

Didier Meseure and Kinan Drak Alsibai

Additional information is available at the end of the chapter

http://dx.doi.org/10.5772/intechopen.79974

Abstract

PiRNAs [P-element-induced wimpy testis (PIWI)-interacting RNAs] represent the most frequent but the least well-investigated subtype of small ncRNAs and are characterized by their interaction with PIWI proteins, a subclass of the Argonaute family. PiRNAs and PIWI proteins maintain integrity of the genomic structure and regulate gene expression in germline and somatic cells. The PIWI-piRNA pathway primarily constitutes a conserved immune-like surveillance process that recognizes self and nonself. This axis controls genome integrity of germline cells and nonaging somatic cells by silencing and suppressing propagation of transposable elements through epigenetic and posttranscriptional mechanisms. However, mounting evidences indicate that the PIWI-piRNA pathway has broader implications in both germinal and somatic cells in various physiological and pathological processes. It modulates mRNAs levels of expression, stability, turnover, and translation and interacts directly with many transcription factors and signaling pathways molecules. PIWI proteins and piRNAs play pivotal roles in germline stem cell maintenance and self-renewal, fertilization and development, genes and proteins expression, genome rearrangement, and homeostasis.

Keywords: piRNA, PIWI proteins, transposable element (TE), transcriptional and posttranscriptional silencing, piRNA cluster, heterochromatin, DNA methylation, ping-pong cycle, nuage

1. Introduction

Cancer is an extremely complex disorder characterized genetically, epigenetically, and histologically by highly heterogeneous proliferative cellular subpopulations, including cancer stem cells (CSCs) and progenies. These cells harbor chromosomal abnormalities, alterations of suppressor genes (TSG) and oncogenes, and aberrant transcriptomic profiles generated by genetic and epigenetic alterations [1, 2]. These cancer cells are in close relationship with a tumor microenvironment (TME), composed of immune and nonimmune stromal cells and modified extracellular matrix. Reciprocal interactions between tumor cells and TME are pivotal in cancer progression, allowing remodeling of TME and reprogramming of cancer cells that develop adaptive strategies to adjust their phenotype to unfavorable environmental conditions. Recently, CSCs were implicated in a new paradigm accounting for tumor heterogeneity [3]. CSCs have the property of self-renewal, lack senescence, maintain an undifferentiated state, and proliferate rapidly. These properties are controlled by epigenetic mechanisms that induce changes in gene expression profiling of tumor cells. Opposite to aging cells that increase genomic and chromosomal instability during adulthood, nonaging immortal cells, such as germline, somatic, and cancer stem cells, harbor a genomic instability triggered by unrepaired mutations with either no or only limited number of genomic alterations [4]. Epigenetic abnormalities are early events in cancer progression, resulting from various environmental injuries, and associate heterogeneity of DNA methylation, posttranscriptional modifications of histones, and deregulation of noncoding RNAs (ncRNAs). Global DNA hypomethylation results in chromosomal instability, overexpression of oncogenes, and reactivation of transposable elements (TEs) [5]. Localized (genes promoters) or wide (>1 Mb) DNA hypermethylation initiates repression of TSGs and modification of epigenetic marks through histone alterations, resulting in occurrence of an aberrantly stemlike state of CSCs. These alterations of the genomic methylation during carcinogenesis allow reprogramming of atypical proliferative cells into highly malignant cells characterized by unlimited proliferation, epithelial-mesenchymal transition (EMT), invasion, and prometastatic properties [6].

Until recently, RNAs were considered as epigenetic regulators and mediators of gene expression, functioning as intermediates of translation in the flow of genetic information from DNA to proteins [7]. Large-scale genomic technologies have provided an astonishing insight into human genome and transcriptome. Next-generation sequencing techniques combined with bioinformatics have revealed that more than 50% of mammalian genomes were composed of TEs and that more than 98% of the human genome was actively transcribed [8]. However, only 1.1% of the genome encodes proteins, and a majority of genes are noncoding RNAs (ncRNAs) [9]. NcRNAs play pivotal roles in developmental and homeostatic processes, and their alterations are implicated in the pathogenesis of many diseases, by modulating expression of numerous genes at epigenetic, transcriptional, and posttranscriptional levels [10]. Most importantly, ncRNAs are frequently deregulated in cancer and have crucial roles in tumor initiation, progression, and metastatic spread. NcRNAs are classified into housekeeper ncRNAs (rRNAs, tRNAs, and snoRNAs) and regulatory ncRNAs. Regulatory ncRNAs are divided into several subfamilies, depending on their size, biogenesis, and biological functions. Small ncRNAs are composed of transcripts shorter than 200 nucleotides (nt), whereas long noncoding RNAs (lncRNAs) comprise transcripts longer than 200 nt [11]. Small ncRNAs also differ by their precursor structure and their mechanisms of biogenesis. They comprise microRNAs (miRNAs), small interfering RNAs (siRNAs), and PIWI-interacting RNA (piRNAs) [12–15].

MiRNAs and siRNAs are generated from double-stranded precursors, whereas piRNAs are processed from long single-stranded precursors. The endoribonuclease Dicer is pivotal in the maturation of miRNAs and siRNAs, but not in the piRNAs processing [16]. Regulatory

http://dx.doi.org/10.5772/intechopen.79974

functions of small ncRNAs are insured by Argonaute (AGO) protein family, which is a very well-conserved master component of RNA silencing complexes in all organisms [17]. The least well-investigated small ncRNAs are the piRNAs, which were first identified in 2006 in mouse and rat germ cells as ncRNAs interacting with PIWI proteins, a subclass of the Argonaute proteins [18–21]. PiRNAs actually constitute the largest and most diverse class of ncRNAs [16]. PiRNAs and PIWI proteins were initially implicated in epigenetic regulation of germline cells and their overexpressions have been more recently observed in various cancers through aberrant DNA methylation.

This review will provide an overview of the PIWI-piRNA pathway, focusing mainly on origin, biochemical properties, biogenesis, functions, and mechanisms of action in germline and somatic tissues. Furthermore, we will discuss emerging implications of piRNAs in carcinogenesis and highlight their potential clinical utilities as diagnostic/prognostic biomarkers and therapeutic tools.

2. The PIWI-piRNA pathway

RNA interference (RNAi) is a widely conserved small-RNA-mediated gene-silencing mechanism involved in crucial homeostatic events of most eukaryotes [22–24]. Small regulatory RNAs of 20–32 nt, such as endogenous siRNAs, miRNAs, and piRNAs, modulate transcriptional and posttranscriptional repression through complementary RNA or DNA recognition by interacting with well-conserved proteins of 95 kDa belonging to the Argonaute family that cleave their targets [25, 27]. AGO proteins include a PAZ domain [P-element-induced wimpy testis (PIWI)-AGO-Zwille] located in the N terminal region, which binds small ncRNAs and a PIWI domain in the C-terminal region that functions as double-strand-specific RNA endonuclease [28]. Based on sequence homology and functional domains in different species, the AGO family of proteins is divided into three subfamilies: AGO proteins (homologous to *Arabidopsis thaliana* AGO1), PIWI proteins (homologous to *Drosophila melanogaster* Piwi), and WAGO (worm-specific Argonaute clade). The number of AGO family proteins varies considerably between species. Eight different proteins were present in humans, whereas 27 proteins were identified in *C. elegans* and only one protein was observed in fission yeast [29]. AGO proteins interact functionally with siRNAs and miRNAs, which are small single-stranded RNAs of 20–22 nt in length processed in a Dicer-dependent manner from double-stranded precursors, to induce posttranscriptional gene silencing in the cytoplasm [30–32]. Conversely, PIWI proteins are implicated in biogenesis of piRNAs and in their main function through transcriptionally and posttranscriptionally repressing TEs in the nucleus and the cytoplasm [33]. PIWI proteins and piRNAs edify ribonucleoproteins named PiRNA-induced silencing complexes (pi-RISCs). Pi-RISCs specificity is determined by piRNA sequence, whereas Argonaute PIWI protein mediates its effector function. PiRNAs associate with PIWI proteins and guide piRISCs to recognize complementary targets and achieve RNA silencing at transcriptional and posttranscriptional levels (**Figure 1**). Cytoplasmic PIWI-piRNA complexes silence their targets posttranscriptionally *via* piRNA-directed cleavage and the "ping-pong" amplification cycle, whereas nuclear PIWI proteins and piRNAs silence gene transcriptionally through epigenetic changes, including DNA methylation, implementation of H3K9me3 repressive marks, interactions with Mael and HP1 proteins, and repression of Pol II.

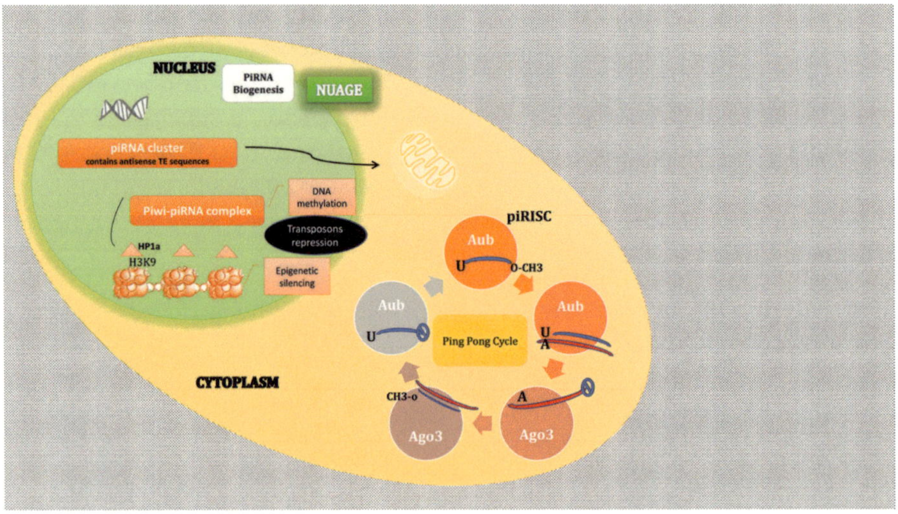

Figure 1. Biogenesis of piRNA and PIWI-piRNA pathways and their function in maintaining genome integrity through transposable element (TE) in germline cells at transcriptional and posttranscriptional level. In Drosophila ovaries, the primary pathway (in the nucleus) operates in both germline and surrounding somatic cells, whereas the "ping-pong" cycle (in the cytoplasm) operates only in germline cells. In the nucleus, PIWI-piRNA complex can regulate HP1, H3K9 methylation, and DNA methylation to influence transposons.

PiRNAs are small single-stranded ncRNAs of 25–33 nt identified in various organisms ranging from sponges to higher vertebrates [34]. Experimental and bioinformatics studies have shown that piRNAs are the most abundant small ncRNAs expressed in mammalian species [35]. They are derived from long single-strand RNA precursors in a Dicer-independent manner. The human genome comprises more than 30,000 piRNAs in which 80% originate from intergenic sequences and 20% from introns and exons of pre-mRNAs [36]. Unlike miRNAs and endosiRNAs, production of piRNAs is not carried out in a precise manner and single strands of long primary precursor transcripts generate numerous piRNAs without a conserved sequence [37]. They comprise Uracil at their 5′ end and methylated 2′-O group at their 3′ end [38, 39]. PiRNAs were first identified in *Drosophila melanogaster* and were named repeat-associated small-interfering RNAs (rasiRNAs) because of their repetitive elements and TEs suppressing activity. RasiRNAs were later found to interact with Argonaute PIWI proteins and ultimately renamed piRNAs in 2006 [40–42]. PiRNAs were further investigated in *Caenorhabditis elegans*, zebrafish, mice, and more recently in humans. Their vast number present in numerous locations of the genome suggests that piRNAs may have potential crucial implications in the control of major biological processes. Indeed, PIWI-piRNA complexes silence TEs and control expression and activity of genes and proteins. They are also instrumental in genome rearrangement, germ stem cell maintenance, reproduction and fertility regulation, embryogenesis, and homeostasis [26, 43–48]. At the opposite of miRNAs and endosiRNAs, piRNAs function only through binding with PIWI proteins and harbor tissue-specific expression in various organs such as prostate and thyroid [49].

PIWI proteins were also initially identified in *Drosophila melanogaster* in which they play crucial roles in germline stem cell maintenance and self-renewal [50]. These proteins contain three functional domains: the PIWI-Argonaute-Zwille (PAZ) domain recognizing the 3′ end of the RNA, the middle domain (MID) providing a binding pocket for the 5′ end of guide strand RNA, and the PIWI domain containing catalytic residues that cleave target transcripts [47]. Unlike proteins of the AGO subclass, PIWI proteins comprise posttranslationally dimethylated arginine-rich motifs that allow interactions with Tudor proteins. These last proteins have pivotal role in functional activities of PIWI proteins by providing a scaffold for edification of higher-order molecular complexes located in *Drosophila* germ cells and mouse testis perinuclear granules named "nuage," similar to P-bodies [51–53]. The PIWI protein family is conserved in numerous organisms, including jellyfish, sponge, *planaria*, zebrafish (Ziwi, Zili), *Caenorhabditis elegans* (Prg1, Prg 2), *Drosophila melanogaster* (Piwi, Aub, Ago3), mouse (MIWI, MILI, MIWI2), and human (PIWIL1, PIWIL2, PIWIL3, PIWIL4) [51] (**Table 1**). PIWIL3 is observed only in human, and its functions are actually largely unknown. PIWI proteins expression is identified in a majority of organs such as liver, lung, heart, brain, pancreas, and kidney [54, 55]. PIWI proteins and piRNAs were first implicated in development, differentiation, and maintenance of germline cells [18, 56, 57]. However, mounting evidence has revealed that the PIWI-piRNA pathway is also instrumental in controlling gene expression both in germinal and somatic cells [58].

This pathway has pivotal roles at all steps of oogenesis and spermatogenesis, but also in somatic cells such as ovary and testis of *Drosophila* [15]. This axis also controls, although at lower levels of expression, numerous biological processes implicated in homeostasis, including brain maturation [59] pancreatic function [55], fat metabolism [60], and regeneration [61]. Indeed, this pathway was initially studied in gonads and implicated in gene silencing of germinal cells [43]. Loss of function studies performed in zebrafish, *Drosophila*, *Caenorhabditis*, and mice have confirmed that the PIWI-piRNA pathway is involved in germline development, spermatogenesis, and maintenance of germline stem cells. Mutations in this pathway resulted in expansive TEs mobility, genomic instability, and sterility [44]. PIWI proteins have nonredundant functions in cell compartments. *Drosophila* PIWI proteins Aub and Ago3 cleaved TEs in the cytoplasm, whereas Piwi inactivated TEs in the nucleus [62, 63]. All mouse PIWI proteins MIWI, MILI, and MIWI2 were expressed during spermatogenesis, whereas only MILI was weakly expressed in female germinal cells [64, 65]. These mouse PIWI proteins not only silenced TEs posttranscriptionally but also inactivated TEs genes transcriptionally through CpG DNA methylation on TEs loci. Homozygous MIWI, MILI, and MIWI2 knockout male mice models were associated with propagation of LINE1 sequences, depleted spermatogenesis, and apoptosis of germinal cells [66]. Particularly, *Drosophila* PIWI mutants were correlated with derepression of TEs, absence of germline stem cell renewal, and depletion of gametes [67–69]. Actually, the main function of this pathway is maintaining germline and somatic genome integrity by silencing TEs at transcriptional and posttranscriptional levels [70]. However, only 20% of piRNAs are localized in TEs and other repeat genomic regions, suggesting that this pathway may have additional biological functions. In germinal cells, the PIWI-piRNA pathway prevents genomic instability of the next generation and sterility. In somatic nonaging cells, this pathway is pivotal in self-renewal, differentiation and maturation of stem cell, embryonic development, and whole body regeneration. In somatic tissues, the PIWI-piRNA pathway is implicated in chromosomal conformation, memory-related synaptic plasticity, transcriptional regulation of mRNAs with deadenylation, and transgenerational inheritance to preserve the memory of self and nonself [71–73].

<table>
<tr><th></th><th>PIWIL1</th><th>PIWIL2</th></tr>
<tr><td>Chromosome
NCBI</td><td>

12q24.33</td><td>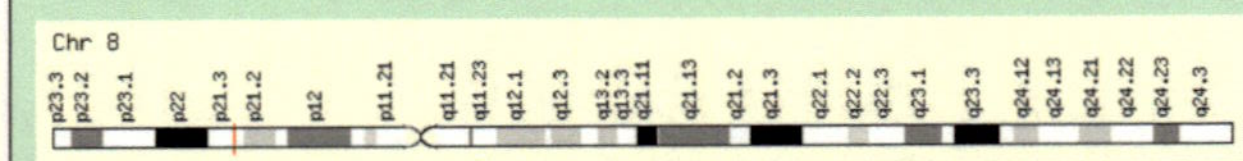

8p21.3</td></tr>
<tr><td>Molecule
*Bamezai et al, 2012</td><td>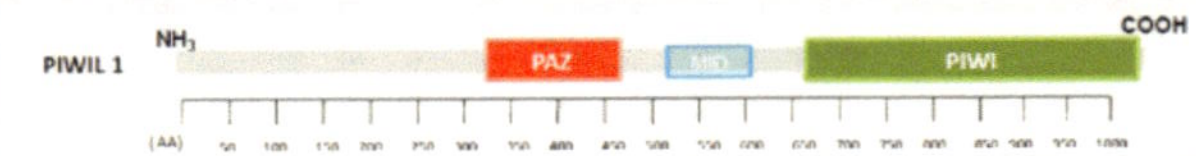
</td><td>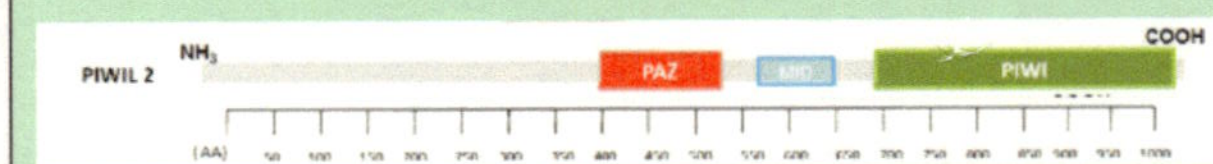
</td></tr>
<tr><td>Network
GeneCards</td><td>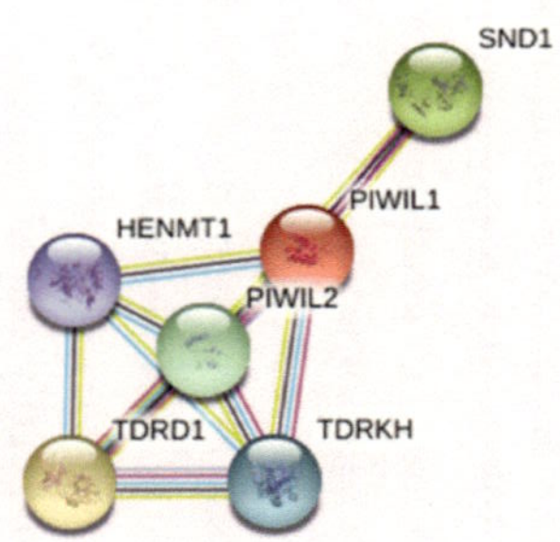
</td><td>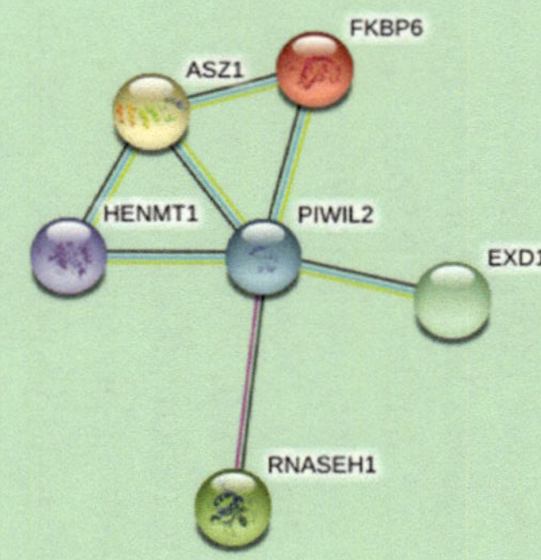
</td></tr>
</table>

	PIWIL3	PIWIL4
Chromosome NCBI	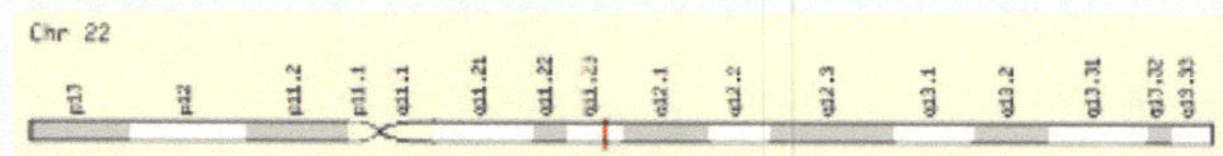 22q11.23	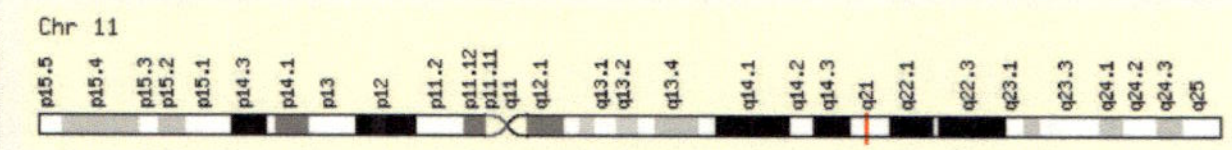 11q21
Molecule Bamezai *et al*, 2012	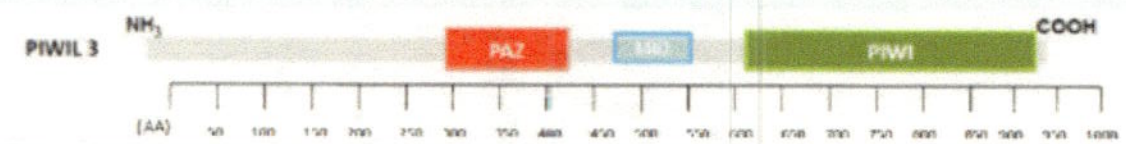	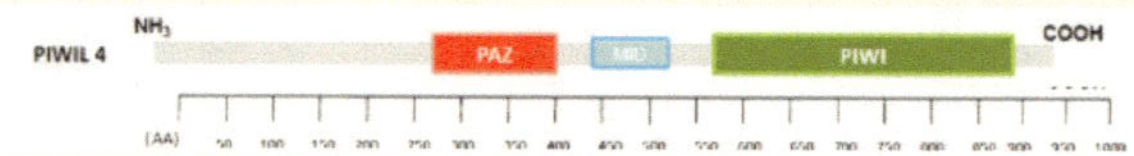
Network GeneCards	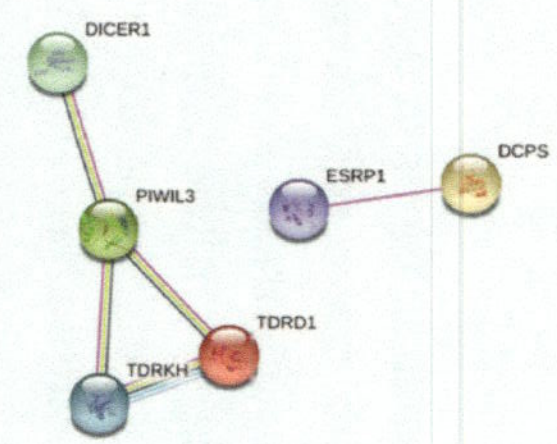	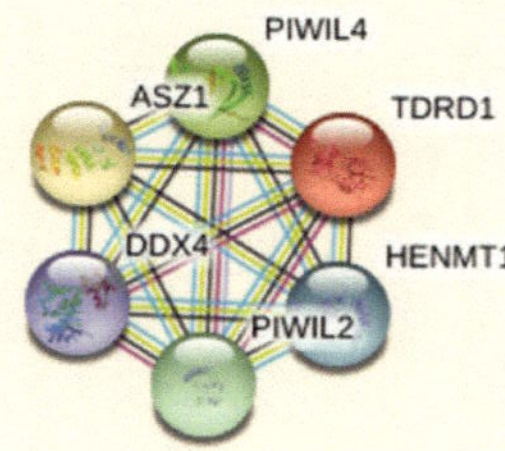

Table 1. This table shows the structure and the network of human PIWI proteins. The human PIWI protein family includes PIWIL1, PIWIL2, PIWIL3, and PIWIL4. The general structure of Argonaute proteins depicting the PAZ domain (red) with the MID domain (blue), and PIWI domain (green). NCBI (http://www.ncbi.nlm.nih.gov) [74].

3. Origin and biogenesis

Understanding of piRNA origin and biogenesis results principally from studies in *Drosophila* and mice [75]. PiRNAs can be classified according to their origin in three subgroups: transposon-derived piRNAs, mRNA-derived piRNAs, and lncRNAs-derived piRNAs. Transposon-derived piRNAs are produced from both genomic strands and generate sense and antisense piRNAs, whereas RNA-derived piRNAs are transcribed from 3′ untranslated regions (UTRs) of mRNAs and lncRNAs-derived piRNAs originate from the entire transcript [76]. Unlike miRNAs and siRNAs, which are derived from stem-loop and double-stranded precursors that are processed by the RNAse III Dicer, piRNAs are predominantly transcribed as large up to 200 kb single-stranded precursors independently from Dicer [43]. Furthermore, piRNAs do not possess secondary structures [77, 78].

The piRNA pathway is composed of PIWI proteins that interact with piRNAs, whose precursors are transcribed from piRNA clusters, cleaved by PIWI proteins, and secondary amplified in the cytoplasm through a sequence-complementary-dependent "ping-pong" cycle. Mature piRNAs are thus derived from two major pathways, the primary pathway and the "ping-pong" cycle that amplifies secondary piRNAs. In germline cells, molecules implicated in biogenesis of the PIWI-piRNA pathway are located at a perinuclear organelle called the "nuage" [79, 80]. Various components of the "nuage" colocalize with mitochondria [81]. In *Drosophila*, the primary pathway was observed in both germline and somatic cells, whereas the "ping-pong" cycle was identified only in germline cells.

3.1. Primary piRNA biogenesis

Deep sequencing of piRNAs recently revealed millions of distinct piRNAs [29]. However, they were usually located to discrete genomic loci, called piRNA clusters [31]. In the primary piRNA biogenesis, piRNAs provide from long single-strand RNA precursors originating from these clusters. These transcriptional units are highly enriched in dysfunctional remnants of TEs and other repetitive elements and are mainly located in pericentromeric and subtelomeric heterochromatin [41, 70, 82]. PiRNA clusters constitute the basis of immunity against TEs dissemination. Primary piRNAs derived from these clusters include uridine (U) at their 5′ nucleic acid and are mostly antisense to TEs mRNA sequences, functioning as guides for PIWI proteins to inactivate TE transcripts through complementary base pairing [40–42]. In the female *Drosophila* germline, these loci are either unidirectionally transcribed (unistrand clusters generating antisense piRNAs) or bidirectionally transcribed (dual-strand clusters generating both sense and antisense piRNAs), producing piRNAs that map to one genomic strand and both strands, respectively [41]. Transcription of unistrand clusters is performed through the canonical polymerase II, whereas dual-strand clusters transcription is generated *via* the noncanonical rhino-deadlock-cutoff (RDC) complexes that are also recruited by PIWI proteins and piRNAs through an intricate feedback loop [34, 83–86]. In female flies, piRNA clusters are expressed in germline cells (oocytes and nurse cells) and somatic cells (follicular cells). Interestingly, germline clusters are transcribed bidirectionally, whereas somatic clusters are transcribed unidirectionally, producing piRNAs antisense to TE coding regions in flies. In mouse spermatogenic cells, one class of piRNA clusters is transcribed during embryonic development and defends the germline against TEs, whereas a second class of clusters is expressed in adolescent mice during the first division of meiosis. The transcription factor A-MYB regulates expression of pachytene piRNA clusters and regulates their transcription through the PIWI-piRNA pathway in mouse [88–97].

In *Drosophila,* nuclear primary transcripts are processed into cytoplasmic mature primary piRNAs (**Figure 1**). These transcripts are resolved of secondary structures by the RNA helicase Armitage and then cleaved by the mitochondria-associated endonuclease Zucchini to generate pre-piRNAs with a characteristic 5′ monophosphate [93–97]. Pre-piRNAs are then loaded on PIWI proteins and their 3′ ends trimmed to a final length by the 3′–5′ exonuclease Nibbler [98, 99]. The 2′ hydroxy group at the 3′ end is then methylated by the small-RNA 2′-O-methyltransferase Hen1 that increases PIWI binding affinity and piRNA stability, while the 5′ end residue of the piRNA incorporated in PIWI shows a strong bias for uridine residues [100–103]. After processed into final length, piRNAs bind PIWI proteins and edify piRNA/PIWI ribonucleoprotein effector complexes (piRISCs) located into the cytoplasmic perinuclear "nuage" [104–106]. PiRISCs migrate back to the nucleus and reach their target genes to epigenetically repress their transcription. Through complementary base pairing of piRNAs and DNA, piRISCs induce transcriptionally heterochromatin formation by establishing a repressive H3K9me3 chromatin state mark on chromatin at target TEs loci and adjacent genes, in order to induce their silencing [107, 108]. H3K9me3 repressive marks are deposited by SETDB1 and Su(var)3–9 methyltransferases and heterochromatin protein 1 (HP1) [109, 110]. In *Drosophila,* the nuclear protein Panoramix is an adaptor allowing interactions between the PIWI-piRNA pathway and the general silencing machinery. Panoramix is implicated with its nuclear partner Asterix, in amplification of the piRNA-dependent TEs silencing [111]. In this way, piRNAs constitute transcriptional regulators that act mainly on TE sequences by recruiting histone methyltransferases, which will lead to establishment of transcriptionally silent heterochromatin [26].

In *Drosophila,* primary piRNAs accumulating in the cytoplasm are amplified by the "ping-pong" cycle [19]. They interact with Ago3 or Aub proteins to form piRNA/Ago or piRNA/Aub complexes, which contain complementary sequences to each other. PiRNA/Ago complexes generate sequences of RNA functioning as substrates for the generation of new piRNAs, which can load Aub proteins. Resulting piRNA/Aub complexes will generate additional RNA substrates to edify new piRNA/Ago3 complexes. The "ping-pong" amplification cycle is mainly observed in early evolutionary species, including *sponges, zebrafish,* and *D. melanogaster* [34].

3.2. Secondary piRNA biogenesis

Secondary piRNAs are generated from mRNA transcripts of active TEs [79]. They are primed in the cytoplasmic "nuage" by primary piRNAs (**Figure 1**) that guide their associated PIWI proteins to cleave target TE transcripts based on sequence complementarity [112]. Cleaved TEs are loaded on another PIWI protein and modified to give rise to multiplied secondary piRNAs in an amplification loop, called the "ping-pong" cycle. This posttranscriptional mechanism associates TEs silencing with piRNAs biogenesis by modifying TEs transcripts to give rise to secondary piRNAs [52]. Cleavage of TEs transcripts by PIWI proteins leads to destruction of TEs message, generation of secondary piRNAs, and concomitant amplification of these defensive sequences targeting active TEs [44]. This process is highly conserved through species and characterized by 5′ U bias of primary piRNAs, 10th adenosine bias of secondary piRNAs, and 10-nt overlap between the 5′ ends of primary and secondary piRNAs [113–115]. The secondary piRNA biogenesis cycle may constitute an adaptive system to TEs propagation by increasing piRNAs production after incorporation of new TEs into piRNA clusters [116, 117].

3.3. Cellular localization and mechanisms of action

PIWI proteins and piRNAs are located in the nucleus and the cytoplasm of cells expressing this pathway. Loading of piRNAs onto PIWI proteins is localized into the cytoplasm, and PIWI-piRNA complexes generated are required for trafficking of PIWI proteins to the nucleus [90]. Several cytoplasmic organelles, including mitochondria and the "nuage," are instrumental in functional activity of the PIWI-piRNA axis by controlling piRNA precursor processing [118]. PIWI-piRNA complexes control gene expression through two different mechanisms of action functioning at transcriptional and posttranscriptional levels.

At transcriptional level, nuclear PIWI-piRNA complexes control TEs and gene expression by promoting epigenetic modifications of the chromatin structure and histone proteins through combining DNA and histone methylation. PIWI proteins and piRNAs regulate expression and activity of three active DNA methyltransferases (DNMT1, DNMT3A, and DNMT3B), which normally repress initiation of transcription through methylation of CpG islands in promoter sites of target genes. When PIWI-piRNA complexes recognize TEs and target transcripts, they directly upregulate expression of these DNA methyltransferases and prevent binding of transcription factors through methylation of promoter regions. PIWIL1 induces overexpression of DNMT1 and DNMT3a [119], and piR-823 upregulates DNMT3A and DNMT3B [120]. PIWIL2 and PIWIL4 promote overexpression of DNMT1, DNMT3A, and DNMT3B, which silence TEs and target genes. Experimental loss of PIWIL2 and PIWIL4 induces downregulation in DNA methylation of promoter regions [121]. The PIWI-piRNA complexes also control methylation of histone lysine residues H3K and H4K through recruiting and interacting with histone methyltransferases (HMTs) such as Suv39H1 and SETDB1, which upregulate the histone H3 lysine 9 methylation (H3K9me). Furthermore, these complexes bind with different isoforms of HP1 and guide them to interact with H3K9me in target regions, which is a gene repressive mark. Accumulated methylation of H3K9 induces a heterochromatin state that allows segregation of chromosomes during cell division and prevents accessibility of TEs and genes to transcription factors [122]. Thus, PIWI-piRNAs complexes promote gene repression by using epigenetic mechanisms that allow HP1α recruitment to TEs loci, heterochromatin edification, and transcription silencing state [123]. PIWIL2 and PIWIL4 increase H3K9 methylation [124]. PIWIL4 recruits SUV39H1 or SETDB1 and promotes H3K9 methylation in promoter region of *CD1A* in monocytes, resulting in recruitment of HP1α and repression of gene transcription [125]. In leukemias, cell cycle-related piRNAs hsa-piR_014637 and hsa_piR_011186 are implicated in edification of molecular complexes combining DNMT1 and HMTs Suv39H1 and EZH2 that induce H3K9 and H3K27 methylation in the *CDKN2B* gene and inhibition of its transcription by DNMT1-induced CpG methylation in promoter region. In *Drosophila*, PIWI proteins also interact with subunits of the polycomb repressive complex 2 (PRC2). They maintain integrity of the ovary germline stem cells genome through preventing binding of PRC2 to HP1α at target gene sites and inhibition of H3K27 di- and trimethylation, a repressive mark upregulated on facultative heterochromatin [126]. Reduction of HP1α interactions with H3K27me3 promotes maintenance of constitutive heterochromatin, which is pivotal for accurate chromatin segregation and repression of developmentally regulated genes [127].

At posttranscriptional level, cytoplasmic PIWI-piRNA complexes principally govern degradation of TEs transcripts through the "ping-pong" amplification cycle. Apart from their implication in repression of TEs transcripts, the cytoplasmic functions of the PIWI proteins are mostly independent of their partner piRNAs. PIWI proteins modulate functions of many intracellular signaling proteins and receptors through degradation of mRNAs, inhibition of translation, and posttranslational modifications. PIWI proteins inhibit gene expression through mRNAs degradation by interacting with deadenylation complexes (Trf4-Air2-Mtr4 polyadenylation complex or CCR4 complex), resulting in shortening of poly-A tails. The PIWIL4-piR30840-Ago4 complex induces degradation of pre-mRNAs through binding to the Trf4-Air2-Mtr4 polyadenylation complex in human T lymphocytes [128]. They repress translation by interacting with translation initiation factors (eIF3a, eIF4E, eIF4F), preventing ribosomal subunits binding to 5′ cap of mRNAs. In mouse, Miwi interacts with eIF4E, while Mili binds to eIF3a, eIF4E, and eIF4F [129]. These proteins also regulate activity and stability of numerous molecules belonging to major signaling pathways by controlling posttranslational modifications such as phosphorylation and ubiquitination. PIWIL2 and PIWIL4 interact with the transcription factor STAT3 and upregulate its phosphorylation and activity. At the opposite, PIWIL4 binds to p53 and prevents serine 15 phosphorylation, inhibiting its functions [130]. PIWI proteins can also upregulate stability of target molecules by preventing their ubiquitination-dependent degradation. Interaction of PIWIL1 with Stathmin 1 inhibits its phosphorylation, resulting in prevention of PIWIL1 degradation by the ubiquitin ligase RLIM. Likewise, PIWIL2 binding to cytokeratin 8 promotes its phosphorylation and upregulates its stability by preventing its ubiquitination-derived degradation [131].

3.4. Biological functions

Up to now, biological functions of piRNAs have been only partially identified, due to the wide variation in piRNA sequences and mechanisms of action over species. However, a great majority of piRNAs are not complementary to mRNAs of target genes and are mainly implicated in epigenetic regulation rather than posttranscriptional modulation of biologic processes. PiRNAs have been implicated in TEs silencing, epigenetic, genes and proteins regulation, genome rearrangement, fertilization, germline and somatic stem cell self-renewal, embryogenesis, and maintenance of homeostasis.

3.4.1. Maintenance of genome stability and integrity

The PIWI-piRNA pathway maintains integrity and stability of the general organization of the genome, including regulation of genes, through recognition of self and nonself and prevention of TEs propagation. During interphase, the genome of eukaryotic cells is organized into various spatial three-dimensional topologically associating domains (TADs) edifying functional subcompartments implicated in pivotal cellular activities [132]. It has been recently observed in somatic cells of Drosophila ovaries PIWI-interacting chromosomal domains overlapping with genomic regions bound by nuclear pore complexes (NPCs). Furthermore, a third of protein-coding genes have been identified in the PIWI-interacting domains. PIWI proteins stochastically interact with nascent transcripts of genes and TEs and scan them through complementarity with piRNAs. Although perfect complementarity allows transcriptional silencing of TEs, imperfect complementarity leads to maintenance of PIWI proteins interactions with transcripts in the mRNPs after their detachment from the sites of transcription until dissociation from mRNPs in the nucleoplasm [133].

3.4.2. Retrotransposons silencing

TEs, first identified in maize by Barbara McClintock in the 1940s, constitute genetic units that can move and propagate within the genome [70]. More recently, genome-sequencing techniques have revealed that TEs occupy 15–22% of the genome of *Drosophila melanogaster* and 55% of the human genome. TEs of the human genome are divided into two major classes. Class II comprises DNA transposons that are generally inactive genetic elements representing less than 2% of the human genome and depending on transposases for their mobilization. They do not need transcription to propagate and use a "cut and paste" mechanism to excise and insert into new genomic sites without increasing their copy number. Conversely, class I retrotransposons are usually active genetic elements propagating through a "copy and paste" mechanism that allows retrotranscription into cDNA by a reverse transcriptase encoded by the retrotransposon and insertion into new genomic sites *via* these RNA transposition intermediates [134]. Retrotransposons are composed of three subclasses: (1) the long interspersed elements 1 and 2 (LINE-1/L1 and LINE-2/L2) are about 6 kb long and encode the two proteins ORF1p and ORF2p. ORF1p is implicated in edification of the retrotransposon particle, and ORF2p allows the enzymatic activities required for retrotransposition such as reverse transcriptase and endonuclease. Analysis of transgenic mice has demonstrated presence of L1 transcripts in gametes, but rare genomic insertion, suggesting posttranscriptional mechanisms allowing preservation of genomic integrity in germline. Conversely, genomic insertions of L1 sequences were mostly identified in somatic tissues during the early phases of embryogenesis [135]; (2) the short interspersed elements (SINEs) belong to the SINE-Alu and SVA classes. Whereas LINEs are autonomous sequences encoding a reverse transcriptase, SINEs are dependent on two proteins encoded by LINEs for their replication and integration [136]. Non-LTR families L1, SVA, and Alu were found to be upregulated in breast, ovarian, colon, and hematological cancers [137]; (3) the third subclass is composed of inactive LTR retrotransposons resulting from ancient germline retroviral infections. Within the human genome, only 80–100 TEs among LINE sequences are competent for the entire retrotransposition activity [8]. In the germline, TEs represent pivotal actors implicated in the shaping of genomes during evolution, and presence of retrotransposition in numerous somatic cells indicates that TEs contribute to edification of mosaicism. TEs have important role in edifying genetic diversity but are also a major source of genetic instability through mutations, chromosomes rearrangements, and epigenetic/genetic deregulations [138].

Although mobilization of class I retrotransposons may be considered as beneficial by promoting biological variability within the genome, existence of an active insertional mutagenesis can induce genomic instability in aging cells, leading to human genetic diseases, degenerative pathologies, and cancer [39]. Class I retrotransposons propagating through their "copy and paste" mechanism result in an increased copy of TEs number, which may become a source of endogenous mutagenesis by producing insertion-mediated deletions with cell cycle arrest and nonhomologous recombination [139]. Gradual release of TEs induces molecular alterations in DNA repair processes, autophagy, chaperones, and ubiquitin-proteasome system [140, 141].

During evolution, organisms have adopted molecular systems to contain expansion of TEs activity. Among them, PIWI proteins and piRNAs constitute a small-RNA-based innate immune-like system mainly expressed in gonads. Upon new expansion, TEs propagate into different regions in the genome, can be trapped into piRNA clusters, and leave traces of their sequences in these TEs traps. By falling into these clusters, novel piRNAs targeting TEs are generated and amplified through the two biogenesis pathways [87]. These pathways are highly conserved in eukaryotes and mainly implicated in protection of the genome integrity and normal gametogenesis by silencing TEs [62]. Within the germline, TEs inactivation is performed by both PIWI-piRNA and siRNA pathways. Propagation of TEs is controlled by the PIWI-piRNA pathway, of which the PIWI proteins are the executive components. The nuclear PIWI proteins allow transcriptional silencing of TEs by recognizing nascent transcripts through perfect complementarity with loaded piRNAs and are assisted by the RNA-binding protein Asterix. Recognition of multiple complementary sites in nascent TE transcripts by Asterix-PIWI-piRNA complexes favors interaction with the adaptor protein Panoramix, resulting in recruitment of the cell silencing machinery that represses TEs transcription. Moreover, introns containing remnants of TEs or genes located in proximity of TEs can be repressed by the PIWI-piRNA axis. Current studies indicated that the high mobility group protein Maelstrom (Mael) may act downstream of Piwi and histone methylation. In mouse, both Mili and Miwi2 promote TEs silencing and a heterochromatin state in mice through DNA and histones methylation. Decreased expression of PIWI proteins and piRNAs is associated with upregulation and propagation of active TEs. However, unlike siRNAs, which are active in both gonadal and somatic aging cells, the PIWI-piRNA pathway predominantly operates in nonaging cells of gonads [142, 143]. This pathway could be part of a mammalian recognition system of coding and non-coding self-genes and non-self-TEs and repeat sequences by using characteristic TEs mobility.

3.4.3. Epigenetic activation

Mounting evidence suggests that PIWI proteins and piRNAs can function as epigenetic activators. In *Drosophila*, Piwi protein increases chromosome 3R telomere-associated sequence (3R-TAS) expression [144].

3.4.4. Genes and proteins regulation

PiRNAs control levels of expression of genes where they are localized. PiR_015520, located in intron 1 of the human melatonin receptor 1A gene (*MTNR1A*), is upregulated in prostate cancer and represses *MTNR1A* gene by directly interacting with its genomic site [145]. PiRNAs also modulate stability of their PIWI partners by promoting direct molecular interactions with specific proteins. During late mouse spermatogenesis, piRNAs regulate ubiquitination of Miwi through its binding to APC/C complex [146]. Furthermore, piRNAs can modify activity and expression of many distant genes. PiRNA-36026 interacts with suppressor proteins Serpin peptidase inhibitor, clade A, member 1 (SERPINA1), and lecithin retinol acyltransferase (LRAT). However, the PIWI-piRNA pathway is also present at lower levels in somatic pluripotent stem cells to differentiated cells [147, 148]. In adult somatic cells of *Drosophila melanogaster*, this pathway is active in ovarian follicle cells, in salivary glands, and in the brain [149]. The pathway is principally observed in stem cells with pluripotent capacities, including mesenchymal and hematopoietic stem cells, but rarely in adult stem cells with limited differentiation capacity [150]. Furthermore, the PIWI-piRNA pathway seems also to regulate protein-coding genes. The first piRNAs identified in *Drosophila melanogaster* were transcribed from the Suppressor of Stellate locus located on the Y chromosome and targeted the protein-coding gene Stellate on the X chromosome [151]. More recently, genome-wide mapping techniques have demonstrated that genic piRNAs derive from TEs and 3′ UTRs of coding genes [56]. Mounting evidence suggests that germline genes could have ancestral implication in regulating stemness. The "nuage" is located in lower metazoan stem cells but restricted to germline cells in upper metazoans [152]. The PIWI-piRNA pathway is expressed in stem cells of metazoans with partial or whole-body regeneration capabilities [153].

3.4.5. Differentiation

PIWI proteins play pivotal roles in cell differentiation during early embryogenesis. In *Drosophila* ovary, self-renewal of differentiated germline stem cells is located in niches composed of different types of cells, including escort cells (ECs). Experimental deregulation of PIWI proteins expression in EC cells was associated with reduction of EC cell population and predominance of undifferentiated germline stem cells. PIWI proteins induced germline cell differentiation by promoting direct interaction between germline stem cells and escort cells through repression of the TGFβ signaling and bone morphogenetic protein (BMP) pathway by preventing edification of Smad complexes. PIWIL2 is the major PIWI protein implicated in cell differentiation through inhibition of the TGFβ signaling pathway. PIWIL2 directly interacts with Smad4 and HSP90 and prevents HSP90-TβR complex formation, resulting in inhibition of the TGFβ signaling pathway. Furthermore, PIWIL2 promotes degradation of TGFβ receptor (TβR) and Smad by upregulating ubiquitination and degradation of TβR by the ubiquitin E3 ligase Smurf2. PIWI proteins contribute to germline stem cells differentiation by repressing c-Fos at post-transcriptional. These proteins promote piRNAs synthesis from 3′ UTR region of c-Fos mRNA, resulting in c-Fos mRNA instability and repression of its translation [151–154].

3.4.6. Cell survival

The PIWI-piRNA axis promotes activation of numerous prosurvival molecules. PIWIL1-induced cell survival by upregulating expression of antiapoptotic molecule FGF8 and down-regulating expression of proapoptotic Bax and p21. In blastema cells of Mexican *axolotl*, experimental defect of PIWIL1 and PIWIL2 promoted apoptosis by suppressing FGF8 expression at transcriptional level and prevented limb regeneration and development. PIWIL2 principally controlled p53 through direct interaction with STAT3 and c-Src by edifying a PIWIL2/STAT3/c-Src complex, resulting in repression of p53 phosphorylation and expression and inhibition of Fas-mediated apoptosis. PIWIL2 - induced activation of STAT3 also upregulated expression of the antiapoptotic Bcl-XL [125].

3.4.7. *Fertilization and development*

Although most attention has been given to the pivotal role of the PIWI-piRNA pathway in germline TEs silencing, mounting evidence has revealed their implication in germline and somatic epigenetic and posttranscriptional regulation of gene expression [151]. This pathway is mainly implicated in the germline biology, including maintenance, differentiation, and function of *Drosophila* and murine GSCs. Furthermore, piRNAs epigenetically activate gene expression with transgenerational epigenetic effects by inducing euchromatin through activation of H3K4me3 and inhibition of H3K27me3 in subtelomeric heterochromatin [56, 73, 152–155]. The PIWI-piRNA axis is implicated in embryonic development, including cell cycle progression, nuclear division, chromatin organization, chromosome integrity during mitosis, control of mRNA translation, and embryonic sex determination [144, 156–158]. Spatial-temporal activation and regulation of PIWI proteins and piRNAs are of pivotal importance during mammalian oogenesis and spermatogenesis, early embryogenesis, organogenesis, and postbirth [159]. PIWIL2 is upregulated in germline cells and appears instrumental in maintaining genome stability, an open state of chromatin and DNA repair via silencing TEs and histones modifications, thus preventing TEs propagation, chromosome rearrangements, oncogenic mutations, and gene dysregulation [160].

3.4.7.1. *Oogenesis*

In *Drosophila*, PIWI-piRNA complexes promote TEs silencing at embryonic germ cell stage, mediate cellular memory of TEs repression, and thus maintain this mechanism in ovaries at the adult stage. In mouse ovary, Miwi upregulation is observed during neonatal stage and its expression is lower in adult ovaries [161]. Human PIWI proteins expression profiling is also variable, depending on the stage of development. PIWIL1 and PIWIL2 are highly upregulated in oocytes of human adult ovary that present a strong activity of TEs, whereas fetal oocytes, whose TEs propagation is lower, overexpress PIWIL2 but not other PIWI proteins [162].

3.4.7.2. *Spermatogenesis*

In mouse, Miwi inactivation occurs during late spermatogenesis and is induced by the anaphase promoting complex (APC)/C-26S proteasomal pathway [159]. Functional destruction box (D-box) is required for Miwi ubiquitination and degradation by (APC)/C system. A genetic analysis in mouse azoospermia showed that mutations in D-box favor Hiwi stabilization in late spermatogenesis. Stabilized mutant Hiwi interacts with RNF8 implicated in histone ubiquitination and prevents its nuclear translocation and ubiquitin ligase activity [163]. Human PIWIL4 function is crucial in accurate spermatogenesis, and genetic polymorphisms of PIWIL4 gene are significantly correlated with defective spermatogenesis associated with spermatogenesis defect and male infertility [164].

3.4.7.3. Organogenesis

In *Drosophila,* PIWI proteins induce ovary tissue morphogenesis through c-Fos inactivation at posttranscriptional level [165]. They are implicated in development of eye color [144]. In silkworm, fempiRNA, a piRNA located on female W-chromosome, is pivotal in sex determination by repressing masculinization mRNA at posttranscriptional level [167]. At early stages of human embryonic lungs development, PIWIL1, PIWIL2, and PIWIL4 levels of expression are strongly upregulated from 6th week to 9th week and then decline [168]. In human, PIWIL1 and PIWIL2 have crucial role in neural polarization and radial migration during maturation of the cerebral cortex region of the brain [59].

3.4.8. Physiological processes

The PIWI-piRNA pathway has pivotal role in numerous physiological processes.

3.4.8.1. Brain plasticity

PIWI proteins and piRNAs are instrumental in synaptic plasticity and stabilization of long-term memory through serotonin-dependent suppression of CREB2 at transcription level that is induced by methylation of CpG islands in the promoter region of the *CREB2* gene [59]. In rodents, several piRNAs are upregulated in hippocampal neurons and Miwi associated with piRNAs control dentritic spine development and morphogenesis [149]. Mili expression is associated with anxiety and locomotory drive [54]. In humans, PIWIL1 controls cortical neuron activity through modulation of microtubule-associated proteins (MAPs) expression [169]. Furthermore, mutations of PIWIL2 and PIWIL4 are significantly correlated with autism [170].

3.4.8.2. Regeneration

PIWI proteins and piRNAs have crucial role in self-renewal, regeneration, and homeostasis. In planarian *Schmidtea mediterranea,* SMEDWI-2 and SMEDWI-3 increase division of adult stem cells to induce regeneration in injured tissues [113]. In jellyfish, Cniwi is upregulated during transdifferentiation of striated muscle into smooth muscle [171]. In humans, PIWI proteins promote hepatocyte regeneration [61] and maintain integrity of retinal cells [130].

3.4.8.3. Metabolism

The PIWI-piRNA pathway controls fat metabolism through repression of TEs, and fat metabolism inactivation is associated with depletion of lipid synthesis and storage [60]. PIWIL2 and PIWIL4 modulate pancreatic β-cells function and insulin secretion. Alterations of their levels of expression were observed in diabetic conditions [55].

http://dx.doi.org/10.5772/intechopen.79974

4. Deregulation of the PIWI/piRNA pathway in pathological nonneoplastic disorders

Mounting evidence has revealed that many transcription factors and signaling molecules interact with the PIWI-piRNA pathway and represent downstream targets of these complexes under pathological conditions. PIWI proteins and piRNAs are deregulated, and their levels of expression are highly altered in various pathological processes. The PIWI-piRNA pathway is pivotal for regeneration after amputation in *Botrylloides leachi* [172]. In Mexican axolotl, PIWIL1 and PIWIL2 transient upregulation in limb blastemal cells induces regeneration of wounded limb [173]. In rat, PIWIL2 expression increases after 24 h of partial hepatectomy, and a set of 72 piRNAs is deregulated during 48 h of posthepatectomy [58]. In rodents, expression of more than 100 piRNAs is deregulated in brain during ischemic condition [174]. In rat, PIWIL2 enhances activity of the autophagic process in diabetic nephropathy by regulating expression of beclin 1 and LC3A study in diabetic rat kidney [175]. Pro-inflammatory cytokines IL1β and TNFα promote PIWIL2 and PIWIL4 upregulation in synovial fibroblasts of rheumatoid arthritis [176].

Acknowledgements

This work was supported by grant INCa-DGOS-4654.

Disclosure: conflict of interest

The authors declare that they have no competing interests.

Author details

Didier Meseure[1,2]* and Kinan Drak Alsibai[3]

*Address all correspondence to: didier.meseure@curie.fr

1 Platform of Experimental Pathology, Curie Institute, Paris, France

2 Department of Diagnostic and Theranostic Medicine, Curie Institute, Paris, France

3 Institut de Pathologie de Paris, Malakoff, France

References

[1] Berdasco M, Esteller M. Aberrant epigenetic landscape in cancer: How cellular identity goes awry. Developmental Cell. 2010;**19**:698-711

[2] Kanwal R, Gupta S. Epigenetic modifications in cancer. Clinical Genetics. 2012;**81**:303-311

[3] Magee JA, Piskounova E, Morrison SJ. Cancer stem cells: Impact, heterogeneity and uncertainty. Cancer Cell. 2012;**21**:283-296

[4] López-Otín C, Blasco MA, Partridge L, Serrano M, Kroemer G. The hallmarks of aging. Cell. 2013;**153**:1194-1217

[5] Yong WS, Hsu FM, Chen PY. Profiling genome-wide DNA methylation. Epigenetics & Chromatin. 2016;**9**:26. DOI: 10.1186/s13072-016-0075-3

[6] Meseure D, Drak Alsibai K, Nicolas A. Pivotal role of pervasive neoplastic and stromal cells reprogramming in circulating tumor cells dissemination and metastatic colonization. Cancer Microenvironment. 2014;**7**:95-115

[7] Amaral PP, Mattick JS. Noncoding RNA in development. Mammalian Genome. 2008;**19**: 454-492

[8] Ponnusamy M, Yan KW, Liu CY, Li PF, Wang K. PIWI family emerging as a decisive factor of cell fate: An overview. European Journal of Cell Biology. 2017;**96**:746-757

[9] Carninci P, Kasukawa T, Katayama S, Gough J, Frith MC, Maeda N, et al. The transcriptional Landscape of the mammalian genome. Science. 2005;**309**:1559-1563. Erratum in: Science 2006;**311**:1713

[10] Lewis BP, Burge CB, Bartel DP. Conserved seed pairing, often flanked by adenosines, indicates that thousands of human genes are microRNA targets. Cell. 2005;**120**:15-20

[11] Iyer MK, Niknafs YS, Malik R, Singhal U, Sahu A, Hosono Y, et al. The landscape of long noncoding RNAs in the human transcriptome. Nature Genetics. 2015;**47**:199-208

[12] Hamilton AJ, Baulcombe DC. A species of small antisense RNA in posttranscriptional gene silencing in plants. Science. 1999;**286**:950-952

[13] Siomi H, Siomi MC. On the road to reading the RNA-interference code. Nature. 2009;**457**:396-404

[14] Ha M, Kim VN. Regulation of microRNA biogenesis. Nature Reviews. Molecular Cell Biology. 2014;**15**:509-524

[15] Aravin A, Gaidatzis D, Pfeffer S, Lagos-Quintana M, Landgraf P, Iovino N, et al. A novel class of small RNAs binds to MILI protein in mouse testes. Nature. 2006;**442**:203-207

[16] Farazi TA, Juranek SA, Tuschl T. The growing catalog of small RNAs and their association with distinct Argonaute/Piwi family members. Development. 2008;**135**:1201-1214

[17] Hock J, Weinmann L, Ender C, Rudel S, Kremmer E, Raabe M, et al. Proteomic and functional analysis of Argonaute-containing mRNA-protein complexes in human cells. EMBO Reports. 2007;**8**:1052-1060

[18] Girard A, Sachidanandam R, Hannon GJ, Carmell MA. A germline-specific class of small RNAs binds mammalian Piwi proteins. Nature. 2006;**442**:199-202

[19] Grivna ST, Beyret E, Wang Z, Lin H. A novel class of small RNAsin mouse spermatogenic cells. Genes & Development. 2006;**20**:1709-1714

[20] Watanabe T, Takeda A, Tsukiyama T, Mise K, Okuno T, Sasaki H, et al. Identification and characterization of two novel classes of small RNAs in the mouse germline: Retrotransposon-derived siRNAs in oocytes and germline small RNAs in testes. Genes & Development. 2006;**20**:1732-1743

[21] Moazed D. Molecular biology. Rejoice RNAi for yeast. Science. 2009;**326**:533-534

[22] Cerutti H, Casas-Mollano JA. On the origin and functions of RNA-mediated silencing: From protists to man. Current Genetics. 2006;**50**:81-99

[23] Ghildiyal M, Zamore PD. Small silencing RNAs: An expanding universe. Nature Reviews. Genetics. 2009;**10**:94-108

[24] Baek M, Gusev Y, Brackett DJ, Nuovo GJ, Schmittgen TD. Systematic evaluation of microRNA processing patterns in tissues, cell lines and tumors. RNA. 2008;**14**:35-42

[25] Carmell MA. The Argonaute family: Tentacles that reach into RNAi, developmental control, stem cell maintenance, and tumorigenesis. Genes & Development. 2002;**16**:2733-2742

[26] Peters L, Meister G. Argonaute proteins: Mediators of RNA silencing. Molecular Cell. 2007;**26**:611-623

[27] Ghildiyal M, Xu J, Seitz H, Weng Z, Zamore PD. Sorting of Drosophila small silencing RNAs partitions microRNA strands into the RNA interference pathway. RNA. 2010;**16**:43-5620

[28] Parker JS, Roe SM, Barford D. Structural insights into mRNA recognition from a PIWI domain-siRNA guide complex. Nature. 2005;**434**:663-666

[29] Yigit E, Batista PJ, Bei Y, Pang KM, Chen CC, Tolia NH, et al. Analysis of the *C. elegans* Argonaute family reveals that distinct Argonautes act sequentially during RNAi. Cell. 2006;**127**:747-757

[30] Ketting RF, Fischer SE, Bernstein E, Sijen T, Hannon GJ, Plasterk RH. Dicer functions in RNA interference and in synthesis of small RNA involved in developmental timing in *C. elegans*. Genes. 2001;**15**:2654-2659

[31] Kawamura Y, Saito K, Kin T, Ono Y, Asai K, Sunohara T, et al. Drosophila endogenous small RNAs bind to Argonaute 2 in somatic cells. Nature. 2008;**453**:793-797

[32] Okamura K, Chung WJ, Lai EC. The long and short of inverted repeat genes in animals: MicroRNAs, mirtrons and hairpin RNAs. Cell Cycle. 2008;**7**:2840-2845

[33] Völler D, Linck L, Bruckmann A, Hauptmann J, Deutzmann R, Meister G, et al. Argonaute family protein expression in normal tissue and cancer entities. PLoS One. 2016;**11**:e0161165. DOI: 10.1371/journal.pone.0161165

[34] Grimson A, Srivastava M, Fahey B, Woodcroft BJ, Chiang HR, King N, et al. Early origins and evolution of microRNAs and Piwi-interacting RNAs in animals. Nature. 2008;**455**:1193-1197

[35] Sarkar A, Maji RK, Saha S, Ghosh Z. piRNAQuest: Searching the piRNAome for silencers. BMC Genomics. 2014;**15**:555. DOI: 10.1186/1471-2164-15-555

[36] Fu A, Jacobs DI, Zhu Y. Epigenome-wide analysis of piRNAs in gene-specific DNA methylation. RNA Biology. 2014;**11**:1301-1312

[37] Betel D, Sheridan R, Marks DS, Sander C. PIWI promotes cell survival by increasing the activity of Pro-survival factors and signaling pathways, piRNA sequence and biogenesis. PLoS Computational Biology. 2007;**3**:e222

[38] Aravin AA, Lagos-Quintana M, Yalcin A, Zavolan M, Marks D, Snyder B, et al. The small RNA profile during *Drosophila melanogaster* development. Developmental Cell. 2003;**5**:337-350

[39] Sturm Á, Ivics Z, Vellai T. The mechanism of ageing: Primary role of transposable elements in genome disintegration. Cellular and Molecular Life Sciences. 2015;**72**: 1839-1847

[40] Saito K, Nishida KM, Mori T, Kawamura Y, Miyoshi K, Nagami T, et al. Specific association of Piwi with rasiRNAs derived from retrotransposon and heterochromatic regions in the Drosophila genome. Genes & Development. 2006;**20**:2214-2222

[41] Brennecke J, Aravin AA, Stark A, Dus M, Kellis M, Sachidanandam R, et al. Discrete small RNA-generating loci as master regulators of transposonactivity in Drosophila. Cell. 2007;**128**:1089-1103

[42] Gunawardane LS, Saito K, Nishida KM, Miyoshi K, Kawamura Y, Nagami T, et al. A slicer-mediated mechanism for repeat-associated siRNA 5′ end formation in Drosophila. Science. 2007;**315**:1587-1590

[43] Siomi MC, Sato K, Pezic D, Aravin AA. PIWI-interacting small RNAs: The vanguard of genome defence. Nature Reviews. Molecular Cell Biology. 2011;**12**:246-258

[44] Iwasaki YW, Siomi MC, Siomi H. PIWI-interacting RNA: Its biogenesis and functions. Annual Review of Biochemistry. 2015;**84**:405-433

[45] Czech B, Malone CD, Zhou R, Stark A, Schlingeheyde C, Dus M, et al. An endogenous small interfering RNA pathway in Drosophila. Nature. 2008;**453**:798-802

[46] O'Donnell KA, Boeke JD. Mighty Piwis defend the germline against genome intruders. Cell. 2007;**129**:37-44

[47] Höck J, Meister G. The Argonaute protein family. Genome Biology. 2008;**9**:210. DOI: 10.1186/gb-2008-9-2-210

[48] Pek JW, Anand A, Kai T. Tudor domain proteins in development. Development. 2012;**139**:2255-2266

[49] Martinez VD, Vucic EA, Thu KL, Hubaux R, Enfield KS, Pikor LA, et al. Unique somatic and malignant expression patterns implicate PIWI-interacting RNAs in cancer-type specific biology. Scientific Reports. 2015;**5**:10423

[50] Aravin AA, Sachidanandam R, Bourc'his D, Schaefer C, Pezic D, Toth KF, et al. A piRNA pathway primed by individual transposons is linked to de novo DNA methylation in mice. Molecular Cell. 2008;**31**:785-799

[51] Sasaki T, Shiohama A, Minoshima S, Shimizu N. Identification of eight members of the Argonaute family in the human genome. Genomics. 2003;**82**:323-330

[52] Houwing S, Berezikov E, Ketting RF. Zili is required for germ cell differentiation and meiosis in zebrafish. The EMBO Journal. 2008;**27**:2702-2711

[53] Wang G, Reinke V. A *C. elegans* Piwi, PRG-1, regulates 21U-RNAs during spermatogenesis. Current Biology. 2008;**18**:861-867

[54] Nandi S, Chandramohan D, Fioriti L, Melnick AM, Hébert JM, Mason CE, Rajasethupathy P, Kandel ER. Roles for small noncoding RNAs in silencing of retrotransposons in the mammalian brain. Proceedings of the National Academy of Sciences of the United States of America. 2016. pii: 201609287. PubMed PMID: 27791114; PubMed Central PMCID: PMC5111663

[55] Henaoui IS, Jacovetti C, Guerra Mollet I, Guay C, Sobel J, Eliasson L, et al. PIWI-interacting RNAs as novel regulators of pancreatic beta cell function. Diabetologia. 2017;**17**:4368-4376

[56] Cox DN, Chao A, Baker J, Chang L, Qiao D, Lin H. A novel class of evolutionarily conserved genes defined by piwi are essential for stem cell self-renewal. Genes & Development. 1998;**12**:3715-3727

[57] Lim RS, Kai T. A piece of the pi(e): The diverse roles of animal piRNAs and their PIWI partners. Seminars in Cell & Developmental Biology. 2015;**47-48**:17-31. DOI: 10.1016/j.semcdb.2015.10.025

[58] Rizzo F, Hashim A, Marchese G, Ravo M, Tarallo R, Nassa G, et al. Timed regulation of P-element-induced wimpy testis-interacting RNA expression during rat liver regeneration. Hepatology. 2014;**60**:798-806

[59] Rajasethupathy P, Antonov I, Sheridan R, Frey S, Sander C, Tuschl T, et al. A role for neuronal piRNAs in the epigenetic control of memory-related synaptic plasticity. Cell. 2012;**149**:693-707

[60] Jones BC, Wood JG, Chang C, Tam AD, Franklin MJ, Siegel ER, et al. A somatic piRNA pathway in the Drosophila fat body ensures metabolic homeostasis and normal lifespan.

Nature Communications. 2016;**7**:13856. DOI: 10.1038/ncomms13856. PubMed PMID: 28000665

[61] Keam SP, Young PE, McCorkindale AL, Dang TH, Clancy JL, Humphreys DT, et al. The human Piwi protein Hiwi2 associates with tRNA-derived piRNAs in somatic cells. Nucleic Acids Research. 2014;**42**:8984-8995

[62] Kalmykova AI, Klenov MS, Gvozdev VA. Argonaute protein PIWI controls mobilization of retrotransposons in the Drosophila male germline. Nucleic Acids Research. 2005;**33**:2052-2059

[63] Sabin LR, Zheng Q, Thekkat P, Yang J, Hannon GJ, Gregory BD, et al. Dicer-2 processes diverse viral RNA species. PLoS One. 2013;**8**(2):e55458. DOI: 10.1371/journal.pone.0055458

[64] Aravin AA, Sachidanandam R, Girard A, Fejes-Toth K, Hannon GJ. Developmentally regulated piRNA clusters implicate MILI in transposon control. Science. 2007;**316**:744-747

[65] Kuramochi-Miyagawa S, Watanabe T, Gotoh K, Totoki Y, Toyoda A, Ikawa M, et al. DNA methylation of retrotransposon genes is regulated by Piwi family members MILI and MIWI2 in murine fetal testes. Genes & Development. 2008;**22**:908-917

[66] Grivna ST, Beyret E, Wang Z, Lin H. A novel class of small RNAs in mouse spermatogenic cells. Genes & Development. 2006;**20**:1709-1714

[67] Lin H, Spradling AC. A novel group of pumilio mutations affects the asymmetric division of germline stem cells in the Drosophila ovary. Development. 1997;**124**:2463-2476

[68] Carmell MA, Girard A, van de Kant HJ, Bourc'his D, Bestor TH, de Rooij DG, et al. MIWI2 is essential for spermatogenesis and repression of transposons in the mouse male germline. Developmental Cell. 2007;**12**:503-514

[69] Das PP, Bagijn MP, Goldstein LD, Woolford JR, Lehrbach NJ, Sapetschnig A, et al. Piwi and piRNAs act upstream of an endogenous siRNA pathway to suppress Tc3 transposon mobility in the *Caenorhabditis elegans* germline. Molecular Cell. 2008;**31**:79-90

[70] Malone CD, Brennecke J, Dus M, Stark A, McCombie WR, Sachidanandam R, et al. Specialized piRNA pathways act in germline and somatic tissues of the Drosophila ovary. Cell. 2009;**137**:522-535

[71] Buckley BA, Burkhart KB, Gu SG, Spracklin G, Kershner A, Fritz H, et al. A nuclear Argonaute promotes multigenerational epigenetic inheritance and germline immortality. Nature. 2012;**489**:447-451

[72] Gu SG, Pak J, Guang S, Maniar JM, Kennedy S, Fire A. Amplification of siRNA in *Caenorhabditis elegans* generates a transgenerational sequence-targeted histone H3 lysine 9 methylation footprint. Nature Genetics. 2012;**44**:157-164

[73] Shirayama M, Seth M, Lee H-C, Gu W, Ishidate T, Conte D. piRNAs initiate an epigenetic memory of nonself RNA in the *C. elegans* germline. Cell. 2012;**150**:65-77

[74] Bamezai S, Rawat VP, Buske C. Concise review: The Piwi-piRNA axis: pivotal beyond transposon silencing. Stem Cells. 2012;**30**(12):2603-2611

[75] Thomson T, Lin H. The biogenesis and function of PIWI proteins and piRNAs: Progress and prospect. Annual Review of Cell and Developmental Biology. 2009;**25**:355-376

[76] Robine N, Lau NC, Balla S, Jin Z, Okamura K, Kuramochi-Miyagawa S, et al. A broadly conserved pathway generates 3′UTR-directed primary piRNAs. Current Biology. 2009;**19**:2066-2076

[77] Bartel DP. MicroRNAs: Target recognition and regulatory functions. Cell. 2009;**136**:215-233

[78] Masi LN, Serdan TD, Levada-Pires AC, Hatanaka E, Silveira LD, Cury-Boaventura MF, et al. Regulation of gene expression by exercise-related micrornas. Cellular Physiology and Biochemistry. 2016;**39**:2381-2397

[79] Lim AK, Kai T. Unique germ-line organelle, nuage, functions to repress selfish genetic elements in *Drosophila melanogaster*. Proceedings of the National Academy of Sciences of the United States of America. 2007;**104**:6714-6719

[80] Voronina E, Seydoux G, Sassone-Corsi P, Nagamori I. RNA granules in germ cells. Cold Spring Harbor Perspectives in Biology. 2011;**3**. pii: a002774. DOI: 10.1101/cshperspect.a002774

[81] Extavour CG. Evolution of the bilaterian germ line: Lineage origin and modulation of specification mechanisms. Integrative and Comparative Biology. 2007;**47**:770-785

[82] Olovnikov IA, Kalmykova AI. piRNA clusters as a main source of small RNAs in the animal germline. Biochemistry (Mosc). 2013;**78**:572-584

[83] Li W, Prazak L, Chatterjee N, Grüninger S, Krug L, Theodorou D, et al. Activation of transposable elements during aging and neuronal decline in Drosophila. Nature Neuroscience. 2013;**16**:529-531

[84] Zhang F, Wang J, Xu J, Zhang Z, Koppetsch BS, Schultz N, et al. UAP56 couples piRNA clusters to the perinuclear transposon silencing machinery. Cell. 2012;**151**:871-884

[85] Mohn F, Sienski G, Handler D, Brennecke J. The rhino-deadlock-cutoff complex licenses noncanonical transcription of dual-strand piRNA clusters in Drosophila. Cell. 2014;**157**:1364-1379

[86] Zhang Z, Wang J, Schultz N, Zhang F, Parhad SS, Tu S, et al. The HP1 homolog rhino anchors a nuclear complex that suppresses piRNA precursor splicing. Cell. 2014;**157**:1353-1363

[87] Tóth KF, Pezic D, Stuwe E, Webster A. The piRNA pathway guards the germline genome against transposable elements. Advances in Experimental Medicine and Biology. 2016;**886**:51-77

[88] Theurkauf WE, Klattenhoff C, Bratu DP, McGinnis-Schultz N, Koppetsch BS, Cook HA. rasiRNAs, DNA damage, and embryonic axis specification. Cold Spring Harbor Symposia on Quantitative Biology. 2006;**71**:171-180

[89] Sienski G, Dönertas D, Brennecke J. Transcriptional silencing of transposons by Piwi and maelstrom and its impact on chromatin state and gene expression. Cell. 2012;**151**:964-980

[90] Saito K, Ishizu H, Komai M, Kotani H, Kawamura Y, Nishida KM, et al. Roles for the Yb body components Armitage and Yb in primary piRNA biogenesis in Drosophila. Genes & Development. 2010;**24**:2493-2498

[91] Czech B, Preall JB, McGinn J, Hannon GJ. A transcriptome wide RNAi screen in the drosophila ovary reveals factors of the germline piRNA pathway. Molecular Cell. 2013;**50**:749-761

[92] Handler D, Meixner K, Pizka M, Lauss K, Schmied C, Gruber FS, et al. The genetic makeup of the Drosophila piRNA pathway. Molecular Cell. 2013;**50**:762-777. DOI: 10.1016/j.molcel.2013.04.031

[93] Vourekas A, Zheng K, Fu Q, Maragkakis M, Alexiou P, Ma J, et al. The RNA helicase MOV10L1 binds piRNA precursors to initiate piRNA processing. Genes & Development. 2015;**29**:617-629

[94] Pane A, Wehr K, Schüpbach T. zucchini and squash encode two putative nucleases required for rasiRNA production in the Drosophila germline. Developmental Cell. 2007;**12**:851-862

[95] Ipsaro JJ, Haase AD, Knott SR, Joshua-Tor L, Hannon GJ. The structural biochemistry of Zucchini implicates it as a nuclease in piRNA biogenesis. Nature. 2012;**491**:279-283

[96] Nishimasu H, Ishizu H, Saito K, Fukuhara S, Kamatani MK, Bonnefond L, et al. Structure and function of Zucchini endoribonuclease in piRNA biogenesis. Nature. 2012;**491**:284-287

[97] Voigt F, Reuter M, Kasaruho A, Schulz EC, Pillai RS, Barabas O. Crystal structure of the primary piRNA biogenesis factor Zucchini reveals similarity to the bacterial PLD endonuclease Nuc. RNA. 2012;**18**:2128-2134

[98] Kawaoka S, Izumi N, Katsuma S, Tomari Y. 3′ end formation ofPIWI-interacting RNAs in vitro. Molecular Cell. 2011;**43**:1015-1022

[99] Feltzin VL, Khaladkar M, Abe M, Parisi M, Hendriks GJ, Kim J, et al. The exonuclease Nibbler regulates age-associated traits and modulates piRNA length in Drosophila. Aging Cell. 2015;**14**:443-452

[100] Horwich MD, Li C, Matranga C, Vagin V, Farley G, Wang P, et al. The Drosophila RNA methyltransferase, DmHen1, modifies germline piRNAs and single-stranded siRNAs in RISC. Current Biology. 2007;**17**:1265-1272

[101] Saito K, Sakaguchi Y, Suzuki T, Suzuki T, Siomi H, Siomi MC. Pimet, the Drosophila homolog of HEN1, mediates 2′-O-methylation of Piwi-interacting RNAsat their 3′ ends. Genes & Development. 2007;**21**:1603-1608

[102] Tian Y, Simanshu DK, Ma JB, Patel DJ. Structural basis for piRNA 2′-O-methylated 3′-end recognition by Piwi PAZ (Piwi/Argonaute/Zwille) domains. Proceedings of the National Academy of Sciences of the United States of America. 2011;**108**:903-910

[103] Montgomery TA, Rim YS, Zhang C, Dowen RH, Phillips CM, Fischer SE, et al. PIWI associated siRNAs and piRNAs specifically require the *Caenorhabditis elegans* HEN1 ortholog henn-1. PLoS Genetics. 2012;**8**:e1002616. DOI: 10.1371/journal.pgen.1002616

[104] Chen C, Jin J, James DA, Adams-Cioaba MA, Park JG, Guo Y, et al. Mouse Piwi interactome identifies binding mechanism of Tdrkh Tudor domain to arginine methylated Miwi. Proceedings of the National Academy of Sciences of the United States of America. 2009;**106**:20336-20341

[105] Handler D, Olivieri D, Novatchkova M, Gruber FS, Meixner K, Mechtler K, et al. A systematic analysis of Drosophila TUDOR domain-containing proteins identifies Vreteno and the Tdrd12 family as essential primary piRNA pathway factors. The EMBO Journal. 2011;**30**:3977-3993

[106] Rouhana L, Vieira AP, Roberts-Galbraith RH, Newmark PA. PRMT5 and the role of symmetrical dimethylarginine in chromatoid bodies of planarian stem cells. Development. 2012;**139**:1083-1094

[107] Huang X, Yuan T, Tschannen M, Sun Z, Jacob H, Du M, et al. Characterization of human plasma-derived exosomal RNAs by deep sequencing. BMC Genomics. 2013;**14**:319

[108] Le Thomas A, Rogers AK, Webster A, Marinov GK, Liao SE, Perkins EM, et al. Piwi induces piRNA-guided transcriptional silencing and establishment of a repressive chromatin state. Genes & Development. 2013;**27**:390-399

[109] Klenov MS, Sokolova OA, Yakushev EY, Stolyarenko AD, Mikhaleva EA, Lavrov SA, et al. Separation of stem cell maintenance and transposon silencing functions of Piwi protein. Proceedings of the National Academy of Sciences of the United States of America. 2011;**108**:18760-18765

[110] Ross RJ, Weiner MM, Lin H. PIWI proteins and PIWI-interacting RNAs in the soma. Nature. 2014;**505**:353-359

[111] Dönertas D, Sienski G, Brennecke J. Drosophila Gtsf1 is an essential component of the Piwi-mediated transcriptional silencing complex. Genes & Development. 2013;**27**:1693-1705

[112] Müller S, Raulefs S, Bruns P, Afonso-Grunz F, Plötner A, Thermann R, et al. Next-generation sequencing reveals novel differentially regulated mRNAs, lncRNAs, miRNAs, sdRNAs and a piRNA in pancreatic cancer. Molecular Cancer. 2015;**14**:94. DOI: 10.1186/s12943-015-0358-5

[113] Palakodeti D, Smielewska M, Lu YC, Yeo GW, Graveley BR. The PIWI proteins SMEDWI-2 and SMEDWI-3 are required for stem cell function and piRNA expression in planarians. RNA. 2008;**14**:1174-1186

[114] Lau NC, Seto AG, Kim J, Kuramochi-Miyagawa S, Nakano T, Bartel DP, et al. Characterization of the piRNA complex from rat testes. Science. 2006;**313**:363-367

[115] Ha H, Song J, Wang S, Kapusta A, Feschotte C, Chen KC, et al. A comprehensive analysis of piRNAs from adult human testis and their relationship with genes and mobile elements. BMC Genomics. 2014;**15**:545

[116] Khurana JS, Wang J, Xu J, Koppetsch BS, Thomson TC, Nowosielska A, et al. Adaptation to P element transposon invasion in *Drosophila melanogaster*. Cell. 2011;**147**:1551-1563

[117] Beyret E, Liu N, Lin H. piRNA biogenesis during adult spermatogenesis in mice is independent of the ping-pong mechanism. Cell Research. 2012;**22**:1429-1439

[118] Honda S, Kirino Y, Maragkakis M, Alexiou P, Ohtaki A, Murali R, et al. Mitochondrial protein BmPAPI modulates the length of mature piRNAs. RNA. 2013;**19**:1405-1418

[119] Chen Z, Che Q, Jiang FZ, Wang HH, Wang FY, Liao Y, et al. Piwil1 causes epigenetic alteration of PTEN gene via upregulation of DNA methyltransferase in type I endometrial cancer. Biochemical and Biophysical Research Communications. 2015;**463**:876-880

[120] Yan Z, Hu HY, Jiang X, Maierhofer V, Neb E, He L, et al. Widespread expression of piRNA-like molecules in somatic tissues. Nucleic Acids Research. 2011;**39**:6596-6607

[121] Kuramochi-Miyagawa S, Kimura T, Ijiri TW, Isobe T, Asada N, Fujita Y, et al. Mili, a mammalian member of piwi family gene, is essential for spermatogenesis. Development. 2004;**131**:839-849

[122] Giauque CC, Bickel SE. Heterochromatin-associated proteins HP1a and Piwi collaborate to maintain the association of achiasmate homologs in Drosophila oocytes. Genetics. 2016;**203**:173-189

[123] Brower-Toland B, Findley SD, Jiang L, Liu L, Yin H, Dus M, et al. Drosophila PIWI associates with chromatin and interacts directly with HP1a. Genes & Development. 2007;**21**:2300-2311

[124] Lu Y, Zhang K, Li C, Yao Y, Tao D, Liu Y, et al. Piwil2 suppresses p53 by inducing phosphorylation of signal transducer and activator of transcription 3 in tumor cells. PLoS One. 2012;**7**:e30999. DOI: 10.1371/journal.pone.0030999

[125] Zhang X, He X, Liu C, Liu J, Hu Q, Pan T, et al. IL-4 inhibits the biogenesis of an epigenetically suppressive PIWI-interacting RNA to upregulate CD1a molecules on monocytes/dendritic cells. Journal of Immunology. 2016;**196**:1591-1603

[126] Peng JC, Valouev A, Liu N, Lin H. Piwi maintains germline stem cells and oogenesis in Drosophila through negative regulation of Polycomb group proteins. Nature Genetics. 2016;**48**:283-291

[127] Boros J, Arnoult N, Stroobant V, Collet JF, Decottignies A. Polycomb repressive complex 2 and H3K27me3 cooperate with H3K9 methylation to maintain heterochromatin protein 1α at chromatin. Molecular and Cellular Biology. 2014;**34**:3662-3674

[128] Zhong F, Zhou N, Wu K, Guo Y, Tan W, Zhang H, et al. A SnoRNA-derived piRNA interacts with human interleukin-4 pre-mRNA and induces its decay in nuclear exosomes. Nucleic Acids Research. 2015;**43**:10474-10491

[129] Unhavaithaya Y, Hao Y, Beyret E, Yin H, Kuramochi-Miyagawa S, Nakano T, et al. MILI, a PIWI-interacting RNA-binding protein, is required for germ line stem cell self-renewal and appears to positively regulate translation. The Journal of Biological Chemistry. 2009;**284**:6507-6519

[130] Sivagurunathan S, Palanisamy K, Arunachalam JP, Chidambaram S. Possible role of HIWI2 in modulating tight junction proteins in retinal pigment epithelial cells through Akt signaling pathway. Molecular and Cellular Biochemistry. 2017;**427**:145-156

[131] Li C, Zhou X, Chen J, Lu Y, Sun Q, Tao D, et al. PIWIL1 destabilizes microtubule by suppressing phosphorylation at Ser16 and RLIM-mediated degradation of Stathmin1. Oncotarget. 2015;**6**:27794-27804

[132] Wachsmuth M, Knoch TA, Rippe K. Dynamic properties of independent chromatin domains measured by correlation spectroscopy in living cells. Epigenetics & Chromatin. 2016;**9**:57. DOI: 10.1186/s13072-016-0093-1

[133] Ilyin AA, Ryazansky SS, Doronin SA, Olenkina OM, Mikhaleva EA, Yakushev EY, et al. Piwi interacts with chromatin at nuclear pores and promiscuously binds nuclear transcripts in Drosophila ovarian somatic cells. Nucleic Acids Research. 2017;**45**:7666-7680

[134] Slotkin RK, Martienssen R. Transposable elements and the epigenetic regulation of the genome. Nature Reviews. Genetics. 2007;**8**:272-285

[135] Huang G, Hu H, Xue X, Shen S, Gao E, Guo G, et al. Altered expression of piRNAs and their relation with clinicopathologic features of breast cancer. Clinical & Translational Oncology. 2013;**15**:563-568

[136] De Cecco M, Criscione SW, Peterson AL, Neretti N, Sedivy JM, Kreiling JA. Transposable elements become active and mobile in the genomes of aging mammalian somatic tissues. Aging (Albany NY). 2013;**5**:867-883

[137] Konkel MK, Batzer MA. A mobile threat to genome stability: The impact of non-LTR retrotransposons upon the human genome. Seminars in Cancer Biology. 2010;**20**:211-221

[138] Kazazian HH Jr. Mobile elements: Drivers of genome evolution. Science. 2004;**303**: 1626-1632

[139] Deragon JM, Capy P. Impact of transposable elements on the human genome. Annals of Medicine. 2000;**32**:264-273

[140] Gorbunova V, Boeke JD, Helfand SL, Sedivy JM. Human genomics. Sleeping dogs of the genome. Science. 2014;**346**:1187-1188

[141] Wood JG, Jones BC, Jiang N, Chang C, Hosier S, Wickremesinghe P, et al. Chromatin-modifying genetic interventions suppress age-associated transposable element activation and extend life span in Drosophila. Proceedings of the National Academy of Sciences of the United States of America. 2016;**113**:11277-11282

[142] Ghildiyal M, Seitz H, Horwich MD, Li C, Du T, Lee S, et al. Endogenous siRNAs derived from transposons and mRNAs in Drosophila somatic cells. Science. 2008;**320**:1077-1081

[143] Law JA, Jacobsen SE. Establishing, maintaining and modifying DNA methylation patterns in plants and animals. Nature Reviews. Genetics. 2010;**11**:204-220

[144] Yin H, Lin H. An epigenetic activation role of Piwi and a Piwi-associated piRNA in *Drosophila melanogaster*. Nature. 2007;**450**:304-308

[145] Esposito T, Magliocca S, Formicola D, Gianfrancesco F. PiR_015520 belongs to Piwi-associated RNAs regulates expression of the human melatonin receptor 1A gene. PLoS One. 2011;**6**:e22727

[146] Cox DN, Chao A, Lin H. Piwi encodes a nucleoplasmic factor whose activity modulates the number and division rate of germline stem cells. Development. 2000;**127**:503-514

[147] Alié A, Leclère L, Jager M, Dayraud C, Chang P, Le Guyader H, et al. Somatic stem cells express Piwi and Vasa genes in an adult ctenophore: Ancient association of "germline genes" with stemness. Developmental Biology. 2011;**350**:183-197

[148] Juliano CE, Swartz SZ, Wessel GM. A conserved germline multipotency program. Development. 2010;**137**:4113-4126

[149] Lee EJ, Banerjee S, Zhou H, Jammalamadaka A, Arcila M, Manjunath BS, et al. Identification of piRNAs in the central nervous system. RNA. 2011;**17**:1090-1099

[150] Nolde MJ, Cheng EC, Guo S, Lin H. Piwi genes are dispensable for normal hematopoiesis in mice. PLoS One. 2013;**8**(8):e71950. DOI: 10.1371/journal.pone.0071950

[151] Peng JC, Lin H. Beyond transposons: The epigenetic and somatic functions of the Piwi-piRNA mechanism. Current Opinion in Cell Biology. 2013;**25**:190-194

[152] Szakmary A, Cox DN, Wang Z, Lin H. Regulatory relationship among piwi, pumilio, and bag-of-marbles in Drosophila germline stem cell self-renewal and differentiation. Current Biology. 2005;**15**:171-178

[153] Pek JW, Kai T. Non-coding RNAs enter mitosis: Functions, conservation and implications. Cell Division. 2011;**6**:6. DOI: 10.1186/1747-1028-6-6

[154] Ashe A, Sapetschnig A, Weick EM, Mitchell J, Bagijn MP, Cording AC, et al. piRNAs can trigger a multigenerational epigenetic memory in the germline of *C. elegans*. Cell. 2012;**150**:88-99

[155] Stuwe E, Tóth KF, Aravin AA. Small but sturdy: Small RNAs in cellular memory and epigenetics. Genes & Development. 2014;**28**:423-431

[156] Khurana JS, Xu J, Weng Z, Theurkauf WE. Distinct functions for the Drosophila piRNA pathway in genome maintenance and telomere protection. PLoS Genetics. 2010;**6**:e1001246. DOI: 10.1371/journal.pgen.1001246

[157] Kawaoka S, Izumi N, Katsuma S, Tomari Y. 3′ end formation of PIWI-interacting RNAs in vitro. Molecular Cell. 2011;**43**:1015-1022

[158] Schwager EE, Meng Y, Extavour CG. Vasa and piwi are required for mitotic integrity in early embryogenesis in the spider *Parasteatoda tepidariorum*. Developmental Biology. 2015;**402**:276-290

[159] Zhao S, Gou LT, Zhang M, Zu LD, Hua MM, Hua Y, et al. piRNA-triggered MIWI ubiquitination and removal by APC/C in late spermatogenesis. Developmental Cell. 2013;**24**:13-25

[160] Yin DT, Wang Q, Chen L, Liu MY, Han C, Yan Q, et al. Germline stem cell gene PIWIL2 mediates DNA repair through relaxation of chromatin. PLoS One. 2011;**6**(11):e27154. DOI: 10.1371/journal.pone.0027154

[161] Ding X, Guan H, Li H. Characterization of a piRNA binding protein Miwi in mouse oocytes. Theriogenology. 2013;**79**:610-5.e1. DOI: 10.1016/j.theriogenology.2012.11.013

[162] Roovers EF, Rosenkranz D, Mahdipour M, Han CT, He N, Chuva de Sousa Lopes SM, et al. Piwi proteins and piRNAs in mammalian oocytes and early embryos. Cell Reports. 2015;**10**:2069-2082

[163] Gou LT, Kang JY, Dai P, Wang X, Li F, Zhao S, et al. Ubiquitination-deficient mutations in human Piwi cause male infertility by impairing histone-to-protamine exchange during spermiogenesis. Cell. 2017;**169**:1090-1104

[164] Kamaliyan Z, Pouriamanesh S, Amin-Beidokhti M, Rezagholizadeh A, Mirfakhraie R. HIWI2 rs508485 polymorphism is associated with non-obstructive Azoospermia in Iranian patients. Reports of Biochemistry and Molecular Biology. 2017;**5**:108-111

[165] Klein JD, Qu C, Yang X, Fan Y, Tang C, Peng JC. c-Fos repression by Piwi regulates drosophila ovarian germline formation and tissue morphogenesis. PLoS Genetics. 2016;**12**(9):e1006281. DOI: 10.1371/journal.pgen.1006281

[166] Vagin VV, Sigova A, Li C, Seitz H, Gvozdev V, Zamore PD. A distinct small RNA pathway silences selfishgenetic elements in the germline. Science. 2006;**313**:320-324

[167] Kiuchi T, Koga H, Kawamoto M, Shoji K, Sakai H, Arai Y, et al. A single female-specific piRNA is the primary determiner of sex in the silkworm. Nature. 2014;**509**:633-636

[168] Navarro A, Tejero R, Vinolas N, Cordeiro A, Marrades RM, Fuster D, et al. The significance of PIWI family expression in human lung embryogenesis and non-small cell lung cancer. Oncotarget. 2015;**6**:31544-31556

[169] Zhao PP, Yao MJ, Chang SY, Gou LT, Liu MF, Qiu ZL, et al. Novel function of PIWIL1 in neuronal polarization and migration via regulation of microtubule-associated proteins. Molecular Brain. 2015;**8**:39-46

[170] Iossifov I, O'Roak BJ, Sanders SJ, Ronemus M, Krumm N, Levy D, et al. The contribution of de novo coding mutations to autism spectrum disorder. Nature. 2014;**515**:216-221

[171] Seipel K, Yanze N, Schmid V. The germ line and somatic stem cell gene Cniwi in the jellyfish *Podocoryne carnea*. The International Journal of Developmental Biology. 2004;**48**:1-7

[172] Rinkevich Y, Rosner A, Rabinowitz C, Lapidot Z, Moiseeva E, Rinkevich B. Piwi positive cells that line the vasculature epithelium underlie whole body regeneration in a basal chordate. Developmental Biology. 2010;**345**:94-104

[173] Zhu W, Pao GM, Satoh A, Cummings G, Monaghan JR, Harkins TT, et al. Activation of germline-specific genes is required for limb regeneration in the Mexican axolotl. Developmental Biology. 2012;**370**:42-51

[174] Dharap A, Nakka VP, Vemuganti R. Altered expression of PIWI RNA in the rat brain after transient focal ischemia. Stroke. 2011;**42**:1105-1109

[175] Wu W, Zhang M, Liu Q, Xue L, Li Y, OU S. Piwil 2 gene transfection changes the autophagy status in a rat model of diabetic nephropathy. International Journal of Clinical and Experimental Pathology. 2015;**8**:10734-10742

[176] Plestilova L, Neidhart M, Russo G, Frank-Bertoncelj M, Ospelt C, Ciurea A, et al. Expression and regulation of PIWIL-proteins and PIWI-interacting RNAs in rheumatoid arthritis. PLoS One. 2016;**11**:e0166920

Chapter 13

Part 2: Deregulated Expressions of PIWI Proteins and piRNAs as New Candidate Biomarkers and Potential Therapeutic Tools in Cancer

Didier Meseure and Kinan Drak Alsibai

Additional information is available at the end of the chapter

http://dx.doi.org/10.5772/intechopen.81738

Abstract

Epigenetic abnormalities are early events in carcinogenesis and associate heterogeneity of DNA methylation, modifications of histones, and deregulation of noncoding RNAs. Aberrant expressions of PIWI proteins and piRNAs were recently observed in numerous subtypes of malignant tumors and were implicated in occurrence of most cancer hallmarks such as cell proliferation, genomic stability, apoptosis inhibition, invasion, and metastatic spread. However, this pathway is a new emerging research field, and further investigations are necessary to elucidate their oncogenic or tumor-suppressing status. Since the aberrant expression of this pathway may induce stemness, analysis of relationship between PIWI proteins, piRNAs, and cancer stem cells may open new avenues in cancer research. The objective of this review is to provide a broad overview of the emerging implication of PIWI proteins and piRNAs in carcinogenesis and their potential clinical interest as diagnostic and prognostic biomarkers and therapeutic tools.

Keywords: piRNA, PIWI proteins, transposable element (TE), transcriptional and post-transcriptional silencing, piRNA cluster, heterochromatin, DNA methylation, ping-pong cycle, Nuage

1. Deregulation of the PIWI-piRNA pathway in cancer

Recent evidences have suggested that activation of the PIWI-piRNA pathway has important implications in carcinogenesis (**Table 1**). PIWI proteins and piRNAs are deregulated in numerous human cancers such as breast cancer, prostate cancer, hepatic cancer, gastric cancer, malignant melanoma, and lung cancer. Experimental analyses have demonstrated that

PIWI proteins	PiRNAs	Cancer type	Expression	Cancer hallmarks	Mechanisms in cancer	References
PIWIL1		Breast cancer	Overexpression	Proliferation and stemness	Cell-cycle regulation	[10]
PIWIL1		Lung cancer	Overexpression	CSC self-renewal	Maintenance of CSCs	[9]
PIWIL1		Ovary cancer	Overexpression	Migration and invasion		[72]
PIWIL1		Gastric cancer	Overexpression	Proliferation	Cell-cycle regulation	[8, 22, 62]
PIWIL1		Colon cancer	Overexpression	Proliferation and stemness	Interactions with OCT4 and SOX2	[17]
PIWIL1		Liver cancer	Overexpression	Proliferation		[12]
PIWIL1		Cervix cancer	Overexpression	Migration and invasion		[5]
PIWIL1		Seminoma	Overexpression	Proliferation		[21]
PIWIL1		Glioblastoma	Overexpression	Proliferation, migration, and invasion	Overexpression of MMP2 and MMP9	[11]
PIWIL1		Sarcomas	Overexpression	Genomic instability	Decreasing of differentiation and inhibition of TSGs (p15, p21, and p27)	[73]
PIWIL2		Colon cancer	Overexpression	Proliferation, migration, and invasion	MMP9 transcriptional activation	[14, 24]
PIWIL2		Breast cancer	Overexpression	Proliferation, apoptosis inhibition, migration, and invasion	Maintenance of CSCs, proliferation via STAT3/cyclin D1 activation, and apoptosis inhibition via STAT3/Bcl-xL activation	[16, 26]
PIWIL2		Liver cancer	Overexpression	Apoptosis inhibition	PIWIL2/STAT3/c-Src activation and repression of p53 transcription	[12, 13]
PIWIL2		Cervix cancer	Overexpression	Apoptosis inhibition	PIWIL2/STAT3/c-Src activation and repression of p53 transcription	[29]
PIWIL2		Ovary cancer	Overexpression	Genomic instability	Chromatin decondensation, cell survival, and DNA repair	[65]
PIWIL4		Colon cancer	Overexpression	Migration and invasion		[74]
PIWIL4		Lung cancer	Overexpression	Proliferation	Increasing of cyclin A and cdk2 expression	[68]

PIWI proteins	PiRNAs	Cancer type	Expression	Cancer hallmarks	Mechanisms in cancer	References
PIWIL4		Cervix cancer	Overexpression	Proliferation, migration, and invasion	Impairing apoptosis via p14ARF/p53 pathway	[34]
	piR-651	Breast, colon, gastric, and lung cancers	Overexpression	Proliferation, invasion, and metastasis	Cyclin D1 and CDK4 overexpression	[28, 43, 69, 70]
	piR-932	Breast cancer	Overexpression	Stemness, EMT, and invasion	TGF-β pathway activation and hypermethylation of the TSG Latexin	[48]
	piR-823	Gastric cancer	Underexpression	Chromatin regulation and angiogenesis	Hypomethylation via decreased DNMT3A/B and decreased pro-angiogenic activity	[5]
	piR-823	Multiple myeloma	Overexpression	DNA methylation and angiogenesis	Hypomethylation via decreased DNMT3A/B and decreased pro-angiogenic activity	[44]
	piR-Hep1	Liver cancer	Overexpression	Migration and invasion	Akt phosphorylation and apoptosis inhibition via STAT3/Bcl-xL pathway	[41, 71]
	piR-017061	Pancreatic cancer	Underexpression			[42]
	piR-L-163	Lung cancer	Underexpression	Proliferation and invasion	Cell-cycle regulation	[38]
	piR ABC	Bladder cancer	Underexpression	Inhibition of proliferation apoptosis promotion	TNFSF4 regulation	[75]

Table 1. Role of PIWI and/or piRNA in carcinogenesis. This table shows recent evidences that deregulation (overexpression and/ or underexpression) of the PIWI-piRNA and PIWI proteins pathway in different cancer types has important implications in cancer hallmarks.

PIWI proteins and piRNAs altered levels of expression are associated with major hallmarks of cancer, including genomic chronic proliferation, differentiation, and survival [1]. Emerging roles of PIWI-piRNA-mediated epigenetic alterations in cancer seem related to promotion of a stem-like state of tumor cells through aberrant DNA methylation. Experimental studies in *Drosophila* have confirmed that PIWI proteins and piRNAs functions are upstream modulators of cell cycle progression and proliferation through control of DNA synthesis, maintain of chromatin structure, and assembly of mitotic spindle. In *Drosophila*, maternal alterations of the three PIWI proteins, Piwi, Aub, and Ago3, were associated with chromatin disorganization, asynchronous nuclear division, abnormal nuclear morphology, and cell cycle arrest. Analysis of this pathway in cancer is still in its infancy, and investigation is currently underway to determine whether ectopic expression of piRNAs and PIWI proteins could have a driver role. Expression status and roles of the PIWI proteins and piRNAs remain poorly understood in cancer, and most of the current data about association of this pathway with carcinogenesis result from clinic-pathological reports. Aberrant expressions of these genes have been associated with various hallmarks of cancer and prognostic/predictive factors [2, 3].

1.1. Deregulation of PIWI proteins in cancer

Based mainly on both *in vivo* and *in vitro* functional studies combined with clinico-pathologic analysis, mounting evidence has identified all four human PIWI proteins as new molecular players in carcinogenesis [4]. Alterations of their expression levels are actually better investigated than piRNAs, with PIWI proteins overexpressed in germline and somatic malignant tumors. Deregulated PIWI proteins observed in somatic malignant tumors can be included in the cancer/testis antigens (CTAs) class and could thus be pertinent targets of immunotherapy [5]. They are linked to most of cancer hallmarks, suggesting an oncogenic role [6]. The four PIWI proteins PIWIL1/HIWI, PIWIL2/HILI, PIWIL3, and PIWIL4/HIWI2 were thus found to be involved in cancer cell proliferation, apoptosis, invasion, and metastasis and could represent diagnostic and prognostic biomarkers.

1.1.1. Proliferation, apoptosis, and stemness

PIWIL1: PIWIL1-induced DNA hypermethylation generates genetic and epigenetic changes promoting carcinogenesis [7]. PIWIL1 was upregulated in seminoma. In gastric cancer, PIWIL1 progressive increased expression was observed from preneoplastic lesions to invasive cancer, whereas inhibition of PIWIL1 suppressed growth of tumor epithelial cells and induced cell-cycle arrest at the G2/M phase. PIWIL1 induced cyclin B1 overexpression and promoted cell cycle progression from G2 to M phase [8]. In sarcomas, PIWIL1 upregulation was associated with repression of tumor differentiation and indefinite proliferation and inversely correlated with TSGs expression, including p15, p21, and p27. Lung cancer stem cells overexpressed PIWIL1 whose knockdown was associated with inhibition of sphere formation ability and colony forming capacity in nude mice [9]. The promoting role of PIWIL1 in cell proliferation was also demonstrated in breast cancer [10]. In glioma, PIWIL1 repression was associated with inhibition of cell proliferation by promoting apoptosis and increasing cell cycle arrest. PIWIL1 induced cell cycle progression through increasing cyclin D1 and cyclin B levels of expression and repressed expression of p21, an inhibitor of cyclin D1 and CDK4, whereas its suppression through siRNAs blocked S phase entry *via* p21 upregulation and cyclin D1 downregulation [11]. In hepatocellular carcinoma, PIWIL1 was underexpressed leading to reduction of proliferation and migration of cancer cells [12].

PIWIL2: PIWIL2-induced Stat3/Bcl-XL pathway activation promoted oncogenesis [13]. In breast cancer, PIWIL2 played an anti-apoptotic role and increased cancer stem cells proliferation by promoting the STAT3/Bcl-xL pathway [14]. PIWIL2 activated the STAT3/cyclin D1 pathway. The splice isoform PL2L60 of PIWIL2 was overexpressed in tumors and was implicated in G0/1 to S phase transition and cancer cells proliferation through activation of NF-κB and upregulation of STAT3 and Bcl-2 [15, 16]. In this way, PIWIL2 can directly bind to STAT3 protein *via* its PAZ domain and edify a PIWIL2/STAT3/c-Src triple protein-protein complex. Then, c-Src-induced phosphorylated STAT3 is translocated to the nucleus and binds to p53 promoter, inhibiting its transcription and reducing apoptosis in cancer cells. PIWIL2 silencing could reduce tumor proliferation and colony formation but increase apoptosis *in vitro*. PIWIL2 inactivation in colon cancer cells reduced proliferation and colony formation [14]. A significant positive correlation at mRNA level was observed between PIWIL1 and OCT4, as well as PIWIL2 and SOX2 in colon cancer cells [17]. In gastric cancer, PIWIL2 upregulated the expression of CDK2, CDK4, cyclin A, and c-Myc-dependent cyclin D1. In lung carcinoma, PIWIL2 enhanced CDK2 and cyclin A expression and initiated mitotic phase [22]. In hepatocellular and cervical carcinoma, PIWIL2 promoted c-Myc expression through interaction of NME2 kinase to c-Myc promoter region [41]. PIWIL1 and PIWIL2 modulated microtubules activity to increase malignant cell proliferation and invasion. PIWIL1 upregulated Stathmin1 expression, a pivotal cytosolic phosphoprotein implicated in edification of the mitotic spindle and segregation of chromosomes [43]. Direct interaction between Stathmin1 and PIWIL1 prevented its inhibition by CaMKII and its degradation by E3 ubiquitin ligase RLIM. PIWIL2 promoted interaction of the tubulin folding cofactor B (TBCB) with HSP90 and repressed binding of TBCB to the E3 ubiquitin ligase adaptor protein gigaxonin, resulting in tubulin polymerization and microtubule edification during cell division [31].

PIWIL4: PIWIL4 also plays an oncogenic role in cervical cancer by inhibiting apoptosis through H3K9 methylation at the *p14-ARF* locus [18]. In glioma, PIWIL4 upregulated cyclin D1 expression and repressed the inhibitor of cyclin D1/CDK4 p16. PIWIL4 also promoted cyclin D1 expression through activation of STAT3, Bcl2, and Bcl-xL, and its expression was modulated by miR384 [46]. In cervical cancer, PIWIL4 promoted survival of cancer cells by inhibiting expression of p14-ARF and p53 and prevented apoptosis by inactivating the p14-ARF/p53 axis. Furthermore, PIWIL4 downregulated PTEN expression and activity at the transcriptional level, through DNMT1-induced hypermethylation of the *PTEN* gene promoter. PTEN inhibits numerous survival factors such as STAT3 and Akt/PI3K and through dephosphorylation induced by its phosphatase activity [1].

Furthermore, overexpression of PIWI proteins contributes to carcinogenesis by decreasing differentiation and promoting cancer stemness. PIWIL1 and PIWIL2 are implicated in maintenance and proliferation of lung and breast cancer stem cells, respectively [15].

1.1.2. Genomic integrity

Cancer cells overexpressing PIWI proteins often have genomic alterations partially resulting from reactivation of TEs. In sarcomas, PIWIL1, and p15, p21 and p27 expression levels are inversely correlated [7]. PIWI proteins maintain genome integrity in tumor cells by using epigenetic mechanisms. PIWIL1-induced carcinomas were characterized by global DNA hypermethylation at nonpromoter CpG regions. In mouse embryonic fibroblasts, PIWIL2 was implicated in DNA repair by promoting histone acetylation, chromatin relaxation, and DNA damage response [1].

1.1.3. EMT, invasion, and metastasis

Reciprocal interactions between PIWI proteins and transcription factors involved in EMT have been identified [19]. In colon carcinomas, PIWI proteins and piRNA 34736 activate genes implicated in EMT [20].

The four PIWI proteins identified in humans are variably overexpressed in numerous cancers and implicated in invasion and metastatic spread. The first example of PIWI protein deregulated in carcinogenesis was that of PIWIL1 overexpression in seminomas [21]. Since then, increased levels of PIWIL1 have been detected in lung, breast, gastrointestinal tract, pancreas, liver, cervical, ovarian, and endometrial carcinomas. In most of these malignant tumors, PIWIL1 overexpression is associated with large tumor size, high histologic grade, advanced stage, and poorer prognosis [22–24]. Compared with PIWIL1 and PIWIL2, PIWIL3 and PIWIL4 have been studied in few cancers.

PIWIL1 upregulation was implicated in carcinogenesis by promoting expression of CSC transcription factors OCT4, NANOG, and BMI1 with increased self-renewal properties and resistance to chemotherapies [25]. In breast cancer, PIWIL1 modified cell cycle progression through regulating levels of TGF-β receptors, CDK4, CDK6, and CDK8 [26]. During gastric carcinogenesis, PIWIL1 levels of expression were gradually increased in normal tissues, atrophic gastritis, intestinal metaplasia, and invasive carcinoma [25]. In colon cancer, PIWIL1 overexpression induced global DNA methylation and proliferation. In hepatocellular carcinomas, PIWIL1 overexpression promoted invasion and metastatic spread, whereas PIWIL1 inactivation decreased metastatic evolution [27]. In glioblastomas, PIWIL1 inactivation reduced migration by inactivating expression of MMP2 and MMP9 [1]. Conversely, PIWIL1 had an inhibiting effect on invasion in ovarian cancer [28].

PIWIL2 upregulation was identified in various stages of cervical low-grade and high-grade squamous, intraepithelial lesions, and invasive cervical carcinomas and had the potential to be used as a complementary biomarker for p16 [29]. In colon cancer, PIWIL2 overexpression was significantly correlated with more aggressive clinical and pathological parameters, including lymph node and distant metastasis and poor prognosis [30]. PIWIL2 overexpression induced invasion through activation of MMP9, whereas PIWIL2 inhibition decreased migration via E-cadherin upregulation and TWIST, vimentin, and N-cadherin downregulation [30]. Experimental PIWIL2 knockdown was associated with inhibition of colon cancer cells invasion through MMP9 downregulation. In breast cancer, PIWIL2 upregulation in cancer stem cells was correlated with large tumor size, high histological grade, estrogen receptor expression, proliferation marker Ki67, advanced tumor stage, and lymph node metastasis [26]. In prostate cancer, PIWIL2 overexpression was associated with deregulation of EMT factors. In breast cancer, PIWIL2 and piRNA 932 overexpression promoted EMT in CD44+/CD24− CSCs [31].

PIWIL4 was upregulated in breast and renal cell cancers, as compared with normal tissue [32,33]. PIWIL4 can also induce cervical cancer cell invasion through inhibiting p14/ARF and p53 expression [34].

1.2. Deregulation of piRNAs in cancer

Compared to PIWI proteins, few literatures are available about piRNAs in carcinogenesis. Despite numerous piRNAs are generated in the human genome, only a small number is consistently expressed in normal and tumor tissues [1]. With the previously established evidence on PIWI proteins deregulation in cancer, the hypothesis that piRNAs are also aberrantly expressed in various cancers is very likely because piRNAs are pivotal part of the pi-RISC effector complexes that allow recognition of TEs. Deregulation of piRNAs that target mRNA transcripts containing TE-derived sequences could induce alterations of TSGs and oncogenes [15]. Deregulation of both PIWI proteins and molecules implicated in biogenesis of the PIWI-piRNA are also in favor of a driver role for piRNAs in cancer. Recent evidence suggests that piRNAs control transcriptional and post-transcriptional genes regulation in cancer cells through epigenetic mechanisms associating global DNA hypomethylation, gene-specific DNA hypermethylation, and histones hypoacetylation, which result in oncogenes activation, TSG repression, genomic silencing, and induction of a stem-like state [35, 36]. Furthermore, piRNAs drive carcinogenesis by using nonepigenetic mechanisms such as cell cycle deregulation, proliferation, and invasion [37, 38]. Differential expressions of few piRNAs were reported between malignant tumors and normal tissues. Transcriptomic analysis of 6260 human piRNAs from 11 types of tumors and normal tissues has revealed that among the 20,831 known piRNAs, only 522 piRNAs were expressed in tumor tissues in a cancer-type specific manner, whereas 273 piRNAs were expressed in normal tissues [39].

The first examples of piRNAs deregulated in cancer were piR-651, piR-823, and piR-932. PiR-651 was aberrantly overexpressed in numerous tumors, such as breast, gastric, colon, and lung cancers, compared to normal tissues. By using small RNA sequencing techniques in breast cancer, over 100 deregulated piRNAs were identified in tumors compared to normal tissues, including piR-34736, piR-36249, piR-35407, piR-36318, piR-34377, piR-36743, piR-36026, and piR-31106. Among them, piR-36743, piR-36026, and piR-31106 were overexpressed, whereas piR-34736, piR-36249, piR-35407, piR-36318, and piR-34377 were underexpressed [33]. Furthermore, estrogen deficiency and the estrogen receptor ERβ controlled metastatic spread of breast cancer cells by directly modulating piRNA expression. Eight piRNAs were significantly differentially expressed between breast tumors and normal tissue, with overexpression of piR-4987, piR-20365, piR-20485, and piR-20582 correlated with lymph node metastasis [33]. In lung cancer, 555 piRNAs were differentially expressed between lung adenocarcinomas and squamous cell carcinomas and normal bronchial tissue. The most frequently underexpressed piRNA in lung cancer was piR-L-163, localized in intron 10 of the *LAMC2* gene. PiR-651 promoted cyclin D1 and CDK4 overexpression, resulting in G1 phase entry [38]. In gastric cancer, piR-823 underexpression and piR-651 overexpression were initially identified by using piRNA microarray and PCR compared to matched nonmalignant tissues. PiR-651 overexpression induced transition from G2 to M phase [40]. A total of 156 piRNAs, including piR-32105, piR-58099, and piR-59056, were significantly differentially expressed by using small RNA sequencing from a series of 320 carcinomas and 38 nonmalignant tissues [1]. In liver cancer, piR-Hep1 was overexpressed compared to normal liver tissues. PiR-Hep1 levels of expression positively correlated with

PIWIL2, suggesting that piR-Hep1 may interact with PIWIL2 to induce carcinogenesis [41]. In pancreatic cancer, piR-017061, localized within HBII-296A snoRNA, was shown to be underexpressed compared to normal pancreatic tissues [42]. In leukemias, piR-823 inhibited p16 expression through activation of DNMT3A and DNMT3B, resulting in upregulation of cyclin D1 and CDK4. PiR_011186A downregulated p15 expression and thus promoted cyclin D1 and CDK4 upregulation through edification of a molecular complex combining DNMT1, EZH2, and Suv39H1 [44]. In kidney cancer, study of 24 tumors revealed that 19 piRNAs were differentially expressed between clear cell renal cell carcinomas and benign tissues. Furthermore, 46 piRNAs were differentially expressed between primary and metastatic carcinomas. Among piRNAs deregulated in renal cell metastatic carcinomas, three overexpressed piRNAs (piR-32051, piR-39894, and piR-43607) were localized on the same piRNA cluster on chromosome 17 [43]. In multiple myeloma, piR-823 was significantly underexpressed in tumors, compared to normal tissues [44]. Furthermore, numerous PiRNAs, including piR-Hep1, piR-651, piR-823, piR-932, piR-L-163, piR-4987, piR-20365, piR-20485, piR-20582 and piR-ABC, were associated with hallmarks of cancer and could be pertinent diagnostic and prognostic biomarkers [40, 41, 45].

1.2.1. Proliferation, survival, and apoptosis

In bladder cancer, piRNAs microarray study identified 106 overexpressed piRNAs and 91 underexpressed piRNAs, including piRNA DQ594040. PiRABC repressed cancer cells proliferation and increased cell apoptosis through TNFSF4 protein overexpression. In multiple myeloma, PiR-823 was upregulated and implicated in proliferation, apoptosis, cell-cycle regulation, and angiogenesis related to DNMT3A and p16-INK4A repression and associated with advanced clinical stage [44]. PiRNA-823 repression promoted deregulation of cell cycle regulators and apoptosis-related proteins and inhibited pro-angiogenic activity [44]. In glioblastoma, piR-598 promoted cancer cells survival and proliferation [46]. In lung carcinoma, piR-55490 underexpression was associated with increased proliferation through reduction of piRNA-induced 3′UTR mTOR mRNA binding and degradation. Furthermore, inhibition of piR-L-163 enhanced DNA synthesis and promoted tumor cells survival and proliferation [38, 47].

1.2.2. Invasion and metastatic spread

In breast cancer, piR-4987 upregulation was significantly associated with lymph node metastasis and PIWIL2-piR-932 complex enhanced EMT through Latexin methylation of its promoter region [31, 48, 49]. In clear cell renal cell carcinoma, 19 piRNAs were differentially expressed between tumor and normal tissues and 46 piRNAs were associated with metastasis and poor survival [43]. PiRNA microarray analysis revealed 235 piRNAs upregulated and 369 piRNAs downregulated in malignant tissue from 106 patient samples. RT-qPCR analysis confirmed that piR-57125 was downregulated in metastatic tumors, whereas piR-30924 and piR-38756 were upregulated [50]. In gastric cancer, downregulated blood levels of piR-823 were correlated with stage, lymph node, and distant metastasis, suggesting that piRNAs may be pertinent blood biomarkers [51]. Furthermore, PiR-651 inhibition promoted growth suppression and cell cycle arrest at the G2/M stage [40]. In hepatocellular carcinoma (HCC), deep sequencing analysis identified oncogenic piR-Hep1 upregulation associated with PIWIL2 overexpression. Inhibition of piR-Hep1 repressed Akt phosphorylation, motility, and invasion [41]. In pancreatic ductal adenocarcinoma, piR-017061 located within HBII-296A snoRNA was found to be downregulated in tumors compared to normal tissues [42]. In lung cancer, PiR-651 overexpression promoted survival, invasion, and metastatic progression through cyclin D1 and CDK4 overexpression [70]. In colon cancer, PiR-823 upregulation was correlated with lymph node and distant metastasis, whereas in gastric cancer, piR-823 over-expression suppressed tumor cell growth [1, 69].

1.3. Mechanisms of the PIWI-piRNA pathway deregulation in cancer

Few studies have identified molecular mechanisms implicating PIWI proteins and piRNAs reactivation in carcinogenesis [52, 53]. The first hypothesis concerning deregulation of the PIWI-piRNA pathway in cancer results from its control of TEs propagation. In normal germline stem cells, piRNAs cooperate with abundant PIWI proteins to regulate TEs inactivation *via* DNA methylation. During embryonic development, these cells containing high levels of PIWI proteins develop into various somatic tissues in which PIWI proteins are not normally expressed. In cancer, recent studies have demonstrated that methylation of TEs was decreased, whereas TEs transcripts and proteins were upregulated and correlated with higher metastasis frequency [54, 55]. Prolonged exposure to detrimental environmental factors may promote TEs mobilization with disruption of various TSGs *via* TEs

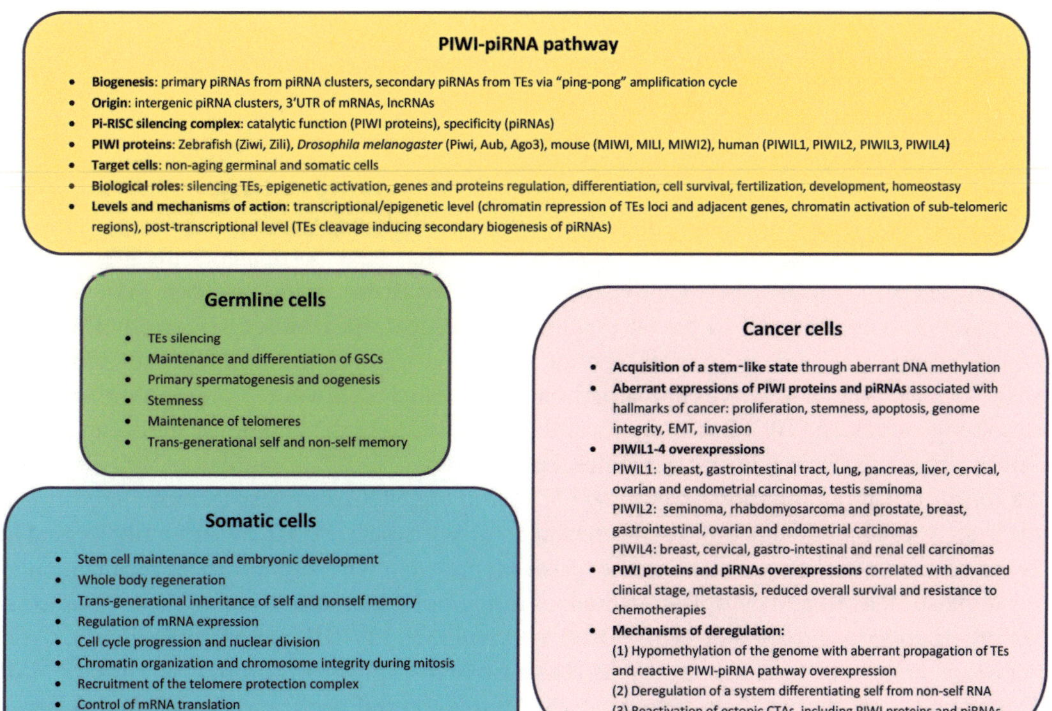

Figure 1. The PIWI-piRNA pathway in germline and somatic cells, and mechanisms of the PIWI-piRNA pathway deregulation in cancer cells.

integration [56]. A widespread hypomethylation of the genome during progression could induce both aberrant reactivation and propagation of TEs with reactive ectopic overexpression of the PIWI-piRNA pathway. High levels of piRNAs and PIWI proteins observed in somatic cancers could promote abnormal extensive DNA methylation, resulting in silencing of TSGs and acquisition of an aberrant stem-like state [57, 58]. The hypothesis that piRNAs guide PIWIL1 and PIWIL2 to transcriptionally silence TSGs is consolidated by analysis of their orthologs in mice, in which MILI and MIWI2 mutants fail to establish DNA methylation of TEs [59]. Furthermore, by suppressing expression of particular TEs, aberrant reactivation of the PIWI-piRNA axis in cancer may induce genomic and chromosomal instability. Thus, this pathway could be implicated in cancer not only in repression of TEs but also in inactivation of various mRNAs not exposed to germline cells. The second hypothesis is the participation of this pathway in a putative biological system that can differentiate self from nonself RNA [60]. The third hypothesis is the ectopic and coordinated reactivation and overexpression in cancer of cancer/testis antigens (CTA) normally restricted to the germline and including molecules of the PIWI-piRNA axis. As observed in normal gametogenesis, CTAs promote immortalization, implantation, and migration that could correspond in carcinogenesis to transformation, invasion, and metastasis, respectively (**Figure 1**). In this ectopic context, reactivation of the PIWI-piRNA pathway could confer properties of proliferative germline stem cells to cancer cells [61–68].

2. PIWIs and piRNAs as potential biomarkers and therapeutic tools

2.1. PIWI proteins as biomarkers

PIWIL1 overexpression was correlated with poor 5-year survival in malignant glioma, hepatocellular, gastric cancer, colon cancer, pancreatic ductal adenocarcinoma, and ovarian cancer [1]. In hepatocarcinoma, PIWIL1 high levels of expression were correlated with larger tumor size, intrahepatic metastasis, overall survival, and recurrence-free survival [41]. PIWIL2 upregulation was observed in breast cancer, colon cancer, gastrointestinal stromal tumors, renal cell carcinoma, and endometrial carcinoma. In colorectal cancer, PIWIL2 overexpression was associated with aggressive clinico-pathological characteristics and poorer prognostic [30]. In esophageal cancer, PIWIL2 was significantly correlated to high histological grade, advanced clinical stage, and poorer clinical outcomes. Furthermore, PIWIL2 upregulation was identified in a cancer cell subpopulation expressing OCT4 and NANOG stemness factors, suggesting a role in stem-cell maintenance and self-renewal. In breast cancer, PIWIL2 was mainly located in cancer stem cells and showed different expression patterns, with slight nuclear localization in preneoplastic lesions and cytoplasmic/nucleus topography in invasive and metastatic cancers. Its upregulation was correlated with poorer survival [33]. PIWIL2 can promote invasion and metastasis among various cancers by increasing c-Myc expression through binding of NM/NM23 nucleoside diphosphate kinase 2 (NME2) to G4-motif region within c-Myc. In hepatocellular carcinoma, nuclear co-expression of PIWIL2 and PIWIL4 had a worse prognostic phenotype [71]. In ovarian cancer, PIWIL2 overexpression was associated with cisplatin resistance and poorer prognostic [28]. PIWIL3 and PIWIL4 overexpression were associated with worse prognosis in breast cancer. In gastric cancer, PIWIL1-4 upregulation was significantly associated with TNM stage histological grade and lymph node metastasis [15, 40, 62].

2.2. PiRNAS as biomarkers

PiRNAs are increasingly investigated as diagnostic and prognostic biomarkers [69]. PiRNAs are small RNAs of only 24–32 nt in length which can pass through cell membrane. Like miRNAs, piRNAs are not degraded in circulation and possess the capability to resist incubation and storage conditions used in laboratories. They are thus detectable in patient samples like blood plasma and serum, saliva, sputum, and urine. Furthermore, piRNAs possess higher sensitivity and specificity compared to an existing lncRNA and miRNA-based biomarkers. They are highly expressed in germinal tissues, but also although at lower levels in various somatic normal tissues, saliva, and plasma-derived exosomes. Furthermore, piRNAs present higher specificity and sensitivity, when compared to miRNAs. In breast cancer, 8 piRNAs were identified as independent prognostic markers and associated with overall survival [63]. Moreover, piR-4987 upregulation in peripheral blood was correlated with lymph node metastasis [48]. In gastric cancer, piR-651 high level of expression in peripheral blood was associated with poor differentiation, advanced TNM stage, and metastasis. A three-piRNA signature, including piR-59056, piR-54878, and piR-62701, could separate patients by risk of recurrence [5]. PiR-651 was also upregulated in other cancers, including lung, colon, and breast tumors. In clear cell renal cell carcinoma, piR-30924 and piR-57125 were shown to be independent prognostic predictors in nonmetastatic patients. In colon cancer, piR59056, piR-54878, and piR-62701 were associated with recurrence-free survival. Few studies have investigated on interactions between piRNAs and PIWI proteins in cancer [70, 72–75]. Actually, it remains unclear whether PIWI proteins independently possess cytoplasmic pro-oncogenic properties that promote cancer cell proliferation, invasion, and metastasis or PIWI proteins located in the nucleus epigenetically modulate numerous biological processes by edifying ribonucleoproteic complexes with piRNAs.

2.3. PIWI proteins and piRNAs as therapeutic tools

Recent studies have revealed the potential for piRNAs as therapeutic tools that target PIWI proteins and oncoproteins [71]. Mounting evidence has revealed that upregulation of PIWI proteins was negatively correlated with patient survival, and that downregulation of PIWI proteins could reduce the number of G2/M phase cells and enhance expression of p53 protein, thus inhibiting proliferation and promoting apoptosis. In addition, PIWI proteins could increase resistance to chemotherapy drugs such as cisplatin, and their downregulation could increase sensitivity of cancer cells to cisplatin. In this way, synthetic piRNAs targeting PIWI genes are potential pertinent tools in transcriptional silencing during cancer progression, and PIWI antibodies targeting PIWI proteins constitute another approach to antagonize cancer cells proliferation at post-transcriptional level. Synthetic piRNAs could also block synthesis of cancer-related proteins by binding to mRNAs. Compared to miRNAs, piRNAs do not require enzymes processing and have better specificity to targets. In a mouse model, artificial piRNAs could be obtained through expression of sense and antisense transcripts, resulting in epigenetic silencing of target genes [1, 72]. Furthermore, piRNAs could be considered as tools to regulate expression levels of TSGs. A lncRNA (*GAS5*)/SnoRNA-derived piRNA enhanced activation of *TRAIL* gene by site-specifically recruiting MLL/COMPASS-like complexes with induction of H3K4 methylation and H3K27 demethylation, resulting in transcriptional activation of *TRAIL* and inhibition of tumor growth [64].

3. Conclusion and perspectives

PiRNAs and PIWI proteins were first recognized more than a decade ago and are coming into attention with development of high-throughput sequencing technologies and bioinformatics methods. This pathway is considered as a conserved immune-like surveillance process to suppress propagation of TEs in germline cells and various types of somatic mostly nonaging cells. Furthermore, the PIWI-piRNA pathway seems to be implicated in maintain of the genome organization, epigenetic modifications of genes expression, and identification of self and non-self genes that are trans-generationally inherited. Moreover, this axis could be implicated in dual DNA/RNA-level regulation of genes expression. Nevertheless, there is still lack of complete understanding of the functions and interactions of piRNAs and PIWI proteins. Therefore, the complicated biogenesis and functions of piRNAs need further elucidation to improve our understanding of the implication of these molecules in cancer. Since discovery of the unexpected role of this pathway in seminoma, aberrant levels of expression of these molecules have been observed across numerous malignant tumors, though further research is needed to elucidate their oncogenic or tumor-suppressing status. Growing evidence suggests that the PIWI-piRNA pathway modulates occurrence of most of cancer hallmarks. PIWI proteins and piRNAs could be pertinent diagnostic/prognostic biomarkers in cancer and therapeutic tools in targeted therapies. However, the potential driver role of a deregulated PIWI-piRNA pathway in cancer needs to be further evaluated. Furthermore, it remains unclear whether PIWI proteins regulate cancer cell proliferation, apoptosis, metastasis, and invasion in the cytoplasm independently or PIWI proteins perform epigenetic control of homeostasis by taken to the nucleus with piRNAs. Most importantly, since the aberrant expression of this pathway may induce stemness, analysis of relationship between PIWI proteins, piRNAs, and cancer stem cells may open new avenues in future investigations.

Acknowledgements

This work was supported by grant INCa-DGOS-4654.

Disclosure/conflict of interest

The authors declare that they have no competing interests.

Author details

Didier Meseure[1,2]* and Kinan Drak Alsibai[3]

*Address all correspondence to: didier.meseure@curie.fr

1 Platform of Experimental Pathology, Curie Institute, Paris, France

2 Department of Diagnostic and Theranostic Medicine, Curie Institute, Paris, France

3 Institut de Pathologie de Paris, Malakoff, France

References

[1] Ng KW, Anderson C, Marshall EA, Minatel BC, Enfield KS, Saprunoff HL, et al. Piwi-interacting RNAs in cancer: Emerging functions and clinical utility. Molecular Cancer. 2016;**15**:5

[2] Hanahan D, Weinberg RA. The hallmarks of cancer. Cell. 2000;**100**:57-70

[3] Hanahan D, Weinberg RA. Hallmarks of cancer: The next generation. Cell. 2011;**144**: 646-674

[4] Litwin M, Szczepańska-Buda A, Piotrowska A, Dzięgiel P, Witkiewicz W. The meaning of PIWI proteins in cancer development. Oncology Letters. 2017;**13**:3354-3362

[5] Cheng J, Deng H, Xiao B, Zhou H, Zhou F, Shen Z, et al. piR-823, a novel non-coding small RNA, demonstrates in vitro and in vivo tumor suppressive activity in human gastric cancer cells. Cancer Letters. 2012;**315**:12-17

[6] Suzuki R, Honda S, Kirino Y. PIWI Expression and Function in Cancer. Frontiers in Genetics. 2012;**3**:204-212

[7] Siddiqi S, Terry M, Matushansky I. Hiwi mediated tumorigenesis is associated with DNA hypermethylation. PLoS One. 2012;**7**:e33711

[8] Liu X, Sun Y, Guo J, Ma H, Li J, Dong B, et al. Expression of Hiwi gene in human gastric cancer was associated with proliferation of cancer cells. International Journal of Cancer. 2006;**118**:1922-1929

[9] Liang D, Yang Y, Liu Y. The role of Hiwi gene in the maintenance of lung cancer stem cell populations. Neoplasma. DOI: 10.4149/neo_2014_022

[10] Wang DW, Wang ZH, Wang LL, Song Y, Zhang GZ. Overexpression of hiwi promotes growth of human breast cancer cells. Asian Pacific Journal of Cancer Prevention. 2014; **15**:7553-7558

[11] Wang X, Tong X, Gao H, Yan X, Xu X, Sun S, et al. Silencing HIWI suppresses the growth, invasion and migration of glioma cells. International Journal of Oncology. 2014;**45**: 2385-2392

[12] Xie Y, Yang Y, Ji D, Zhang D, Yao X, Zhang X. Hiwi downregulation, mediated by shRNA, reduces the proliferation and migration of human hepatocellular carcinoma cells. Molecular Medicine Reports. 2015;**11**:1455-1461

[13] Lee JH, Schutte D, Wulf G, Fuzesi L, Radzun HJ, Schweyer S, et al. Stem-cell protein Piwil2 is widely expressed in tumors and inhibits apoptosis through activation of Stat3/Bcl-XL pathway. Human Molecular Genetics. 2006;**15**:201-211

[14] Oh SJ, Kim SM, Kim YO, Chang HK. Clinicopathologic implications of PIWIL2 expression in colorectal cancer. Korean Journal of Pathology. 2012;**46**:318-323

[15] Ye Y, Yin DT, Chen L, Zhou Q, Shen R, He G, et al. Identification of Piwil2-like (PL2L) proteins that promote tumorigenesis. PLoS One. 2010;**5**(10):e13406. DOI: 10.1371/journal.pone.0013406

[16] Lee JH, Jung C, Javadian-Elyaderani P, Schweyer S, Schütte D, Shoukier M, et al. Pathways of proliferation and antiapoptosis driven in breast cancer stem cells by stem cell protein piwil2. Cancer Research. 2010;**70**:4569-4579

[17] Litwin M, Dubis J, Arczyńska K, Piotrowska A, Frydlewicz A, Karczewski M, et al. Correlation of HIWI and HILI expression with cancer stem cell markers in colorectal cancer. Anticancer Research. 2015;**35**:3317-3324

[18] Sugimoto K, Kage H, Aki N, Sano A, Kitagawa H, Nagase T, et al. The induction of H3K9 methylation by PIWIL4 at the p16Ink4a locus. Biochemical and Biophysical Research Communications. 2007;**359**:497-502

[19] Kessenbrock K, Plaks V, Werb Z. Matrix metalloproteinases: Regulators of the tumor microenvironment. Cell. 2010;**141**:52-67

[20] Botchkina IL, Rowehl RA, Rivadeneira DE, Karpeh MS Jr, Crawford H, Dufour A, et al. Phenotypic subpopulations of metastatic colon cancer stem cells: Genomic analysis. Cancer Genomics Proteomics. 2009;**6**:19-29

[21] Qiao D, Zeeman A-M, Deng W, Looijenga LHJ, Lin H. Molecular characterization of hiwi, a human member of the piwi gene family whose overexpression is correlated to seminomas. Oncogene. 2002;**21**:3988-3999

[22] Liu XY, Sun Y, Guo JP, et al. Expression of hiwi gene in human gastric cancer was associated with proliferation of cancer cells. International Journal of Cancer. 2006;**118**:1922-1929

[23] Grochola LF, Greither T, Taubert H, et al. The stem cell-associatedHiwi gene in human adenocarcinoma of the pancreas: Expressionand risk of tumour-related death. British Journal of Cancer. 2008;**99**:1083-1088

[24] Zeng Y, Qu LK, Meng L, Liu CY, Dong B, Xing XF, et al. HIWI expression profile in cancer cells and its prognostic value for patients with colorectal cancer. Chinese Medical Journal. 2011;**124**:2144-2149

[25] Liu Y. MicroRNAs and PIWI-interacting RNAs in oncology. Oncology Letters. 2016; **12**:2289-2292

[26] Zhang H, Ren Y, Xu H, Pang D, Duan C, Liu C. The expression of stem cell protein Piwil2 and piR-932 in breast cancer. Surgical Oncology. 2013;**22**:217-223

[27] Zhao YM, Zhou JM, Wang LR, He HW, Wang XL, Tao ZH, et al. HIWI is associated with prognosis in patients with hepatocellular carcinoma after curative resection. Cancer. 2012;**118**:2708-2717

[28] Lim SL, Ricciardelli C, Oehler MK, Tan IM, Russell D, Grutzner F. Overexpression of piRNA pathway genes in epithelial ovarian cancer. PLoS One. 2014;**9**:e99687

[29] He G, Chen L, Ye Y, Xiao Y, Hua K, Jarjoura D, et al. Piwil2 expressed in various stages of cervical neoplasia is a potential complementary marker for p16. American Journal of Translational Research. 2010;**2**:156-169

[30] Li D, Sun X, Yan D, Huang J, Luo Q, Tang H, et al. Piwil2 modulates the proliferation and metastasis of colon cancer via regulation of matrix metallopeptidase 9 transcriptional activity. Experimental Biology and Medicine (Maywood, N.J.). 2012;**237**:1231-1240

[31] Tan H, Liao H, Zhao L, Lu Y, Jiang S, Tao D, et al. HILI destabilizes microtubules by suppressing phosphorylation and Gigaxonin-mediated degradation of TBCB. Scientific Reports. 2017;**7**:46376

[32] Al-Janabi O, Wach S, Nolte E, Weigelt K, Rau TT, Stöhr C, et al. Piwi-like 1 and 4 gene transcript levels are associated with clinicopathological parameters in renal cell carcinomas. Biochimica et Biophysica Acta. 2014;**1842**:686-690

[33] Hashim A, Rizzo F, Marchese G, Ravo M, Tarallo R, Nassa G, et al. RNA sequencing identifies specific PIWI-interacting small non-coding RNA expression patterns in breast cancer. Oncotarget. 2014;**5**:9901-9910

[34] Su C, Ren ZJ, Wang F, Liu M, Li X, Tang H. PIWIL4 regulates cervical cancer cell line growth and is involved in down-regulating the expression of p14ARF and p53. FEBS Letters. 2012;**586**:1356-1362

[35] Feldman N, Gerson A, Fang J, Li E, Zhang Y, Shinkai Y, et al. G9a-mediated irreversible epigenetic inactivation of Oct-3/4 during early embryogenesis. Nature Cell Biology. 2006;**8**:188-194

[36] Baylin SB. DNA methylation and gene silencing in cancer. Nature Clinical Practice. Oncology. 2005;**2**:S4-S11

[37] Fu A, Jacobs DI, Zhu Y. Epigenome-wide analysis of piRNAs in gene-specific DNA methylation. RNA Biology. 2014;**11**:1301-1312

[38] Mei Y, Wang Y, Kumari P, Shetty AC, Clark D, Gable T, et al. A piRNA-like small RNA interacts with and modulates p-ERM proteins in human somatic cells. Nature Communications. 2015;**6**:7316. DOI: 10.1038/ncomms8316

[39] Martinez VD, Vucic EA, Thu KL, Hubaux R, Enfield KS, Pikor LA, et al. Unique somatic and malignant expression patterns implicate PIWI-interacting RNAs in cancer-type specific biology. Scientific Reports. 2015;**5**:10423

[40] Cheng J, Guo JM, Xiao BX, Miao Y, Jiang Z, Zhou H, et al. piRNA, the new non-coding RNA, is aberrantly expressed in human cancer cells. Clinica Chimica Acta. 2011;**412**: 1621-1625

[41] Law PT, Qin H, Ching AK, Lai KP, Co NN, He M, et al. Deep sequencing of small RNA transcriptome reveals novel non-coding RNAs in hepatocellular carcinoma. Journal of Hepatology. 2013;**58**:1165-1173

[42] Müller S, Raulefs S, Bruns P, Afonso-Grunz F, Plötner A, Thermann R, et al. Next-generation sequencing reveals novel differentially regulated mRNAs, lncRNAs, miRNAs, sdRNAs and a piRNA in pancreatic cancer. Molecular Cancer. 2015;**14**:94. DOI: 10.1186/s12943-015-0358-5. Erratum in: Molecular Cancer. 2015;**14**:144

[43] Li Y, Wu X, Gao H, Jin JM, Li AX, Kim YS, et al. Piwi-Interacting RNAs (piRNAs) Are Dysregulated in Renal Cell Carcinoma and Associated with Tumor Metastasis and Cancer-Specific Survival. Molecular Medicine. 2015;**21**:381-388

[44] Yan H, Wu Q-L, Sun C-Y, Ai L-S, Deng J, Zhang L, et al. piRNA-823 contributes to tumorigenesis by regulating de novo DNA methylation and angiogenesis in multiple myeloma. Leukemia. 2014;**29**:196-206

[45] Schupbach T, Wieschaus E. Female sterile mutations on the second chromosome of Drosophila melanogaster. II. Mutations blocking oogenesis or altering egg morphology. Genetics. 1991;**129**:1119-1136

[46] Jacobs DI, Qin Q, Lerro MC, Fu A, Dubrow R, Claus EB, et al. PIWI-interacting RNAs in Gliomagenesis: Evidence from Post-GWAS and Functional Analyses. Cancer Epidemiology, Biomarkers & Prevention. 2016;**25**:1073-1080

[47] Bartel DP. MicroRNAs: Genomics, biogenesis, mechanism, and function. Cell. 2004; **116**:281-297

[48] Huang G, Hu H, Xue X, Shen S, Gao E, Guo G, et al. Altered expression of piRNAs and their relation with clinicopathologic features of breast cancer. Clinical & Translational Oncology. 2013;**15**:563-568

[49] Luteijn MJ, Ketting RF. PIWI-interacting RNAs: From generation to transgenerational epigenetics. Nature Reviews Genetics. 2013;**14**:523-534

[50] Busch J, Ralla B, Jung M, Wotschofsky Z, Trujillo-Arribas E, Schwabe P, et al. Piwi interacting RNAs as novel prognostic markers in clear cell renal cell carcinomas. Journal of Experimental & Clinical Cancer Research. 2015;**34**:61

[51] Cui L, Lou Y, Zhang X, Zhou H, Deng H, Song H, et al. Detection of circulating tumor cells in peripheral blood from patients with gastric cancer using piRNAs as marks. Clinical Biochemistry. 2011;**44**:1050-1057

[52] Heyns M, Kovalchuk O. Non-coding RNAs including miRNAs, piRNAs, and tRNAs in human cancer. Oncotarget. 2015;**6**:23055-23057

[53] Moyano M, Stefani G. piRNA involvement in genome stability and human cancer. Hematological Oncology. 2015;**8**:38. DOI: 10.1186/s13045-015-0133-5

[54] Iskow RC, McCabe MT, Mills RE, Torene S, Pittard WS, Neuwald AF, et al. Natural mutagenesis of human genomes by endogenous retrotransposons. Cell. 2010;**141**:1253-1261

[55] Baba Y, Huttenhower C, Nosho K, Tanaka N, Shima K, Hazra A, et al. Epigenomic diversity of colorectal cancer indicated by LINE-1 methylation in a database of 869 tumors. Molecular Cancer. 2010;**9**:125

[56] Miki Y, Nishisho I, Horii A, Miyoshi Y, Utsunomiya J, Kinzler KW, et al. Disruption of the APC gene by a retrotransposal insertion of L1 sequence in a colon cancer. Cancer Research. 1992;**52**:643-645

[57] Pattamadilok J, Huapai N, Rattanatanyong P, Vasurattana A, Triratanachat S, Tresukosol D, et al. LINE-1 hypomethylation level as a potential prognostic factor for epithelial ovarian cancer. Journal of Gynecological Cancer. 2008;**18**:711-717

[58] Ogino S, Nosho K, Kirkner GJ, Kawasaki T, Chan AT, Schernhammer ES, et al. A cohort study of tumoral LINE-1 hypomethylation and prognosis in colon cancer. Journal of the National Cancer Institute. 2008;**100**:1734-1738

[59] Carreira PE, Richardson SR, Faulkner GJ. L1 retrotransposons, cancer stem cells and oncogenesis. The FEBS Journal. 2014;**281**:63-73

[60] Conine CC, Moresco JJ, Gu W, Shirayama M, Conte D, Yates JR, et al. Argonautes promote male fertility and provide a paternal memory of germline gene expression in C. elegans. Cell. 2013;**155**:1532-1544

[61] Simpson AJ, Caballero OL, Jungbluth A, Chen YT, Old LJ. Cancer/testis antigens, gametogenesis and cancer. Nature Reviews. Cancer. 2005;**5**:615-625

[62] Wang Y, Liu Y, Shen X, Zhang X, Chen X, Yang C, et al. The PIWI protein acts as a predictive marker for human gastric cancer. International Journal of Clinical and Experimental Pathology. 2012;**5**:315-325

[63] Houwing S, Kamminga LM, Berezikov E, Cronembold D, Girard A, van den Elst H, et al. A role for Piwi and piRNAs in germ cell maintenance and transposon silencing in Zebrafish. Cell. 2007;**129**:69-82

[64] He X, Chen X, Zhang X, Duan X, Pan T, Hu Q, et al. An Lnc RNA (GAS5)/SnoRNA-derived piRNA induces activation of TRAIL gene by site-specifically recruiting MLL/COMPASS-like complexes. Nucleic Acids Research. 2015;**43**:3712-3725

[65] Klein JD, Qu C, Yang X, Fan Y, Tang C, Peng JC. C-Fos repression by Piwi regulates drosophila ovarian germline formation and tissue morphogenesis. PLoS Genetics. 2016;**12**:e1006281

[66] Sun H, Li D, Chen S, Liu Y, Liao X, Deng W, et al. Zili inhibits transforming growth factor-beta signaling by interacting with Smad4. The Journal of Biological Chemistry. 2010;**285**:4243-4250

[67] Yakushev EY, Mikhaleva EA, Abramov YA, Sokolova OA, Zyrianova IM, Gvozdev VA, et al. The role of Piwi nuclear localization in the differentiation and proliferation of germline stem cells. Molecular Biology. 2016 (Mosk);**50**:713-720

[68] Qu X, Liu J, Zhong X, Li X, Zhang Q. PIWIL2 promotes progression of non small cell lung cancer by inducing CDK2 and Cyclin A expression. Journal of Translational Medicine. 2015;**13**:301-312

[69] Weng W, Liu N, Toiyama Y, Kusunoki M, Nagasaka T, Fujiwara T, et al. Novel evidence for a PIWI-interacting RNA (piRNA) as an oncogenic mediator of disease progression, and a potential prognostic biomarker in colorectal cancer. Molecular Cancer. 2018;**17**:16. DOI: 10.1186/s12943-018-0767-3

[70] Zhang SJ, Yao J, Shen BZ, Li GB, Kong SS, Bi DD, et al. Role of piwi-interacting RNA-651 in the carcinogenesis of non-small cell lung cancer. Oncology Letters. 2018;**15**:940-946

[71] Xiao Z, Shen J, Zhang L, Li M, Hu W, Cho C. Therapeutic targeting of noncoding RNAs in hepatocellular carcinoma: Recent progress and future prospects. Oncology Letters. 2018;**15**:3395-3402

[72] Singh G, Roy J, Rout P, Mallick B. Genome-wide profiling of the PIWI-interacting RNA-mRNA regulatory networks in epithelial ovarian cancers. PLoS One. 2018;**13**:e0190485. DOI: 10.1371/journal.pone.0190485

[73] Taubert H, Greither T, Kaushal D, Würl P, Bache M, Bartel F, et al. Expression of the stem cell self-renewal gene Hiwi and risk of tumour-related death in patients with soft-tissue sarcoma. Oncogene. 2007;**26**:1098-1100

[74] Li L, Yu C, Gao H, Li Y. Argonaute proteins: Potential biomarkers for human colon cancer. BMC Cancer. 2010;**10**:38. DOI: 10.1186/1471-2407-10-38

[75] Chu H, Hui G, Yuan L, Shi D, Wang Y, Du M, et al. Identification of novel piRNAs in bladder cancer. Cancer Letters. 2015;**356**:561-567